Gas Hydrates 1

Series Editor
Allain Dollet

Gas Hydrates 1

Fundamentals, Characterization and Modeling

Edited by

Daniel Broseta
Livio Ruffine
Arnaud Desmedt

iSTE

WILEY

First published 2017 in Great Britain and the United States by ISTE Ltd and John Wiley & Sons, Inc.

ISTE Ltd
27-37 St George's Road
London SW19 4EU
UK

www.iste.co.uk

John Wiley & Sons, Inc.
111 River Street
Hoboken, NJ 07030
USA

www.wiley.com

Library of Congress Control Number: 2017936797

British Library Cataloguing-in-Publication Data
A CIP record for this book is available from the British Library
ISBN 978-1-84821-969-4

Cover image: DIC image of cyclopentane hydrate formed on a water drop (with diameter in the mm range) on hydrophilic glass, showing a faceted crust over the water, surrounded by a fine-grained halo, on the substrate under the guest phase (see Chapter 3). Pixel coloring by intensity (dark to light shades) highlights the delicate beauty of hydrate crystals revealed by high resolution microscopy.

Contents

Preface

Clathrate hydrates are crystalline inclusion compounds resulting from the hydrogen bonding of water (host) molecules enclosing relatively small (guest) molecules, such as hydrogen, noble gases, carbon dioxide, hydrogen sulfide, methane and other low-molecular-weight hydrocarbons. They form and remain stable at low temperatures – often well below ambient temperature – and high pressures – ranging from a few bar to hundreds of bar, depending on the guest molecule. Long considered either an academic curiosity or a nuisance for oil and gas producers confronted with pipeline blockage, they are now being investigated for applications as diverse as hydrogen or methane storage, gas separation, cold storage and transport, water treatment, etc. The ubiquitous presence of natural gas hydrates not only in the permafrost, but also in deep marine sediments, has been identified, and their role in past and present environmental changes and other geohazards, as well as their potential as an energy source, are under intense scrutiny.

These perspectives are motivating an ever-increasing research effort in the area of gas hydrates, which addresses both fundamental issues and applications. Gas hydrates exhibit fascinating yet poorly understood phenomena. Perhaps the most fascinating feature exhibited by gas hydrates is *self-preservation,* or the existence of long-lived metastable states in some conditions far from stable thermodynamic equilibrium. Strong departures from equilibrium are also noted in gas hydrate compositions, depending on their formation and kinetic pathways. A proper understanding of these two effects could serve in developing gas storage and selective molecular-capture processes. The *memory effect*, or the ability of gas hydrates to reform rapidly in an aqueous solution where gas hydrates have been freshly melted, is another puzzling phenomenon. Gas hydrates are likely to be soon exploited for storing gas (guest) molecules or for separating or capturing some of them selectively; yet, the occupancy rates of the different hydrate crystal cavities by the various guest molecules are not fully understood. Very little is known as well on hydrate formation and their stability in the extreme conditions (e.g. low or high pressures) such as on extraterrestrial bodies like comets and planets. How hydrates interact with substrates is a topic of prime interest for understanding not only the behavior of hydrates in sediments, but also why some mesoporous particles act as hydrate promoters. Nucleation and growth processes are still unsettled issues, together with the mechanisms by which additives (co-guest molecules, surfactants, polymers, particles, etc.) promote or inhibit hydrate formation. Depending on the application, these additives are needed to either accelerate or slow down the crystallization process; but their selection is still carried out on a very empirical basis. This book gathers contributions from scientists who actively work in complementary areas of gas hydrate research. They have been meeting and exchanging views regularly over the past few years at a national (French) level, and recently at a European level, within the COST Action MIGRATE (Marine gas hydrate – an indigenous resource of natural gas for Europe). This book is somehow the written expression of those meetings and exchanges. It is divided into two volumes: the first (and present) volume is devoted to the fundamentals, characterization and modeling of gas hydrates, whereas the second volume will focus on gas hydrates in their natural environment and for industrial applications.

The present volume starts with an extensive presentation of the experimental tools capable of probing small spatial and temporal scales: neutron scattering (Chapter 1), spectroscopy (Chapter 2) and optical microscopy (Chapter 3). In addition to providing fundamental insights into structural and dynamical properties, these tools have allowed considerable progress in the understanding of the molecular and mesoscopic mechanisms governing hydrate formation and growth. Moving to larger scales, the calorimetric tools used to measure heat and related thermodynamic properties are described in Chapter 4. Chapter 5 provides a comprehensive view of the thermodynamic modelling of solid-fluid equilibria, from pure solid phases to gas semiclathrate hydrates. Finally, Chapter 6 presents a novel approach coupling thermodynamics and kinetics to describe the non-equilibrium effects occurring during hydrate formation, with a focus on the evolution of the composition of the hydrate phase. Most of these chapters extend their scope to semi-clathrates, in which gas or small molecules still occupy the crystal cavities, but the cavities themselves consist of water and organic species, such as quaternary ammonium salts, strong acids or bases. These semiclathrates hold great promise from a practical point of view, because the temperature and pressure conditions of their formation and stability are closer to the ambient than their hydrate counterparts.

Volume 2 addresses geoscience issues and potential industrial applications. It deals with marine gas hydrates through a multidisciplinary lens, integrating both field studies and laboratory work and analyses, with a focus on the instrumentations and methods used to investigate the dynamics of natural deposits. This is followed by the description of the geochemical models used for investigating the temporal and spatial behavior of hydrate deposits. Finally, potential industrial applications of clathrate and semiclathrate hydrates are also presented in that volume.

To conclude, we would like to warmly thank all the contributors to the present volume for taking the time to write concise and clear introductions to their fields.

Daniel BROSETA
Livio RUFFINE
Arnaud DESMEDT
April 2017

1

Neutron Scattering of Clathrate and Semiclathrate Hydrates

1.1. Introduction

Neutron scattering is a standard tool when dealing with the microscopic properties of the condensed matter at the atomic level. This comes from the fact that the neutron matches with the distances and energy scales, and thus with the microscopic properties of most solids and liquids. Neutrons, with wavelengths in the order of angstroms, are capable of probing molecular structures and motions and increasingly find applications in a wide array of scientific fields, including biochemistry, biology, biotechnology, cultural heritage materials, earth and environmental sciences, engineering, material sciences, mineralogy, molecular chemistry, solid state and soft matter physics.

The striking features of neutrons can be summarized as follows. Neutrons are neutral particles. They interact with other nuclei rather than with electronic clouds. They have (de Broglie) wavelengths in the range of interatomic distances. They have an intrinsic magnetic moment (a spin) that interacts with the unpaired electrons of magnetic atoms. Their mass is in the atomic mass range. They carry, thus, similar energies and momentum than those of condensed matter, and more specifically of gas hydrates.

As gas hydrates are mainly constituted of light elements (H, O, C, etc.), *in situ* neutron scattering appears as a technique particularly suited to their study. In the case of diffraction (i.e. structural properties), while the identification of these light atoms by X-ray diffraction requires the presence of heavy atoms and is therefore extremely complicated, neutron diffraction (NP) is highly sensitive to them due to the interaction of the neutrons with nuclei rather than with electron clouds. Moreover, most of the matter is

---

Chapter written by Arnaud DESMEDT, Laura BEDOURET, Jacques OLLIVIER and Claire PETUYA.

"transparent" to neutron beams. Such a feature provides advantages for studying gas hydrates when a heavy sample environment is required (e.g. high pressure, low temperature). For instance, X-ray powder diffraction studies are usually restricted to small sample volumes, as large sample volumes would be associated with a strong absorption and unwanted scattering from the pressure cell. *Neutron techniques* allow studies of bulk processes *in situ* in representative volumes, hence with high statistical precision and accuracy [STA 03, HEN 00, GEN 04, FAL 11]. Furthermore, although alteration of some types of ionic clathrate hydrates (or semiclathrates), such as the splitting of the tetra-alkylammonium cations into alkyl radicals [BED 91, BED 96], by X-ray irradiation has been reported, neutrons do not damage sample.

Finally, future developments in gas hydrate science will be based on the understanding, at a fundamental level, of the factors governing the specific properties of gas hydrates. In this respect, the investigation of gas hydrate dynamics is a prerequisite. At a fundamental level, host–guest interactions and coupling effects, as well as anharmonicity, play an important role. These phenomena take place over a broad timescale, typically ranging from femtoseconds to microseconds. Investigating the dynamics (intramolecular vibrations, Brownian dynamics, etc.) of gas hydrates thus requires various complementary techniques, such as NMR or Raman spectroscopy, and indeed inelastic and quasi-elastic neutron scattering (QENS), especially when it comes to encapsulating light elements such as hydrogen or methane in water-rich structures.

In this chapter, the recent contributions of neutron scattering techniques in gas hydrate research are reviewed. After an introduction to neutron scattering techniques and theory, an overview of the accessible information (structural and dynamical properties) by means of neutron scattering is provided. Then, selected examples are presented, which illustrate the invaluable information provided by neutron scattering. Some of these examples are directly related to existing or possible applications of gas hydrates.

1.2. Neutron scattering

Both nuclear and magnetic neutron interactions are weak: strong but at very short length scale for the nuclear interaction and at larger scale for the magnetic interaction. In that respect, the probed sample can be considered as transparent to the neutron beam. This highly non-destructive character combined with the large penetration depth, both allowed because of the weak scattering, is one of the main advantages of this probe.

Nuclear scattering deals with nuclear scale interaction and hence presents no wave vector dependent form factor attenuation allowing to offer high momentum transfers for diffraction or specific techniques such as deep inelastic neutron scattering (also known as neutron Compton scattering).

Neutron spectroscopic techniques range from the diffraction of large objects using small-angle scattering, usually made with long incident wavelengths (cold neutrons), to direct imaging through contrast variation (neutron tomography), usually made with short wavelengths (hot neutrons) and going through ordinary diffraction and inelastic scattering in the intermediate wavelength range.

In that respect, neutron scattering complements without necessarily overlapping the other available spectroscopic techniques such as nuclear magnetic resonance (NMR). If one naturally thinks about X-ray for structure determination, neutrons are very competitive for inelastic scattering and even essential for magnetic scattering both in the diffraction and inelastic modes.

The main drawback that contrasts with the numerous advantages comes from the intrinsic relative flux limitation of neutron sources, and thus, this type of spectroscopy can only be performed at dedicated large-scale facilities.

1.2.1. *A basic ideal scattering experiment*

In a generic experiment (Figure 1.1), a beam of monochromated neutrons with single energy (E_i) is directed on a sample. The scattered neutrons are collected along direction (angles θ and ϕ) and analyzed by energy difference with the incident energy by using a detector, covering a solid angle $\Delta\Omega$ of the sphere, which measures the analyzed neutron intensity. The measured intensity in the solid angle spanned by the detector and in a final energy interval ΔE_f in this simple *gedanken* experiment reads:

$$I = \Phi\eta \frac{d^2\sigma}{d\Omega\, dE_f} \Delta\Omega\Delta E_f \qquad [1.1]$$

where Φ stands for the incident flux at the incident energy and η is the efficiency of the detector. The quantity between the identified terms is the double differential scattering cross-section, a surface per unit of energy, which characterizes the interaction of the neutron with the sample or the surface that the sample opposes to the incident beam. Since the intensity has the dimension of count/s, the double differential scattering cross-section can be seen as the ratio of the scattered flux in the given detector per unit energy over the incident flux.

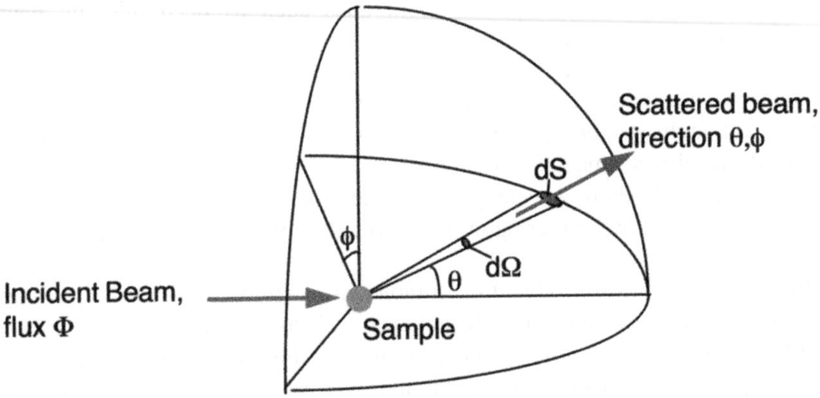

Figure 1.1. *Sketch of an ideal scattering experiment. An incident neutron beam of monochromatic energy E_i and wave vector $\mathbf{k}_i$ is scattered with energy E_f and wave vector $\mathbf{k}_f$. For a color version of this figure, see www.iste.co.uk/broseta/hydrates1.zip*

1.2.2. *Neutron scattering theory*

The mathematical development of the neutron scattering technique comes from the more general scattering theory. The interaction of the neutron with a single nucleus is first examined and then the generalization of the theory for an assembly of scatterers is developed. From scattering theory to its application to neutron scattering, the aim is to convince that the scattering of neutrons by the nuclei or by the spins of an ensemble of atoms provides information on the structure and motions of the atoms, i.e. information on the sample under investigation at the atomic level.

To study the scattering of a single neutron by one nucleus of the sample at the atomic level, one has to consider the incoming neutron as a plane wave, whose square modulus gives the probability of finding the neutron at a given position in space (this probability is a constant for a plane wave). Considering a point-like interaction between the neutron plane wave and the nucleus, the nucleus size in interaction being far smaller than the neutron wavelength and atomic distances, the scattered neutron wave is then described as an isotropic spherical wave whose intensity is proportional to $1/r$ and the strength of the interaction between the neutron and the nucleus of interest, called the scattering length ***b***.

The scattering length ***b*** is specific (and tabulated, see [SEA 92]) to each nucleus and does not vary with the atomic number in a correlated way. It can be positive, meaning repulsive interaction, negative (attractive) and can be complex and energy dependent, which means that the target can absorb the neutron (absorption proportional to the incident wavelength in the thermal neutron range).

Going from the scattering by a single nucleus to the scattering by a macroscopic assembly of nuclei as found in a condensed medium is a matter of properly summing all scattered waves under well-defined approximations, which are generally fulfilled in neutron scattering experiments.

The first obvious approximation is that the scattered waves are weak and thus leave the incident plane wave unperturbed over the coherence volume. This allows retaining only the first term of the Born series of the Lipmann–Schwinger equation. This simplification is known as the Born approximation.

For the sample (but not for the sample nuclei whose states are left unchanged), the neutron is a perturbation, and then the scattering can be treated within perturbation theory. At the quantum level, during the interaction of the neutron with a nucleus of the sample, the sample is changing from an initial state λ_i to a final state λ_f (which are left undefined so far but depends on the system under consideration) via the interaction potential $V(r)$. The point-like interaction potential $V_j(r)$ is known as the Fermi pseudo-potential and reads, with normalization factors:

$$V_j(\boldsymbol{r}) = \frac{2\pi\hbar^2}{m_n} b_j \delta(\boldsymbol{r} - \boldsymbol{r}_j) \qquad [1.2]$$

where m_n is the neutron mass (1.675×10^{-27} kg) for an atom with scattering length b_j at position $\boldsymbol{r}_j$. Conservation laws tell us that the change in energy $\hbar\omega = E_i - E_f$ and momentum $\boldsymbol{Q} = \boldsymbol{k}_i - \boldsymbol{k}_f$ of the probe (the neutron) should be reflected in a similar change in the quantum state of the target.

In the language of quantum mechanics, the summation over the scattered waves is equal to a sum over the final quantum states λ_f while averaging over the statistically weighted initial states of the system. The statistics of the initial state p_{λ_i} is taken as a Boltzmann distribution, from which other statistics can be deduced depending on the quantum nature of the target (bosons or fermions).

The calculation is rather cumbersome but robust within the framework of the approximations valid in ordinary neutron scattering. The theory is extensively described in dedicated books [SEA 92, LOV 84]. One ends up then with the definition of the double differential cross-section, the quantity directly measured in a scattering experiment:

$$\frac{d^2\sigma}{d\Omega dE_f} = \frac{k_f}{k_i} S(\boldsymbol{Q}, \omega) \qquad [1.3]$$

The scattering function $S(\boldsymbol{Q}, \omega)$ or dynamical structure factor reads:

$$S(\boldsymbol{Q}, \omega) = \frac{1}{2\pi\hbar} \sum_{jj'} b_j b_{j'} \int_{-\infty}^{\infty} dt\, e^{-i\omega t} \langle e^{-i\boldsymbol{Q}.\boldsymbol{r}_j(0)} e^{i\boldsymbol{Q}.\boldsymbol{r}_{j'}(t)} \rangle \qquad [1.4]$$

It is a function that depends solely on **Q** and ω for the neutron and that contains all the probed dynamics of the target in the double sum over the Fourier transform of the thermal average of the expectation value of the product of some Heisenberg operator $e^{i\boldsymbol{Q}.\boldsymbol{r}_j(t)}$ (which do not commute, except in the classical limit), i.e. the term between angular brackets.

1.2.3. *Correlation functions*

At this stage, one can be puzzled about the significance of the time Fourier transform of the operator expectation value where all summations and averages over the quantum states of the sample have been condensed in this compact but not very meaningful expression.

Fortunately, van Hove [HOV 54] derived the above expression in terms of intuitive number density pair correlation functions between the atom labeled j at position $\boldsymbol{r}_j(0)$ at initial time ($t = 0$) and another atom j' at position $\boldsymbol{r}_j'(t)$ at time t.

To simplify the discussion, the nucleus-dependent scattering length is usually reduced to a single value $b_j = b$ and removed from the summation in [1.4]. The number density operator reduces here to the sum over the atoms of the sample of the point-like probability of finding a scatterer at position $\boldsymbol{r}_j(t)$ at time t:

$$\rho(\boldsymbol{r}, t) = \sum_j \delta\left(\boldsymbol{r} - \boldsymbol{r}_j(t)\right) \qquad [1.5]$$

Defining the autocorrelation in space and time of the number density operator:

$$G(\boldsymbol{r}, t) = \int d^3\boldsymbol{r}' \langle \rho(\boldsymbol{r} - \boldsymbol{r}', 0) \rho(\boldsymbol{r}', t) \rangle \qquad [1.6]$$

and using the space Fourier transform of a delta function:

$$\frac{1}{(2\pi)^3} \int d^3\boldsymbol{r}\, e^{-i\boldsymbol{Q}.\boldsymbol{r}} \delta(\boldsymbol{r} - \boldsymbol{r}_j(t)) = e^{i\boldsymbol{Q}.\boldsymbol{r}_j(t)} \qquad [1.7]$$

leads to:

$$G(\boldsymbol{r}, t) = \frac{1}{(2\pi)^3} \int d^3\boldsymbol{Q}\, e^{-i\boldsymbol{Q}.\boldsymbol{r}} \sum_{jj'} \langle e^{-i\boldsymbol{Q}.\boldsymbol{r}_j(0)} e^{i\boldsymbol{Q}.\boldsymbol{r}_j'(t)} \rangle \qquad [1.8]$$

that is, once plugged into [1.4]:

$$S(Q,\omega) = \frac{b^2}{2\pi\hbar}\int d^3r \int dt\, e^{i(Q.r-\omega t)}\, G(r,t) \qquad [1.9]$$

The scattering function is thus the time and space Fourier transform of the number density or pair correlation function G$(\boldsymbol{r},t)$. In the classical approximation (usually when $kT \gg \hbar\omega$), $G(\boldsymbol{r},t)$ can be defined as the probability that, given a particle at the origin of time t = 0, any particle is within the elementary volume dr around the position $\boldsymbol{r}$ at time t.

$G(\boldsymbol{r},t)$ is a physical quantity that can be derived from analytical models or from molecular dynamics (MD) calculations. With such models, a direct comparison with the results of a scattering experiment, i.e. $S(\boldsymbol{Q},\omega)$, is thus possible. It has to be noted, however, that the sample can contain several atoms species and thus different scattering lengths b_j that render the interpretation more difficult even if the trend is the same. In particular, reversing the Fourier transform to deduce the correlation functions from the scattering function is not straightforward because information is lost in the superposition of contributions of different scattering lengths.

The partial space Fourier transform, the term under the sum in [1.4] or [1.8], is also called the intermediate scattering function and one can go back and forth from $S(\boldsymbol{Q},\omega)$ to $G(\boldsymbol{r}.t)$ via the intermediate step $I(\boldsymbol{Q},t)$:

$$I(\boldsymbol{Q}.t) = \sum_{jj'}\langle e^{-i\boldsymbol{Q}.r_j(0)} e^{i\boldsymbol{Q}.r'_j(t)}\rangle = \frac{1}{(2\pi)^3}\int d^3\boldsymbol{r}\, e^{i\boldsymbol{Q}.\boldsymbol{r}} G(\boldsymbol{r},t) \qquad [1.10]$$

As a final remark of this section, it is important to note that the expression [1.9] summarizes the direct relationship existing between classical or *ab initio* MD simulations and neutron scattering experiments: the atomic trajectories, computed by means of simulations, can be "simply Fourier transformed" in time and in space, to calculate a simulation-derived scattering law, directly comparable with the experimental one, as it will be illustrated in this chapter.

1.2.4. *Coherent and incoherent scattering*

When coming back to the scattering length one has to note an additional difficulty, which can be in fact taken as an advantage. The scattering length b_j of the nucleus are randomly distributed on the chemical species. This random distribution is due to the isotopic nature of the nuclei. Since it does not affect the chemical properties, the different nuclei isotopes in the sample, each with a different scattering length, are randomly distributed. A second "randomization" comes from the spin interactions in the system neutron plus nucleus during the interaction. The neutron spin from an unpolarized neutron beam couples randomly with the nucleus spin during the interaction and since the scattering length depends on the way the spins couple with each other, it leads to another source of randomization of the effective scattering length in the sample under study.

There is, however, a convenient way of dealing with this problem at an ensemble average level without taking care of the microscopic (atomic) level of the problem.

Starting from equation [1.4], one can rewrite the expression:

$$S(\boldsymbol{Q},\omega) = \sum_{jj'} b_j b_{j'}^* S_{jj'}(\boldsymbol{Q},\omega) \qquad [1.11]$$

to emphasize the scattering length visualization. Keeping the monoatomic sample, the indices carry on the atomic positions but not on the different scattering lengths due to isotopes or spin-interactions. Because the difference in energy or momentum of the scattered neutrons does not change when interacting with different isotopes or neutron–nucleus spin interactions[1], one can take the average over the scattering lengths:

$$S(\boldsymbol{Q},\omega) = \sum_{jj'} \overline{b_j b_{j'}^*}\, S_{jj'}(\boldsymbol{Q},\omega) \qquad [1.12]$$

and because the nuclear spin and the isotopes are not correlated between sites j:

$$\overline{b_j b_{j'}^*} = \bar{b}_j\, \overline{b_{j'}^*} \qquad if\ j \neq j' \qquad [1.13]$$

$$\overline{b_j b_{j'}^*} = \left|\overline{b_j^2}\right| = \overline{b_j^2} \quad if\ j = j' \qquad [1.14]$$

or, in a compact notation with transfer of the $j' = j$ term on the left-hand term on the right-hand side of the equality:

$$\overline{b_j b_{j'}^*} = \bar{b}_j\, \overline{b_{j'}^*} + \left(\overline{b^2} - \bar{b}_j^{\,2}\right)\delta_{jj'} \qquad [1.15]$$

which, plugged into equation [1.12], gives:

$$S(\boldsymbol{Q},\omega) = \sum_{jj'} \bar{b}_j\, \overline{b_{j'}^*}\, S_{jj'}(\boldsymbol{Q},\omega) + \sum_j \left(\overline{b_j^{\,2}} - \bar{b}_j^{\,2}\right) S_{jj}(\boldsymbol{Q},\omega) \qquad [1.16]$$

The first term represents the sum of all possible pairs of atoms (j, j'), each term representing a two-body correlation between the positions of atom j at time 0 and atom j' at time t. It contains interference terms in the scattering and gives rise to *coherent scattering*. The second term represents the sum of all atoms j at one time, each term representing a correlation between the position of atom j at time 0 and that of the same atom at time t, that is, *incoherent scattering*, sensitive to single-particle

1 There are noticeable but scarce counterexamples, the para-/ortho-hydrogen (H_2) is the most known.

spatiotemporal correlations. Incoherent scattering arises thus from the random distribution of scattering lengths about the mean, as schematized in Figure 1.2. As shown by equation [1.16], both scattering occur at the same time during an experiment, whereas they have substantially different origin. For nuclei carrying coherent and incoherent scattering lengths or cross-sections, the Bragg peaks from coherent scattering are superimposed to an unstructured incoherent background reflecting the summation of the two contributions. In inelastic scattering, if the summation still holds, it is not always simple to discriminate between both contributions, at least in simple experiments without polarization analysis.

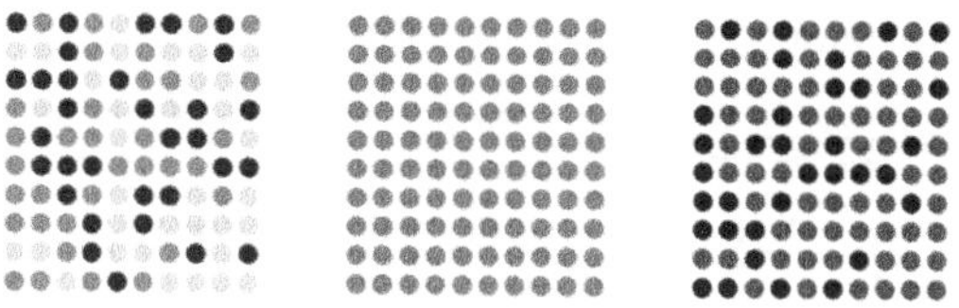

Figure 1.2. *Sketch of the separation coherent/incoherent. A single atomic sample with three isotopes (problem reduced to the static part of the isotopes here) is split in its average scattering length $\overline{b}$ on all nuclei plus the fluctuations around the mean value for each nucleus ($\overline{b_j^2} - \overline{b_j}^2$) . For a color version of this figure, see www.iste.co.uk/broseta/hydrates1.zip*

Fortunately, there are some degrees of freedom with the scattering lengths themselves. Some atomic species are dominated by a single isotope and, moreover, some have no nuclear spin. In the case, for example, with ^{12}C or ^{16}O, $b_j = b$ for all j, $\overline{b^2} = \overline{b}^2$ and thus the incoherent scattering term vanishes. They are then pure coherent scatterers. On the other hand, atoms as simple as the hydrogen have weak coherent scattering length but are dominated by the spin incoherence and thus the scattering is dominated by the incoherent scattering and the coherent scattering is vanishingly negligible. Substituting the ^{1}H isotope with the ^{2}H (deuterium) isotope is largely used (although it might become expensive in certain cases) to inverse completely the problem: in this case, the coherent scattering largely dominates. Depending on the information one wants to get out of the experiment, ensemble average of individual properties or collective atoms interdependent positions and dynamics, one can switch between these two sides of the same coin. In intermediate cases, where the system under study is neither a pure coherent nor an incoherent scatterer, it is still possible to make an isotope substitution to label the contribution of specific atoms in the scattered intensity either in the coherent or incoherent spectra. In this case, the shift in energy due to the isotopic substitution of the excitations that are proportional to the square root of the masses can also be directly measured at the same time on light atoms.

To better separate the two contributions, the expression in equation [1.16] can be written as:

$$S(Q,\omega) = S_{coh}(\boldsymbol{Q},\omega) + S_{inc}(\boldsymbol{Q},\omega) \quad [1.17]$$

and depending on the dominant contribution, one can get rid of one of the terms on the right-hand side.

Within the crude simplification of a monoatomic sample, $\bar{b}_j^{\,2} \equiv \bar{b}^2$, $\overline{b_j^2} = \overline{b^2}$, the indices over the positions drop off because they are all the same, the scattering function can be expressed in terms of coherent and incoherent scattering *cross-sections*: $\sigma_{coh} = 4\pi\bar{b}^2$ and $\sigma_{inc} = 4\pi(\overline{b^2} - \bar{b}^2)$:

$$S_{coh}(Q,\omega) = \frac{\sigma_{coh}}{4\pi}\frac{1}{2\pi\hbar}\sum_{jj'}\int dt\, e^{-i\omega t}\langle e^{-iQ.r_j(0)}e^{iQ.r_{j'}(t)}\rangle \quad [1.18]$$

$$S_{inc}(Q,\omega) = \frac{\sigma_{inc}}{4\pi}\frac{1}{2\pi\hbar}\sum_{j}\int dt\, e^{-i\omega t}\langle e^{-iQ.r_j(0)}e^{iQ.r_j(t)}\rangle \quad [1.19]$$

from which the coherent and incoherent intermediate scattering functions (equation [1.10]) and correlation functions (equation [1.8]) can be deduced. The splitting in equations [1.18] and [1.19] emphasizes that in the former we have a sum over the correlation functions of all pairs of atoms (*j*,*j'*) at different times, whereas in the latter we have a summation over all same atoms *j* at different times.

	c (%)	b_c (fm)	b_i (fm)	σ_c	σ_i	σ_s	σ_a
H		–3.74		1.757	80.26	82.02	0.33
^{1}H	99.9885	–3.74	25.2	1.758	79.9	81.67	0.33
^{2}H	0.0115	6.674	4.03	5.6	2.04	7.64	0.0
C		6.55	0.0	5.55	0	5.55	0.0
^{12}C	98.93	6.65	0.0	5.56	0	5.56	0.0
^{13}C	1.07	6.19	–0.25	4.81	0.034	4.81	0.0
Na	100	3.63	3.69	1.66	1.62	3.22	0.53
V		–0.44		0.02	5.08	5.1	5.08
^{50}V	0.25	7.6		7.3	0.5	7.8	60.0
^{51}V	99.75	–0.4	6.35	0.02	5.07	5.09	4.9

Table 1.1. *Examples of scattering lengths (b in fm) and cross-section (σ in barns) for some selected chemical species [SEA 92]*

Natural hydrogen (H) is made essentially of ^{1}H and thus is a pure incoherent scatterer. The incoherent cross-section is so large that, at any time in an organic sample, containing thus a large proportion of hydrogen, the hydrogen dominates the scattering unless an isotope substitution with deuterium is made. Carbon (C) is a pure coherent scatterer: no nuclear spin and an almost single isotope. In contrast to this, the vanadium (V) is essentially an (spin) incoherent scatterer. Its spatial flatness response is often used for instrument calibration. Many others atoms are less selective in their affinity, e.g. sodium (Na) has one isotope and almost equal incoherent and coherent scattering cross-sections. A large benefit could be obtained from the high sensitivity of hydrogen in clathrate hydrate: contrast between the scattering response of the cages and the guest molecules could be reached by selectively deuterating the cages and/or the guest molecules.

1.2.5. A *simple example of scattering*

A simple example of incoherent scattering and self-correlation function is given by following a particle experiencing Brownian motion in a fluid. The canonical example of this situation is the argon atom in liquid argon [SKO 72] that exhibits both coherent and incoherent scattering. The particle has to obey the Fick law of diffusion due to thermal agitation and successive strokes on the particle in the medium.

In the classical limit, the self-correlation function $G_s(\boldsymbol{r},t)$ is a solution of the Fick equation (second Fick's law or standard diffusion equation) with the self-diffusion coefficient D_s:

$$G_s(\boldsymbol{r},t) = \left(\frac{1}{4\pi D_s t}\right)^{3/2} e^{-\frac{r}{4D_s t}} \qquad [1.20]$$

from which can be deduced the self-intermediate scattering function by space Fourier transform:

$$I_s(Q,t) = e^{-D_s Q^2 t} \qquad [1.21]$$

and thus the incoherent scattering function, the quantity directly measured in a neutron scattering experiment, by time Fourier transform is:

$$S_{inc}(Q,\omega) = \frac{1}{\pi}\frac{D_s Q^2}{\omega^2 + (D_s Q^2)^2} \qquad [1.22]$$

The measured incoherent spectrum of an ensemble of individual particles obeying the Fick law is a Lorentzian depending on $Q = |\boldsymbol{Q}|$ and ω.

Some properties of the scattering functions arise from this simple example. First, $\int S_{inc}(Q,\omega)d\omega = I_s(Q,t=0) = 1$. The incoherent scattering function is normalized whatever Q, and thus the energy-integrated scattering appears as a flat-in-Q, unstructured, intensity[2] as expected from the summation over unrelated

---

2 The Debye–Waller term that comes from atomic vibrations in front of the scattering functions is neglected here.

contributions. In terms of correlation function, the probability of finding the particle at initial position r at time $t = 0$ should be 1: $G_s(r, t = 0) = \delta(r)$ and at infinite time, the probability of finding the particle at the same place r (in the elementary volume dr to be more precise) should be the inverse of the explored volume V; in the liquid case, the entire volume of the sample is given as: $G_s(r, t \to \infty) = 1/V$.

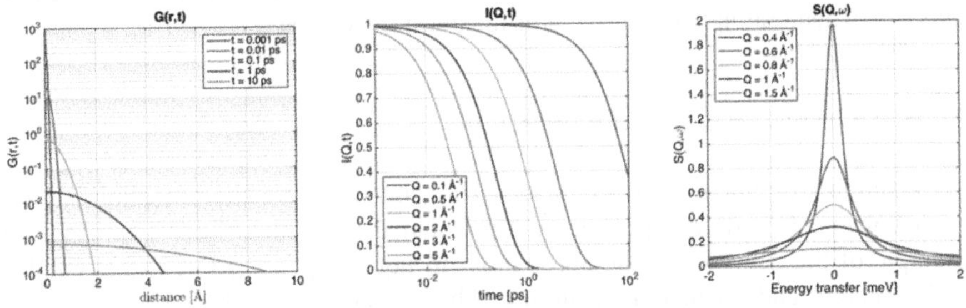

Figure 1.3. *Self-correlation function, intermediate scattering function and dynamical structure factor for the simple case of a self-diffusion of a particle with diffusion coefficient D = 10^{-5} cm²/s. For a color version of this figure, see www.iste.co.uk/broseta/hydrates1.zip*

In the opposite case, the coherent scattering should show the liquid-like structure factor when integrated over the energy: $\int S_{coh}(Q,\omega)d\omega = S(Q)$ and $G_d(r, t \to \infty) = \rho$, the limit at infinite time of the pair correlation function (d stands for distinct) is the average particle density ρ. The coherent scattering functions are more difficult to obtain, even for a simple liquid: a derivation can be found in [HAN 06]. Neglecting the viscosity, thus transverse terms, one ends up with the intermediate scattering function from the number density pair correlation:

$$I_d\,(Q,t) = \langle \rho_Q(t)\rho_{-Q}(0)\rangle = \left(\frac{\gamma-1}{\gamma}\right)e^{-D_T Q^2 t} + \frac{1}{\gamma}e^{-\Gamma Q^2 t}\cos(v_s Q t) \qquad [1.23]$$

where $\gamma = C_P/C_V$ is the specific heat ratio. The expression [1.23] shows an exponential decay, as in the self-diffusion, but with a thermal transport coefficient D_T different from the self-diffusion coefficient, and a damped oscillatory function, whose frequency depends on the speed of sound v_s in the liquid. The time Fourier transform is analytically simple:

$$S_{coh}(Q,\omega) = \frac{S(Q)}{2}\left[\left(\frac{\gamma-1}{\gamma}\right)\frac{2D_T Q^2}{\omega^2+(D_T Q^2)^2} + \frac{1}{\gamma}\left(\frac{\Gamma Q^2}{(\omega+v_s Q)^2+(\Gamma Q^2)^2} + \frac{\Gamma Q^2}{(\omega-v_s Q)^2+(\Gamma Q^2)^2}\right)\right] \qquad [1.24]$$

In addition to the Rayleigh line centered at $\omega = 0$, a collective excitation, the (longitudinal) sound wave, a density wave, occurs in the liquid. Equation [1.24] is an approximation at low Q, in particular, the central Lorentzian width does not show up the so-called "De Gennes narrowing", an increase in the (pair) correlation times (narrowing of the Lorentzian width) at the maximum of S(Q) corresponding to first neighbor distances due to favorable local arrangement. This can be overcome in a first approximation by replacing the Lorentzian width by $D_s Q^2 / S(Q)$.

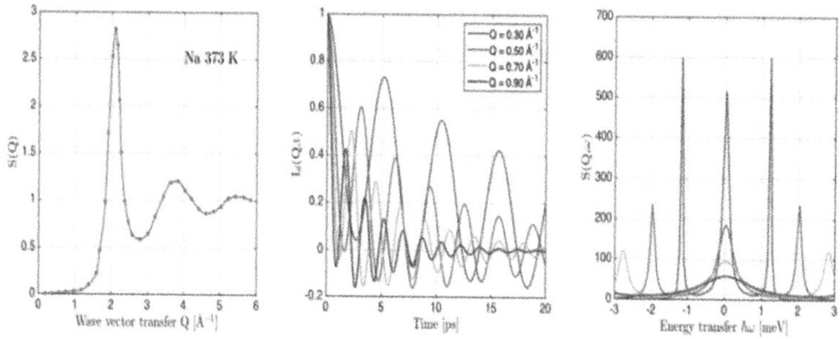

Figure 1.4. *Coherent structure factor of liquid sodium (left). Sketches of intermediate and coherent scattering functions for the coherent case. For a color version of this figure, see www.iste.co.uk/broseta/hydrates1.zip*

The example of monoatomic liquids illustrates the power and weaknesses of inelastic neutron scattering. Self-correlations and pair-correlations give access to different properties of the liquid. Thus, through isotope substitution (when possible) or polarization analysis, the neutron beam behaves as a dual probe.

When dealing with diffusion and relaxation in incoherent scattering, many meaningful microscopic models exist or are easy to imagine; the passage to the scattering law that can be compared to the measurements being rather straightforward [BEE 86]. This holds for all systems where incoherent cross-sections are dominant in the scattering. Even in the monoatomic liquid case, deriving the coherent scattering from the particle density correlations is a difficult step and cannot be usually derived from scratch each time a new compound is being studied. Fortunately, neutron scattering has allowed for many correlation functions or scattering functions of practical interest to be derived. Within approximations or "hacking" of known models, it is possible to adapt each new case to the known theoretical material. In the case of more complex materials, e.g. molecular materials with several chemical species where the superposition of scattering from heterogeneous scattering lengths apply, understandable analytical models are not always possible. In particular, it is not possible to Fourier transform the scattering function $S(Q,\omega)$ to get the correlation functions $G(r,t)$ as it is feasible, although rarely used, in a mononuclear system.

A frequent approach for a complex system involves MD calculations. The quantities of interest can be inferred from the molecular dynamic trajectories and velocities and, moreover, a specific labeling of the atoms in the system under study *in silico* allows to extract more accurately the role of the constituents in a molecular system. MD is another way of obtaining the emergent physics at a macroscopic scale from simpler modeling, although not that simple with *ab initio* calculations, at a microscopic level. From the recorded trajectories, the mean-square displacement can be directly computed from the ensemble average of the distances of a tagged particle:

$$\langle u^2(t)\rangle = \frac{1}{N}\langle \Sigma_j |r_j(t) - r_j(0)|^2 \rangle \qquad [1.25]$$

In the long time limit, the macroscopic self-diffusion coefficient D_S, the same as in the example above, can be computed as:

$$D_s = \lim_{t\to\infty} \frac{\langle u^2(t)\rangle}{6t} \qquad [1.26]$$

It is also possible to compute various kinds of correlation functions. In particular, the velocity autocorrelation function defined as the projection of the velocity at time t onto the initial velocity averaged over the initial conditions:

$$C(t) = \frac{1}{3}\langle \boldsymbol{v}(t).\boldsymbol{v}(0)\rangle \qquad [1.27]$$

has important properties. At $t = 0$, $C(t) = 1/3\,\langle v^2(0)\rangle = k_B T/m$ because of the equipartition of the energy, giving access to the temperature of the system under simulation. Another important equation is the relation between the velocity autocorrelation function and the self-diffusion coefficient:

$$D_s = \int_0^\infty C(t)dt \qquad [1.28]$$

This equation constitutes an example of a Green–Kubo formula relating a macroscopic property in terms of the time integral over a microscopic time–correlation function. From the trajectories, the van Hove correlations can also be computed, and thus the intermediate and scattering functions, which can be directly confronted to neutron measurements.

1.3. Probing structural and dynamical properties of gas hydrates

The host substructure of a clathrate hydrate is constituted with a slightly distorted tetrahedral distribution of water molecules. An ideal clathrate hydrate includes a distribution of the H-bonds within the water substructure, respecting the ice rule, i.e. each water molecule accepts and donates two hydrogen bonds. The dynamic and static proton disorder in the H-bond subsystem is typical of most polymorphous crystalline forms of ice. It appears because of the delocalization of the protons between the neighboring oxygen atoms and because of reorientations of water molecules. Moreover, guest molecules encapsulated within the water cages exhibit a high degree of disorder (e.g. inter- or intracage diffusion, reorientations, rattling), triggered by the host–guest interactions. Clathrate hydrates can then be considered as glassy crystals with respect to the large variety of molecular motions. Neutron scattering techniques constitute the appropriate tools for proposing comprehensive models revealing this dynamical and/or statistical disorder met in clathrate hydrates [DES 12]. This section provides an overview of the dynamical and structural properties met in clathrate hydrates. Focused examples are provided in the following sections.

1.3.1. *Structures*

A primary use of neutron scattering probably resides in the ability of this technique to locate light elements like hydrogen, or more specifically deuterium (see the coherent cross-section of deuterium in Table 1.1), compared to X-ray scattering. Neutron powder diffraction performed on clathrate hydrates possessing deuterated water cages give access to the proton distribution around the oxygen atoms constituting the cages. Using deuterated guest molecules (such as methane, i.e. rich in hydrogen), neutron powder diffraction provides information about the distribution of guest molecules within the cage in terms of orientational distribution or cage occupancies. The readers may find numerous structures of clathrate hydrate resolved by means of neutron powder diffraction in the literature (e.g. [SLO 08]).

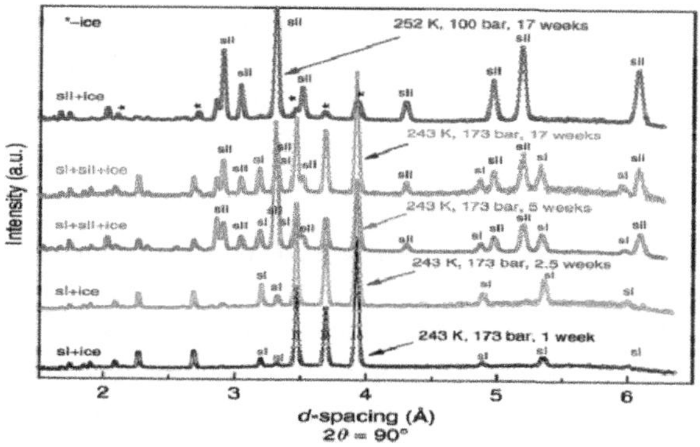

Figure 1.5. *Neutron diffraction data obtained on carbon monoxide clathrate hydrate at 260 K and at indicated pressures for various formation conditions (temperature, pressure and duration of P-T exposure) indicated on the figure (according to [ZHU 14]). For a color version of this figure, see www.iste.co.uk/broseta/hydrates1.zip*

Among the numerous studies dealing with structure refinement by means of NP, let us consider a few examples. A full structural refinement of NP data leads to the location of the hydrogen atom positions with high precision [HEN 00], which is especially suited for determination of the position and quantity of hydrogen (deuterium) (eg. [MUL 08]). In addition, in the case of single crystals, this technique allows the refinement of multiple cage occupancy and orientation ordering in cages (e.g. pure H_2 hydrates [LOK 04], binary H_2/THF hydrates [HES 06]).

Some remarkable results for clathrate science have been obtained by means of neutron powder diffraction (NPD) in recent years. One example is the very recent discovery of a new ice phase, referred to as ice XVI [KUH 14]. This phase is an s-II clathrate structure with empty cages, obtained after 5 days of vacuum pumping neon clathrate hydrates. Another example is the metastability of the type I structure formed with carbon monoxide guest molecules: the s-I structure is kinetically favored, but transforms into the s-II structure – thermodynamically more stable – after a few weeks [ZHU 14].

Due to the high flux available on neutron diffractometers together with its simultaneous readout over a large 2θ range and to the excellent penetration power of neutrons (allowing the use of high-pressure equipment and cryogenic devices with relative ease), *in situ* NPD represents a useful tool for investigating structural changes when external (e.g. pressure) conditions are variable [MAO 02]. Neutron powder diffraction is useful for clarifying the difference in the thermal vibrations of the same kind of atoms at crystallographically distinct sites [IKE 99]. Moreover, NP experiments have enabled some recent successes both in studying the clathrate hydrate kinetics of formation/decomposition [STA 03, MUR 09, KUH 06a], guest replacement reactions (especially when accompanied by a structural transition) and have also provided useful information on the time dependency of hydrate transformations [HAL 01]. Thus, neutron scattering is widely used for *in situ* experiments and this specificity will be illustrated in the following sections.

1.3.2. *Relaxation of guest molecules and water molecules*

Because of the exceptionally large incoherent neutron scattering cross-section, nearly two orders of magnitude greater than that of any other nucleus (see Table 1.1), incoherent QENS techniques are particularly sensitive to proton and have also emerged as invaluable tools for the study of clathrate hydrates. Indeed, as QENS gives access, in both space and time, to the individual displacements of nuclei, a detailed description of the spatial and time characteristics of diffusive mechanisms – especially in the case of hydrogen – can be obtained at a microscopic level.

The stability of the *water molecules* framework is ensured by the directionality of the H-bond, involving the so-called ice rule. The signature of the relative rigidity of the cages resides in the reorientational process of water molecules, exhibiting quite high activation energy in the order of 30–50 $kJ{\cdot}mol^{-1}$ [SLO 08, DAV 73]. ^2H NMR investigations reveal that the relaxation of water molecules in clathrate hydrates occurs on a microsecond timescale [BAC 01, KIR 03, SLO 08]. This characteristic time is relatively long with respect to the observation timescale (typically ranging from picoseconds to nanoseconds) probed by means of QENS experiments. Thus, the incoherent QENS spectra include an elastic-like scattering due to the water molecules, limiting the interest of this technique for probing the water Brownian motions. The process of orientational ordering and the occurrence of H-bond defects (Bjerrum defects, ionic defects, etc.) within the host network may facilitate the water molecule relaxation. These defects might be created through the existence of short lifetime (of the order of the picoseconds) H-bonds between water molecules and guest molecules [ALA 09, BUC 09, PEF 10]. Another way to generate H-bonds defects in the network of the water substructure is to confine acids within the cages. The acidic guest molecules are then enclosed within correspondingly cationic water cages in which the proton excess is delocalized [MOO 87, DES 13], leading to superprotonic conductors [SHI 10]. In such a case, QENS is the technique to elucidate the mechanism of the protonic conduction and to probe the induced modification of water relaxation. Playing with the energy resolution used in experiments (i.e. with the observation time to probe the system), QENS has enabled to disentangle the dynamics occurring on a large timescale and has, for instance, provided direct experimental evidence about the fundamentals of proton diffusive motions in the water cage framework [DES 04, DES 12, DES 13, BED 14]. This aspect will be detailed in section 1.7 dedicated to ionic clathrate hydrates.

The dynamics of the *guest molecules* encapsulated within water cages is triggered by the cage potential energy surface, i.e. the host–guest interactions. It follows that accessing the cage energy landscape could be done through the analysis of the guest molecules dynamics, i.e. through the analysis of guest Brownian dynamics. The combination of incoherent QENS experiments and MD simulations yields a comprehensive model of the dynamics of guest molecules encapsulated in the cages of clathrate hydrate (e.g. [PEF 10, DES 11]). In a QENS experiment, clathrate hydrates are prepared with a deuterated host lattice so that only the guest molecules contribute to the incoherent scattering. The MD simulations box generally consists of a supercell (composed of several elementary unit cells) and MD simulations are run over several nanoseconds. The MD atomic trajectories can be used to calculate the incoherent neutron scattering laws within the same experimental conditions (i.e. by folding the MD-derived scattering laws with the experimental energy resolution function). Comparison of the MD-derived scattering laws with the experimental ones may lead to the "experimental" validation of the

MD simulations as soon as a correct agreement is observed (see example of the CH_3I-$17D_2O$ s-II clathrate hydrate in Figure 1.6). The MD trajectories can then be safely analyzed to disentangle the various dynamical processes met in the clathrate hydrates [DES 11]. In a given type of cage, guest reorientations are generally modeled with a quasi-isotropic model: the guest molecule can adopt all orientations within the cage and preferred orientations may be adopted in the case of polar molecules [DES 11] or in the case of host–guest H-bonding [BUC 09, PEF 10]. The preferred orientations adopted by the guest molecules are different from cage to cage, in agreement with the disorder existing in the distribution of the water hydrogen bonding in the host network. This various local environment leads to a distribution of anisotropic energy landscapes (showed by the quasi-isotropic reorientations) experienced by the guest molecules. Despite these various local environments, no distribution of characteristic times is observed by means of MD-QENS investigation [DES 11].

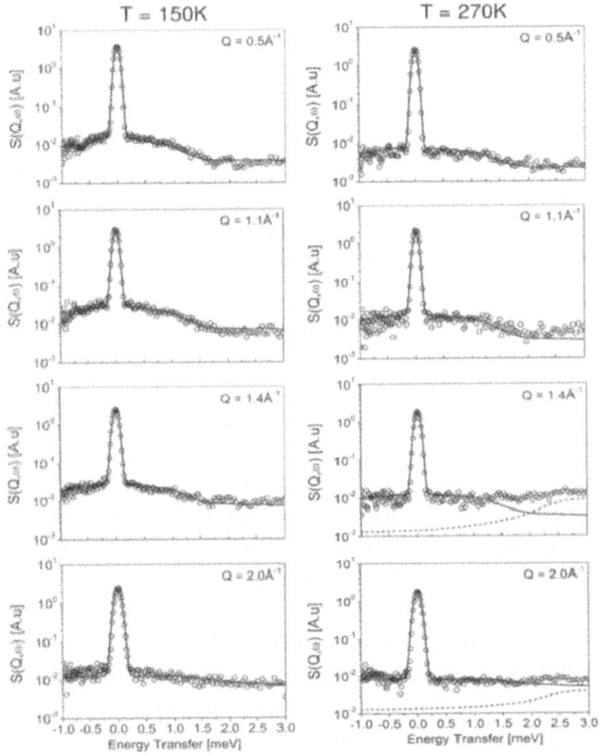

Figure 1.6. *Experimental QENS scattering laws (circles) and MD-derived scattering laws (continuous lines) of the methyl-iodide clathrate hydrate (s-II structure with only the large cage filled with guest molecules). The dashed lines represent inelastic contribution of the host cages (from [DES 11])*

Finally, the main difference in terms of guest molecule relaxation times is observed for guest molecules occupying different types of cages. In other words, the relaxation characteristic time depends on the type of cages occupied by the guest molecule. This confinement differentiation is experimentally observed by means of QENS in the case of the bromomethane clathrate hydrates (s-I structure with both types of cage filled) [PEF 16]. As shown in Figure 1.7, the bromomethane molecule reorientations occur on a longer timescale in small cage than in large cage. This difference is a direct signature of the host–guest interactions, which are more attractive in the small cage than in the large cage [SCH 03]. Such a signature is similar to that observed in the vibrational spectra of methane clathrate hydrate [SUM 97]: the C–H stretching mode of methane confined in large cage is observed at higher frequencies than the one of methane confined in small cage.

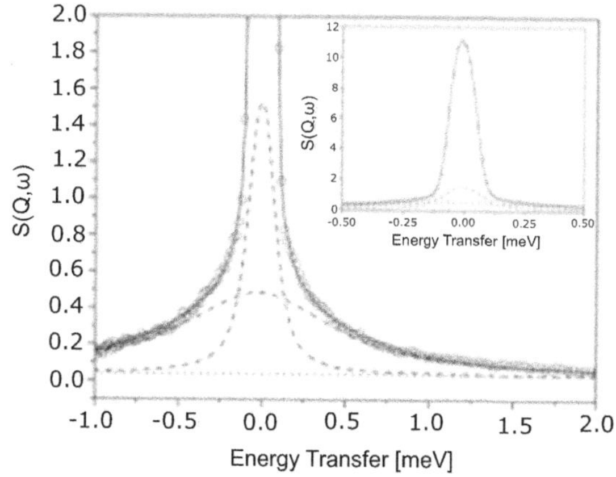

Figure 1.7. *Experimental (points) and fitted (continuous lines) QENS spectra of CH_3Br–5.75D_2O at 50 K for various values of Q. The continuous lines represent the fitted scattering laws and the dashed lines represent the two QENS components. The broader component is associated with the guest molecules located within large cage. Instrument: NEAT at HMI, Q = 2.1 $Å^{-1}$, λ_0 = 5.1 Å, δE = 100 μeV. Adapted from [PEF 16]*

1.3.3. *Excitations and vibrational density of states*

In addition to neutron scattering, Raman scattering or infrared absorption leads to detailed information about the intramolecular vibration modes (i.e. for energy excitations typically greater than 10 meV) (see Chapter 2). The specificity of inelastic incoherent neutron scattering (IINS) with respect to optical spectroscopy probably resides in the investigation of the low frequencies modes met in clathrates hydrates (wavelength and energy of neutron beams are in the range of these excitations). The crucial role played by the knowledge of low frequencies modes is certainly typified by the anomalous behavior of heat transport properties in clathrate hydrates [STO 79, ROS 81, HAN 87, TSE 88, TSE 01, TSE 05]. Moreover, IINS is a highly selective probe for investigating the properties of hydrogen atoms from molecules confined inside the clathrate nanometric cavities. In the case of a relevant concentration of protons, the bands in the spectra due to the host vibrational modes not involving hydrogen are typically quite weak. Another unique feature of IINS is the sensitivity of neutrons to rotational transitions, which are not probed in optical, infrared and Raman spectroscopies. Those transitions involving the conversion from ortho-hydrogen to para-hydrogen and vice versa have been extensively studied, in particular in the context of investigations about hydrogen adsorption in hydrogen storage materials (see section 1.4.3).

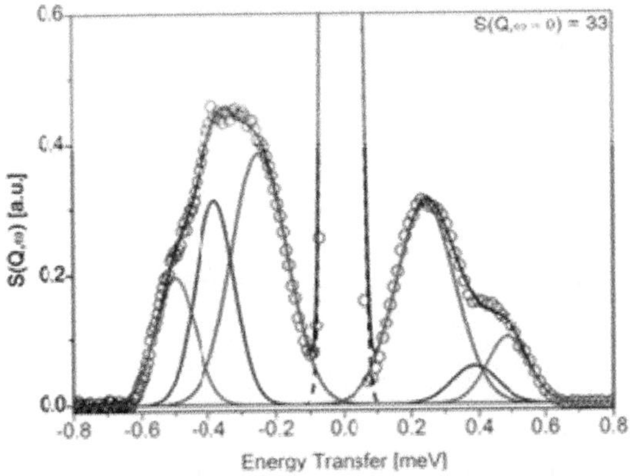

Figure 1.8. *Inelastic neutron scattering spectrum of the iodomethane clathrate hydrate $CH_3I \cdot 17D_2O$ at 1.8 K (NEAT time-of-flight spectrometer @HZB-Berlin with λ_0 = 7.0 Å, ΔE = 50 µeV and <Q> = 1.6 $Å^{-1}$). Points are experimental data and continuous lines are resulting from the fit of three tunneling transitions. Adapted from [PRA 04b]. For a color version of this figure, see www.iste.co.uk/broseta/ hydrates1.zip*

The dynamical properties of s-I methane hydrate has been the subject of numerous works, including MD simulations [GUT 01, ENG 03, SUS 08], and inelastic neutron scattering [TSE 97], inelastic X-ray scattering [BAU 03] and NMR [RIP 88, RIP 04] experiments. Almost free rotation and very anharmonic low-frequency translation (the so-called rattling mode) are observed. The quantitative analysis of the MD at the lowest temperature shows a single-particle quantum rotation in weak rotational potentials differing slightly from cage to cage. This static distribution of the local environment owes to different arrangements of the H bonds in the cage surface [GUT 01]. Details about these various H bonds arrangements have been obtained by analyzing the dynamics of simple model systems of polar guest molecules: high-energy resolution QENS experiments have thus been performed to analyze the rotational tunneling of guest methyl group in the series of methyl halide clathrate hydrates at low temperatures. Investigations have also been performed into the s-I clathrate hydrates (CH_3X–5.75 H_2O with X = F, Cl and Br) formed with fluoromethane [PRA 05], chloromethane [PRA 08] and bromomethane [PRA 07]. Despite the 1-D character of the methyl-halide rotor, complex tunneling spectra are observed due to the distribution of potential energy barriers as observed in the case of the 3-D methane rotor. The iodomethane clathrate hydrate crystallizes in the type II structure with only large cages filled. This "simplification" provides a unique opportunity to explore the nature of the inhomogeneity of potential energy barriers [PRA 04a, PRA 04b]. The neutron scattering spectrum of the iodomethane clathrate hydrate, $CH_3I•17D_2O$ (Figure 1.8) is constituted of three broad inelastic peaks, corresponding to three tunneling transitions. While the broadening can be explained by different distribution of the H-bonds between water molecules (inducing slightly different potentials from cage to cage), the existence of three tunneling transitions is surprising since all iodomethane molecules are cristallographically equivalent (they are all located in $5^{12}6^4$ cages). These peaks are in fact the experimental evidence of three types of adsorption sites existing at the cage surface [PRA 04b].

Large single crystals are generally required for neutron scattering experiments, but obtaining them is a tricky task. Consequently, vibrational densities of states (denoted VDOS) are often recorded on powdered clathrate hydrates to enable the study of their excitations. Such a projection of the lattice modes on the energy transfer axis leads to the loss of the dispersion information. Various options are then offered to interpret the data: use of isotopic substitutions (mainly, deuteration of selected chemical species), use of chemical substitutions (changing the guest molecules) or use of computing science (numerical modeling of the spectra). Examples of inelastic neutron scattering VDOS are shown in Figure 1.9. The translational modes at 7 and 10.5 meV are due to the host substructure [TSE 01] The specific excitations (rattling modes) arising from the encapsulated guest are observed below 5 meV. The guest signature is not observed in the case of the argon clathrate hydrate because of the negligible scattering cross-section of argon, 0.7 barn, to be compared with that of oxygen (4.2 barn) or nitrogen (11.5 barn). This feature allows the extraction of the O_2 or N_2 contribution by differentiating their DOS with that of argon clathrate hydrate (Figure 1.9). Such spectra are composed of two bands attributed to the encapsulation in the two types of cage existing in the s-II clathrate hydrate. MD simulations [SCH 03] confirm this attribution. Additional information is provided by the MD analysis: the guest molecules encapsulated in the large cages exhibit a rattling mode with frequencies lower than the one in the small cages. As in the case of intramolecular vibrations or Brownian motions, these low-frequency signatures underline the importance of host–guest interactions in clathrate hydrates.

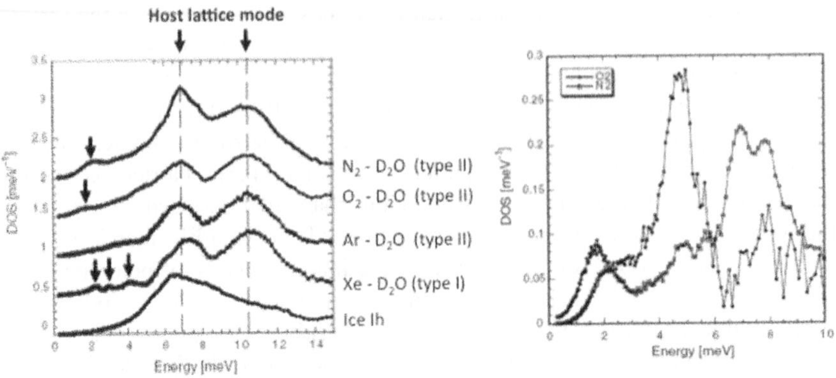

Figure 1.9. *Left: VDOS for various clathrates hydrates (T = 120 K except for O_2, T = 100 K) and hexagonal ice (T = 235 K). Right: Guest molecules contributions to the VDOS for N_2 (T = 120 K) and O_2 (T = 100 K) clathrates hydrates. Adapted from [SCH 03]*

1.4. Selected examples

1.4.1. *Inhibition and formation mechanisms*

The blocking of pipelines by gas hydrate plugs is a source of concern for the oil and gas industry, especially the offshore industry where the conditions for natural gas hydrate formation are met (e.g. [SLO 10]). This industry devotes large capital and financial resources to prevention or remediation strategies by using pipeline insulation or heating and/or the injection of inhibitors. This obviously requires the understanding of fundamental properties of clathrate hydrates, such as thermodynamic stability, formation kinetics and growth mechanisms. This section focuses on the invaluable information provided by NPD.

1.4.1.1. *Thermodynamic inhibition of hydrate formation*

Salts such as NaCl are thermodynamic hydrate inhibitors: in comparison to pure water, higher pressures and/or lower temperatures are required to form methane clathrate hydrates in brine (Figure 1.10). As other examples, dimethylamine (DMA) or ethylamine (EA) also act as inhibitors of CH_4 hydrate formation [YOU 14]. Although these two compounds form type I pure clathrate hydrates, the inclusion of methane molecules as secondary guests under external gaseous CH_4 induces a structural transition into the type II phase. NPD experiments on the mixed methane/EA or DMA clathrates from 10 to 220 K show a significant and irreversible structure transformation at about 200 K. The clathrate structure evolves from a low temperature type II structure to a type I structure (Figure 1.11), an effect attributed to the release of methane just above 190 K, a temperature significantly lower than the usual methane hydrate temperature of dissociation. The P-T phase equilibrium data show that the inhibition effect increases with the increase in DMA or EA mole fraction [YOU 14].

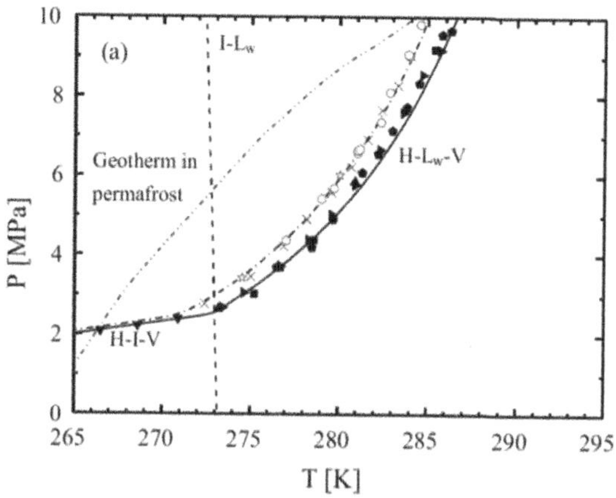

Figure 1.10. *Phase diagrams of the CH_4-H_2O clathrate hydrates systems. H, I, L_W and V represent hydrate, ice, liquid water and vapor phase, respectively. The continuous and dashed-dotted lines show equilibrium pressures of hydrates formation for pure water and for 3.5 wt% NaCl systems, respectively. The dashed and two dashed-dotted lines show melting points of pure water and estimated geotherm in permafrost [KID 09, POH 09] (from [KOM 13]). For a color version of this figure, see www.iste.co.uk/broseta/hydrates1.zip*

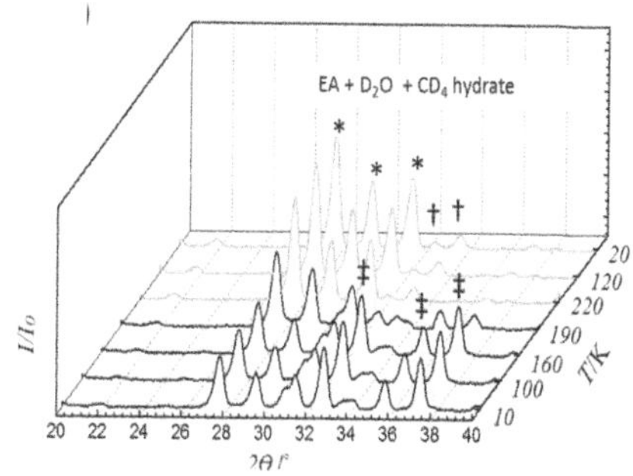

Figure 1.11. *Temperature dependence of the NPD patterns of the hydrate obtained from an EA (x = 0.0286) + D_2O + CD_4 mixture by increasing the temperature from 10 to 220 K followed by decreasing it to 20 K again. The black and light-gray patterns correspond to the s-II and s-I hydrates, respectively; *hexagonal ice; †s-I hydrate; ‡s-II hydrate (from [YOU 14])*

Because of the transparency of high-pressure and cryogenic devices to a neutron beam, NPD represents a useful tool to follow the structural changes that occur under various conditions. This technique has enabled some successes in studying the clathrate hydrate kinetics of both formation and decomposition [STA 03, MUR 09, KUH 06a].

1.4.1.2. *Formation kinetics*

Gas hydrate formation from fluid (aqueous- and guest-rich) phases is a relatively slow process, and a large specific surface area is needed to reach conceivable growth rates in laboratory conditions. However, it has been proved that clathrates can be grown quickly from ice powders. For instance, the formation of deuterium clathrate from ice powder was found to be faster from powdered ice (~10 min) than from water (full conversion not reached yet after 5 days) [LOK 04, LOK 06]. The kinetics of clathrate formation from ice has mostly been studied in the case of methane or carbon dioxide clathrate hydrate. As observed in Figure 1.12, under constant pressure, both CO_2 and CH_4 clathrate formation are temperature-dependent processes, which show a fast initial and then a smoothly decreasing rate [HEN 00, WAN 02]. Methane hydrate formation from ice at low temperature is a much slower process. However, nearly complete conversion from hexagonal ice to both s-I CO_2 and CH_4 hydrate can be reached by slowly increasing the temperature through the melting point of D_2O ice. On the basis of visual observations, in ice grains, a shrinking core model can be postulated, which involve the diffusion of future guest molecules through an outer mantle of hydrate to react with the inner core of ice [STE 01, HWA 90]. This model [JAN 27] considers a sphere of solid phase A (with radius r) that reacts with a mobile phase B (gas, liquid or solid powder) to form a product layer that covers the sphere A as the reaction proceeds. After the exposure of the solid phase A to the mobile species B, an initial layer of product is assumed to be rapidly formed, which may take a measurable amount of time t^*. Then, once a product layer is formed on A after a time t^*, the reaction is supposed to become diffusion controlled. The diffusion process of a particle from the outside during this diffusion-controlled stage can be described as follows:

$$(1-\alpha)^{1/3} = \left(\frac{-(2k)^{1/2}}{r}\right)(t-t^*)^{1/2} + (1-\alpha^*)^{1/3} \qquad [1.29]$$

where k is the diffusion constant, and α and α^* are degrees of reactions at time t and t^* ($<t$), respectively.

NPD data on formation reaction kinetics of carbon dioxide [HEN 00] and methane [WAN 02] clathrates were both successfully analyzed using this shrinking core model. In the case of carbon dioxide, results confirm that, after a time t^* corresponding to ~20% conversion, the rate limiting step of the formation process is the diffusion of CO_2 through the clathrate hydrate layer, characterized by an activation energy of 27.2 kJ/mol [HEN 00]. The conversion of ice into carbon dioxide hydrate was modeled as a two-stage process. First, an initial fast conversion stage occurs. Its rate-limiting step before a layer of hydrate covers the ice particles is the reaction of CO_2 with a quasi-liquid layer (QLL) [TAK 00], a thin mobile phase of water molecules with mobilities in between those of liquid water and crystalline ice [MIZ 87]. Then, a slower second stage takes place, which is controlled by the diffusion of the CO_2 molecules through the layers of hydrate covering the ice particles. As this CO_2 diffusion is characterized by an activation energy lower than the energy needed to break the hydrogen bonding in ice (53.1 kJ/mol [ITA 67]) but greater than that of 21 kJ/mol in liquid water [SLO 08], the formation of the hydrate after the diffusion through the hydrate layer is suggested to occur through a reaction of CO_2 with internal water molecules in the QLL rather than with ice.

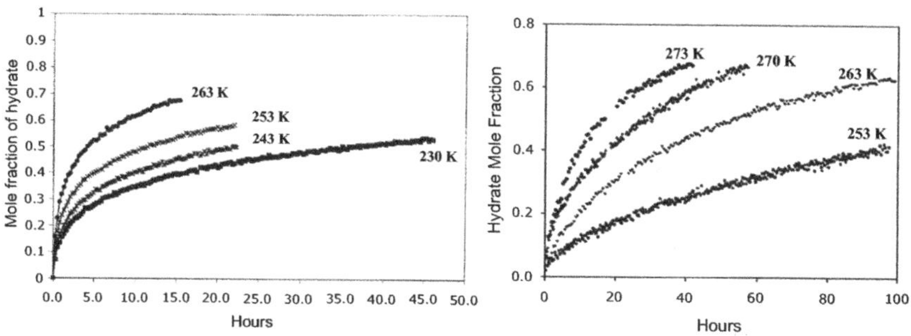

Figure 1.12. *Conversion of deuterated ice into (left) carbon dioxide hydrate at 62 bars (from [HEN 00]) and into (right) methane hydrate at ~70 bar (from [WAN 02]) at various temperatures indicated on the figure. Each data point represents the mole fraction of hydrate refined from a 15 min histogram*

In contrast, the diffusion of methane through a clathrate hydrate layer becomes the rate limiting step of the process at ~10% conversion and was characterized by an activation energy of 61.5 kJ/mol [STE 01]. A more complex shrinking core model [REK 95] describing the clathrate formation by a first initial reaction of methane with the surface of ice particle, followed by the growth of hydrate layer and inner diffusion of methane gas. Finally, the reaction of methane gas with ice at the unreacted ice core was tested. However, if results support the previous conclusion, it does not lead to any improvement of the neutron data fit. Considering the 61.5 kJ/mol activation energy, smaller than what is required for the dissociation of methane hydrate (81.2 kJ/mol [CLA 01] but greater than results in the CO_2 hydrate system involving QLL (27.2 kJ/mol [HEN 00]) and also than the energy needed to break the hydrogen bonding both in ice (53.1 kJ/mol [ITA 67]) and in liquid water (21 kJ/mol [SLO 08]), free liquid water phase does not seem to be required for converting ice into methane hydrate. This hypothesis is supported by results from the monitoring with a neutron spectrometer of the conversion under non-isothermal condition. Indeed, ice, existing above its melting point, is consumed at the same, but opposite, rate as the amount of hydrate increases (Figure 1.13). These results exhibit either the existence of superheated ice or a demonstration of the insulating properties of hydrate layer, which may keep the ice core cooler than the measured temperature of the sample container. As for carbon dioxide hydrate, a two-stage model thus describes the conversion of ice to methane hydrate. The reaction starts quickly at the nucleation stage, and then the hydrate propagates as a hydrate layer that covers the ice particle. Further reaction is limited by the growth of the hydrate layer and inward diffusion of methane molecules through the hydrate layer to the unreacted ice core. However, in contrast with carbon dioxide hydrate formation, the reaction does not seem to involve interactions with a free water phase.

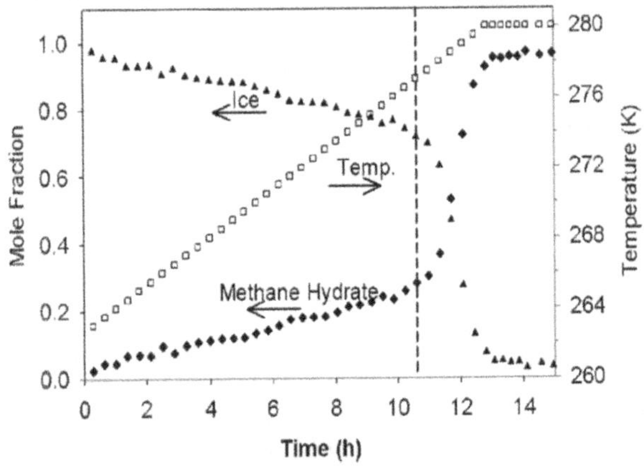

Figure 1.13. *Conversion of deuterated ice to methane hydrate through temperature ramp at 103 bar. Vertical line is the melting point of deuterated ice (from [WAN 02])*

1.4.1.3. *Clathrate formation from ice particles*

In most ice into hydrate conversion kinetics studies at temperatures close to the ice point, a thin gas hydrate film was reported to rapidly spread over the ice surface at the initial stage [HEN 00, STA 03, GEN 04, WAN 02, KUH 06b]. After this initial nucleation and ice surface coating, the only way to continue the subsequent clathration is the transport of gas molecules through the formed hydrate layer to the ice–hydrate interface, and/or the transport of water molecules from the ice core to the outer hydrate–gas interface. Indeed, due to the lower water density in the hydrate frame, excess water molecules diffuse outward to the hydrate surface to react with the ambient gas, which induces an outward expansion of the clathrate hydrate shell during the growth process. This can be modeled by a shrinking core approach, assuming that the reaction starts at the surface and proceeds toward the center of the ice particle. However, these considerations are inferred from the study of crushed ice particles, whose geometry is ill-defined. Moreover, particle size analysis is not taken into account, despite its crucial impact onto formation kinetics.

For instance, during CO_2 clathrate hydrate formation from small ice particles (~10^{-1} μm) below 200 K, hydrate growth does not start with hydrate shell formation [FAL 11]. As the initial hydrate film thickness exceeds the ice particle dimensions, the whole ice volume is already transformed at the initial stage without reaching the permeation-limited regime. CO_2 clathrate hydrate growth quickly consumes the ice grain on which the nucleation has occurred but does not spread to neighboring particles. The overall speed of transformation and the size of the hydrate formed thus depend on the initial ice particle size taking into account the particles' internal discontinuities (cracks, structural defects, grain boundaries) susceptible to hinder the spreading of the clathrate hydrate. As each of these effective particles needs to experience a nucleation event before it can be converted to clathrate (due to the high activation energy of nucleation, the stochastic nature of the nucleation event and the low activation energy of growth), smaller frost particles will tend to transform more slowly. Larger particles with larger surface areas have a higher probability to experience a nucleation event on their surfaces and would then transform faster by the clathrate spreading growth. The reverse tendency will take place when ice particle size exceeds 20–30 μm, allowing for hydrate shell formation around the remaining ice core. At temperatures equal or lower than 200 K, the main limiting factor for forming CO_2 clathrates may thus be the geometry of the starting material as well as the stochastic nature of the nucleation process.

The standard shrinking core model has been modified, mainly to account for the essential outward growth of the transforming particles and necking between them and to specify the initial phase of the clathration reaction on the ice particle surface [RYU 09]. The conversion of individual ice particles into hydrates was thus described by a first ice-surface coating stage in which an initially patchy hydrate film nucleates on and spreads over the surface of much larger ice particles [KAS 08]. The hydrate shell's further growth around each ice core and includes the clathration reaction itself at the inner ice–hydrate as well as the external gas–hydrate interfaces and gas/water mass transport through the hydrate layer. At temperatures between 185 and 272 K, a recent study [FAL 13] of CO_2 hydrate formation from well-defined spherical ice powders has convincingly shown that this process is limited by the CO_2 transport (diffusion) through the hydrate shells growing around the ice cores. The modified shrinking core model gives experimental access to the values of the CO_2 permeation coefficient D in bulk hydrate reported in Figure 1.14, from which different activation energies are found above (~46 kJ/mol) and below 225 K (~19 kJ/mol).

Guests' migration in clathrate hydrate framework requires the opening of the lattice cages, which may be well described by a model of migrating water vacancies

[HAA 09]. Assuming that the transport of water molecules from the ice–gas interface to the particle surface is the hydrate formation limiting step, activation energy is expected to be in the 12 kJ/mol range characterizing the water permeation process [KOM 00]. On the other hand, activation energies for the creation of a water vacancy-interstitial pair in CO_2 hydrate and for molecular CO_2 transport in a cage-to-cage mechanism in presence of a water vacancy are estimated respectively as 39 ± 6 kJ/mol [MAT 11] and between 12.5 kJ/mol (large cage–large cage) and 25 kJ/mol (large cage–small cage) [HAA 09]. The experimental activation energies thus seem to be related to carbon dioxide jumps mediated by available extrinsic water vacancy defects below 225 K, while above 225 K the increased activation energy is either a consequence of the creation of vacancy-interstitial pairs or simply the result of jumps between defect-free cages, which must then be rate limiting due to the small number of extrinsic defects. Therefore, gas molecules can migrate through bulk hydrate via cage-to-cage hopping promoted by vacancies of water molecules in the connecting cage wall. The rate limiting process at temperatures below ~225 K is the cage-to-cage jumping of CO_2 molecules via a "hole-in-the-cage" mechanism involving extrinsic water vacancies in cage walls, while at higher temperature the creation of water vacancy-interstitial pairs become the rate limiting process. These results seem to be generalized to the methane hydrate case at least at temperatures above 225 K. Indeed, the permeation coefficients for CH_4 molecules in the hydrate framework at temperatures above ~225 K can be estimated by MD simulations assuming a water vacancy assisted mechanism for the cage-to-cage hopping [HAA 09]. As seen in Figure 1.14, even if the values are about three times smaller than those experimentally obtained of CO_2, activation energies are quite similar.

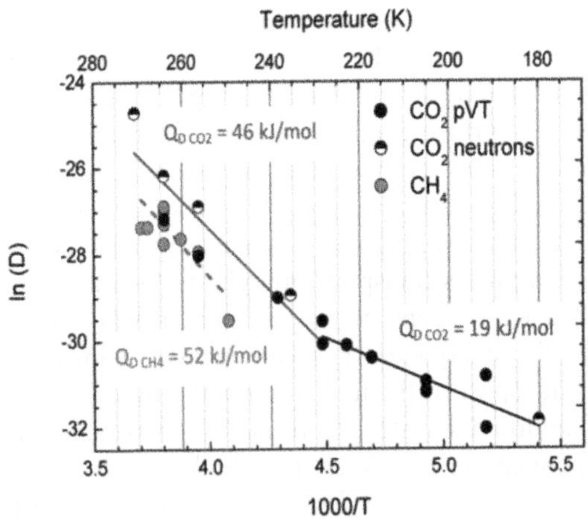

Figure 1.14. *Apparent activation energies for carbon dioxide (red and blue) and methane (green) diffusing through clathrate hydrate lattice deduced from permeation coefficient (denoted D) measurements (from [FAL 13]). For a color version of this figure, see www.iste.co.uk/broseta/hydrates1.zip*

1.4.2. *Guest replacement in gas hydrates*

The large deposits of natural gas hydrate in deep oceans and permafrost have motivated a large research effort [SLO 03, CHA 05, KOH 07, SLO 08]. The amount of methane in these deposits, about 700,000 trillion cubic feet according to some authors [BOS 09], corresponds to an amount of carbon greater than that of all conventional fossil fuels on earth [HAQ 99, MAX 00, COL 02, DEM 10]. It may represent an enormous supply of energy reserve, and a future energy source if the gas can be recovered economically and safely. The recovery of methane from these resources has been extensively studied, leading to actual recovery tests [RYU 09, KID 09, POH 09, CRA 09, LIU 09]. Most recovery techniques involve the promotion of the methane clathrate dissociation by using external stimulation such as thermal treatment [CRA 09], depressurization [LIU 09] and inhibitor addition [DEM 10]. Furthermore, extracting methane from the hydrates and simultaneously replacing it by its greenhouse gas product, carbon dioxide, could be a two-in-one approach. This concept, referred to as *CO_2–CH_4 replacement*, is expected to present benefits for stabilizing the ocean floor and the permafrost zones during the recovery process [OHG 96, BRE 99, KOM 02, LEE 03, YOO 04, OTA 05, PAR 06, GRA 08]. This section is devoted to gas replacement in gas hydrates.

1.4.2.1. *Thermodynamic and structural considerations*

CH_4 and CO_2 clathrate hydrates both crystallize according to a cubic structure s-I, where the guest can be trapped in both small pentagonal dodecahedral (5^{12}) and large tetrakaidecahedral ($5^{12}6^2$) cages [WAN 02, HEN 00]. Indeed, the ellipsoidal shape and size (~6.2 Å in including van der Waals interactions) of the large cavitie allows the confinement of either the methane molecules or the linear carbon dioxide molecule (~5.12 Å long). The small, roughly spherical cavities easily trap the methane molecules and, even if not ideal for a linear guest molecule, provide a tight fit for CO_2 as its van der Waals diameter is roughly equal to the length of the carbon dioxide molecule. Raman scattering [SUM 97] and X-ray diffraction [UDA 01] analysis confirm that encapsulation of CO_2 molecule in large cages is favorable. The methodology of CO_2–CH_4 replacement consists of the injection of high pressure carbon dioxide below sea beds, where replacement of the methane by CO_2 is expected to occur, at least for pressures not exceeding 8 MPa (water depths < 800 m). Indeed, as observed in Figure 1.15 [WAN 02, KOM 13, GOE 06], CH_4 hydrate is less thermodynamically stable than CO_2 hydrate for such pressures or depths. Measurements of methane and carbon dioxide distribution coefficients between gas and hydrate phases also show that the second if preferred in the hydrate phase over methane [OHG 96]. Injection tests with mixtures of CO_2 and N_2 below sea beds were successfully conducted in 2012 by the DOE (Department of Energy, USA).

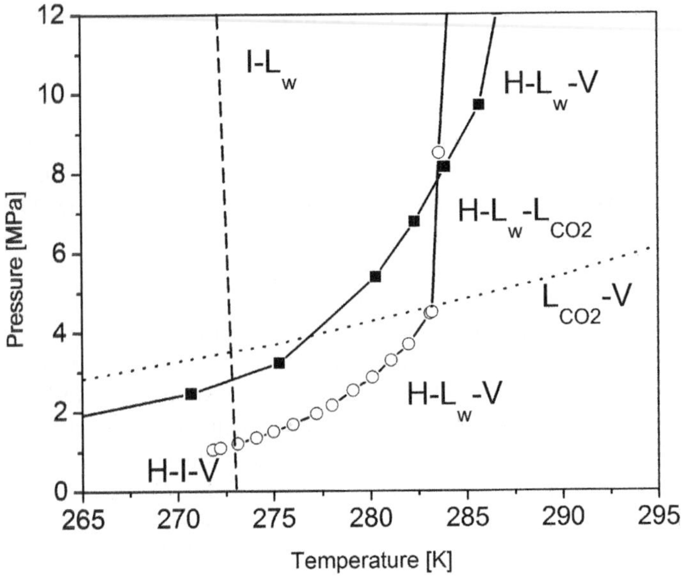

Figure 1.15. *Phase diagrams of the CH_4–H_2O clathrate hydrates (squares) and of the CO_2–H_2O (circles) clathrate hydrates. H, I, L and V represent hydrate, ice, liquid and vapor phase, respectively. Indices w and CO_2 refer to water and carbon dioxide, respectively. The dotted lines show equilibrium curves of CO_2 and the dashed line show melting points of pure water. Adapted from [WAN 02, GOE 06, KOM 13]*

However, partial failures/slowness of this replacement processes may have drastic consequences, as methane is, for instance, a greenhouse gas 20 times more effective in causing global warming than carbon dioxide [HOU 96]. Furthermore, as gas hydrates stability is very sensitive to the gas ratio and the conditions of temperature and pressure, any changes of these factors may lead to regrowth processes and affect the cementing properties of gas hydrates in a sedimentary matrix. Considering such broad implications, a full understanding of the replacement process should be obtained to design safe, efficient and economically feasible large-scale recovery. Kinetic aspects of replacement reaction in clathrate hydrate especially required a particular attention, since results suggest that the replacement reaction could be possibly dominated by kinetics rather than equilibrium thermodynamics [LEE 04].

1.4.2.2. *Kinetics of replacement processes*

Although NPD is a reliable method for studying microscopic occupation by guest molecules, it has not been applied to CH_4–CO_2 replacement reactions yet and most of the available information comes from ^{13}C NMR, Raman spectroscopy or magnetic imaging resonance studies [KOM 13]. The absence of measurable dissociation of methane hydrates during the replacement process [OTA 05] indicates that this reaction does not consist of the individual hydrate formation and dissociation reactions only. The activation energy for CH_4 hydrate decomposition (14.5 kJ/mol [OTA 05]) matches with the hydrate dissociation enthalpy to methane and ice (15–21 kJ/mol [RYD 07]) which implies simultaneous decomposition of both types of cages. Considering the CO_2 molecular diameter not very favorable to the occupation of the small cages compared to CH_4 molecules, a slower replacement rate in the small cages than in the large ones may be attributed to a reoccupation of small cages by methane molecules after its release from the large cages. The activation energy for CO_2 hydrate formation during replacement (73.3 kJ/mol) has been found to be larger than that for the diffusion of CO_2 molecules through the layer of hydrate (27 kJ/mol [HEN 00] and 39 kJ/mol [BOS 09]). Its better agreement with the activation energy for the methane diffusion in an hydrate layer (61.5 kJ/mol [WAN 02]) or for H_2O molecule diffusion in the bulk ice (in the range of 52–70 kJ/mol [LIV 97]) suggests that CO_2 hydrate formation for CH_4–CO_2 replacement reaction could be dominated by the diffusion of these species.

Neutron scattering experiments have led to detailed analysis of the kinetics in the cases of CO_2/Argon and methane/ethane replacement reaction in clathrate hydrate [HAL 01, MUR 10]. A first interesting feature is that hydrate formation on an existing hydrate surface appears faster than the formation of new hydrate. This was assumed to be due to the fact that the initial hydrate can act as a template. Introduction of carbon dioxide molecules at a 62 bar pressure on the s-II Argon clathrate induces its conversion into a mixed Ar/CO_2 s-I clathrate hydrate [HAL 01]. The kinetic rate, in contrast to hydrate formation from ice, indicates that the whole conversion is diffusion-limited (Figure 1.16). The diffusion process is characterized by an activation energy of 58.2 kJ/mol, higher than what is needed to break hydrogen bonding in either water or ice. A study of the s-II Ar hydrate decomposition as a function of the type of gas used and its pressure has reveals that the decomposition is slower under pressure of a non-hydrate forming gas than under the same pressure of CO_2 [HAL 01]. Although the existence of an undetectable intermediate phase cannot entirely be excluded, these observations – the absence of ice peaks in the neutron scattering pattern and the empirical report of a whole conversion dominated by diffusion – suggest that the conversion does not require any intermediate phase with weaker hydrogen-bonding energies. The transformation of the s-II Ar hydrate to the mixed Ar/CO_2 type s-I hydrate was thus assumed to be a direct conversion mechanism without first decomposing into ice. This leads to envision gas hydrate as an equilibrium assemblage in which hydrogen bonds open and close constantly, allowing guest molecules to leave or enter freely during that time.

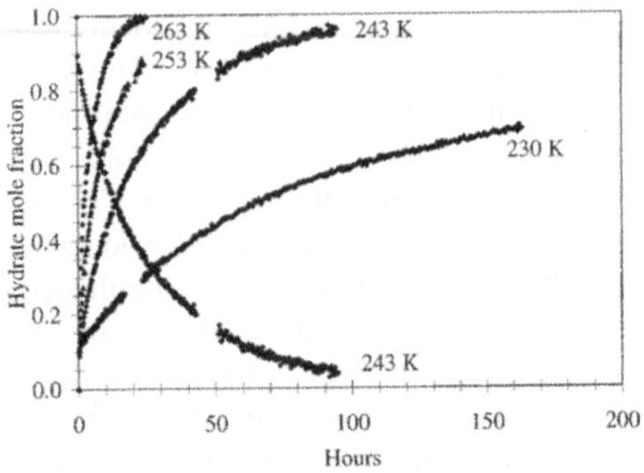

Figure 1.16. *Conversion of s-II Ar hydrate to s-I hydrate after carbon dioxide gas exchange at several temperatures. Upward sloping curves indicate the s-I hydrate formation, whereas the downward curve (for 243 K trial) indicates the s-II Ar hydrate decomposition (from [HAL 01])*

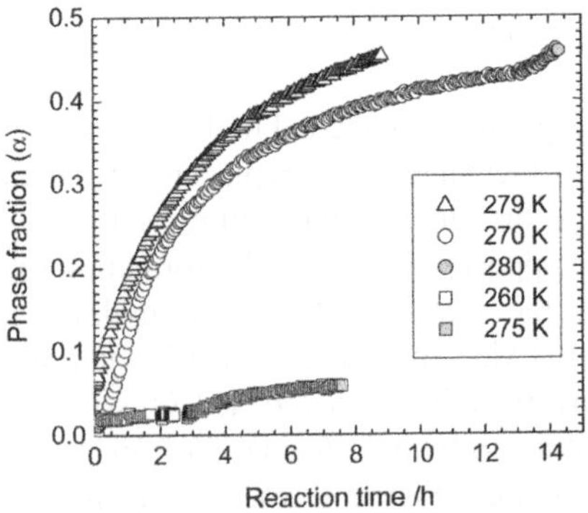

Figure 1.17. *Reaction kinetics of mixed methane/ethane clathrate hydrate formed on deuterated ethane hydrate consolidated (squares) or crushed (circles, triangles) at several temperatures (from [MUR 10])*

Similarly, exposing pure s-I ethane hydrate to free methane gas at 5 Mpa induces an instantaneous formation of mixed s-II hydrate. The formation rate is temperature dependent and shows a nonlinear fast formation followed by a slower diffusion limited growth of the new s-II phase (Figure 1.17) [MUR 10]. Analysis of conversion from crushed and consolidated s-II ethane hydrate indicates that the growth kinetics is highly influenced by the surface area of the starting material. In agreement with the process suggested for hydrate formation from ice powder, the gas replacement in methane/ethane clathrate hydrates has been assumed to be a regrowth process involving the nucleation of new crystallites at the surface of the parent C_2H_6 s-I hydrate with a progressively shrinking core of unreacted material. Thus, mixed s-II

CH_4-C_2H_6 hydrate growth occurs around a shrinking s-I ethane hydrate core. Once a mixed s-II hydrate forms as surface coverage, a (fictitious) spherical s-I ethane hydrate grain shrinks, and its radius decreases due to the inward growth of the s-II hydrate layer. The structural transformation takes place at the replacement front with a rearranged composition of the constituents, and the CH_4-C_2H_6 s-II hydrate layer acts as a diffusion barrier both for the in-bound CH_4 molecules moving toward the replacement front and the out-bound C_2H_6 molecules moving toward the gas phase.

1.4.3. *Hydrogen: from its dynamics properties to its storage capabilities*

As gas hydrates usually show features such as large heat of dissociation and phase change temperatures above the ice point, they are considered promising materials for energy storage due to numerous advantages including intrinsic safety, environmentally benign and quite low processing costs [CHA 05]. Some applications envision new technological opportunities for storage and transportation of natural gases [SLO 08] and, most notably, hydrogen [DYA 99, MAO 02, FLO 04, LEE 05, PEF 12, VEL 14]. Indeed, conventional methods of hydrogen gas storage include compression and liquefaction, which both involve safety risk or demand high energy [SCH 05]. Strategies thus explore the possibility of using solid clathrate hydrate as an environmentally benign as well as non-explosive medium to reversibly capture, concentrate, store and release hydrogen fuel [SUG 09]. All these strategies include concepts in which hydrogen would be stored in its molecular form, and ready for utilization just by depressurization or minimal thermal stimulation. Consequently, numerous studies have been dedicated to improving the stability and storage capacity at near-ambient conditions of hydrogen clathrate hydrate.

In addition to the opportunity for hydrogen storage offered by hydrogen clathrate hydrates, these systems exhibit original and fundamental physicochemical properties. The understanding of the microscopic mechanisms underlying the development of hydrogen storage capabilities requires the detailed knowledge of the dynamic properties of hydrogen clathrate hydrate. Moreover, the fundamental investigations of the inherent quantum mechanical behavior of H_2 molecules “isolated” within cages play a key role in fundamental science. All these applied and fundamental aspects will be detailed in this section.

1.4.3.1. *Thermodynamics characteristics*

Hydrogen clathrate hydrates crystallize according to the s-II cubic structure at pressures between 750 and 3,100 MPa at 295 K or temperature below 145 K at ambient pressure [MAO 02]. NP shows that a maximum of one hydrogen molecule can be trapped in the small cavity in the pure hydrogen hydrate [LOK 04]. Below 50 K, each large cage traps four guests arranged in a tetrahedral geometry with a 2.93 Å D_2–D_2 distance and, increasing the temperature, the occupancy of the large cages gradually decrease from 4 below 70 K to 2 at 160 K. As pure hydrogen hydrate formation requires considerably high pressure at ambient temperature, the co-inclusion of another guest molecule in the hydrate cages is considered in order to significantly reduce the formation pressure. Tetrahydrofuran (THF), which crystallizes in a pure stoichiometric (THF-17H_2O) s-II hydrate when a mix of 5.56 mol% THF with water is freezed [HES 06], has been extensively studied [FLO 04, LEE 05, STR 06, AND 07, HAS 07, NAG 08, SUG 09]. Mixed s-II THF-H_2 clathrate can be formed by applying a pressure of ~100 bar on a grounded THF clathrate hydrate. Phase diagrams for THF concentrations at boundary limits (i.e. 0 mol% THF corresponding to the pure hydrogen clathrate hydrate, 5.56 mol% THF corresponding to the s-II structure with the THF molecule filling all large cages and the hydrogen molecules located in the small cages) are shown in Figure 1.18. Others promoters of s-II mixed hydrogen clathrate hydrate have been identified (phase diagrams reported in [VEL 14]). Cyclopentane increases dissociation temperature compared to THF. Furan and tetrahydrothiophene lead to stabilization at lower temperature and faster absorption of hydrogen than THF systems for the same storage capacities under the tested experimental conditions [VEL 14]. Cyclohexanone (CH) has also been considered but, despite larger size hydrate cages (due to the size of CH), the mixed CH-H_2 hydrate has lower hydrogen occupancy than that formed with THF as a promoter [VEL 14].

Apart from hydrogen storage in s-II clathrate hydrate (pure or mixed with promoters), it is possible to have hydrogen molecules occupy the cages of different hydrate structures that hydrogen cannot usually form. Inclusion of single or two hydrogen molecules in the small cages of s-I CO_2 hydrate while the large ones are filled by CO_2 has been reported [KUM 09]. The opportunity of storing hydrogen in both types of s-I cages has also been demonstrated by initially forming s-I hydrate (with CO_2 or CH_4 molecules as guest species) and then by pressurizing the system with hydrogen gas [GRI 12]. On the other hand, experimental storage of hydrogen along with a larger guest molecule has been in the small s-H cages [VEL 14]. Such storing opportunity leads to an increase in hydrogen storage capacity by 40% compared to binary hydrogen s-II hydrates [VEL 14]. However, the formation of such s-H hydrate requires extreme conditions: about 100 MPa pressure at 273 K. Such pressure prevents from considering practical applications despite its high storage abilities. Moreover, a distinct structure s-VI, which could be capable of storing maximum hydrogen compared to other hydrates due to the size of the cages, has also been recently reported [DU 11]. The mixed tert-butylamine/hydrogen hydrate can then be formed with stability at lower temperature than most of the mixed hydrogen hydrates formed with promoters. It is interesting to note that recent MD simulations outline the existence of interstitial site welcoming hydrogen molecules in between two cages [GRI 14]. Finally, semiclathrate hydrates made up of quaternary ammonium and phosphonium salts also host hydrogen (represented in Figure 1.19). In such cases, the formation conditions are relaxed to lower pressures and higher temperatures (i.e. closer to room temperature condition) as shown in Figure 1.19 [CHA 07].

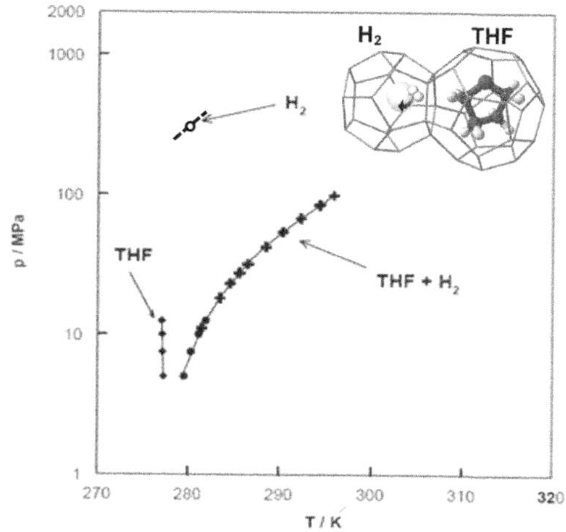

Figure 1.18. *Phase equilibrium data for hydrogen clathrate hydrate (opened circles), THF clathrate hydrate (filled diamonds) and THF-H_2 clathrate hydrate (with 5.56 mol% THF in water; filled circles and crosses). Varying the concentration of THF between 0 and 5.56 mol% in water leads to equilibrium curves of the formed THF-H_2 clathrate hydrate contains in between the equilibrium curves of the pure hydrogen clathrate hydrate and the "THF + H_2" clathrate hydrate [VEL 14]. Adapted from [FLO 04, PEF 12]. For a color version of this figure, see www.iste.co.uk/broseta/hydrates1.zip*

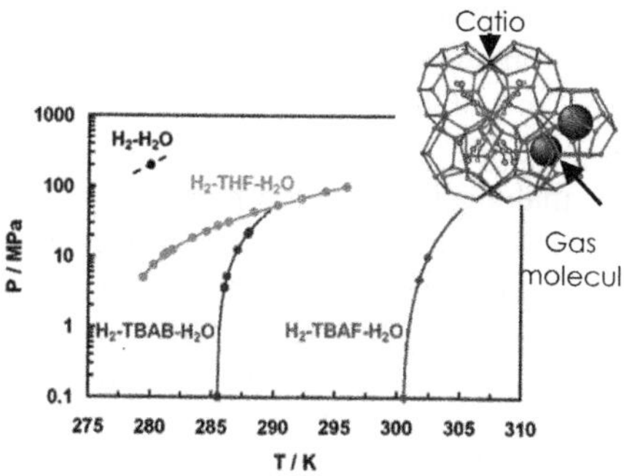

Figure 1.19. *Phase equilibrium data for mixed hydrogen hydrate formed with various semiclathrate promoters. TBAB: tetrabutylammonium bromide. TBAF: tetrabutylammonium fluoride. Adapted from [CHA 07]. For a color version of this figure, see www.iste.co.uk/broseta/hydrates1.zip*

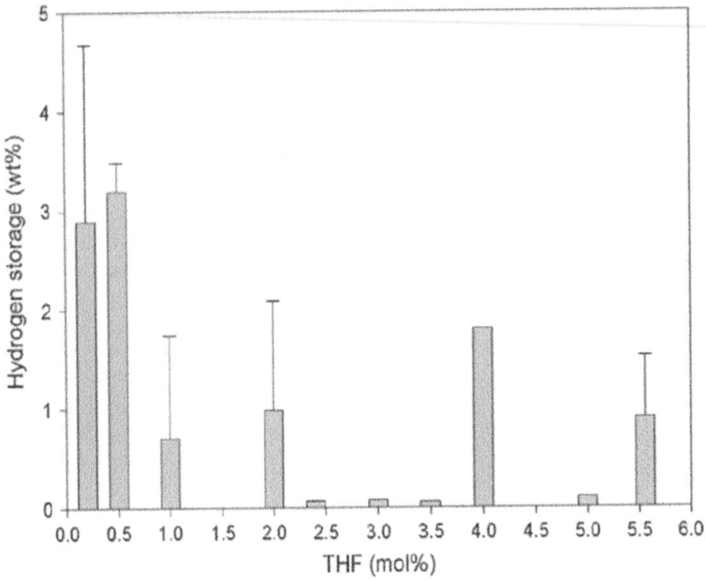

Figure 1.20. *Hydrogen storage capacity as a function of THF concentration (from [VEL 14])*

1.4.3.2. *Cage occupancy*

Guest occupancy is an essential factor for storage in clathrate hydrates. It varies with temperature and pressure, the use or lack of additives, and is mainly dominated by adsorption–desorption equilibrium. The guest occupancy of hydrogen clathrate hydrates takes a particular importance in the case of mixed hydrogen hydrates. Indeed, even if the use of promoters enhances the hydrate formation it also reduces the storage ability. When mixed hydrogen hydrates form in the presence of a promoter, the promoter molecule preferably occupies large cages stabilizing the hydrate structure thereby making it difficult for hydrogen molecules to enter and occupy the large cages. However, at certain low concentrations of promoter, hydrogen molecules have been reported to partially occupy the large cages, thus increasing the hydrogen content [LEE 05, KUM 09]. This phenomenon, referred as the "tuning effect" since the concentration of the promoter is tuned to achieve the maximum hydrogen storage, has first been reported in mixed THF/hydrogen hydrates. A maximum amount of hydrogen of 4.03 wt% was observed by lowering the concentration of THF from stoechiometric composition (5.56 mol%) to 0.15 mol% at 12 MPa [LEE 05]. Data about hydrogen storage capacity as a function of THF concentration compiled from literature data are presented in Figure 1.20. Even if such a tuning effect has also been reported in mixed hydrogen hydrates with various promoters (see various examples in [VEL 14]), the guest occupancy of hydrogen clathrate hydrate remains a conflicting subject [LEE 05, MUL 08, KUM 09, PEF 12, VEL 14].

First, most of the studies reporting such a tuning effect used Raman spectroscopy, for which quantitative measurements of hydrogen content in gas hydrates are known to be a tricky task [STR 09]. Besides, other studies categorically refute the possibility of any tuning effect [HES 06, STR 06, HAS 07, AND 07, MUL 08] and report only a maximum of around 1 wt% corresponding to hydrogen stored in small cages [STR 06, AND 07]. For instance, the direct determination by neutron scattering of hydrogen occupancy in the small cavities of the binary deuterated hydrogen/THF hydrate provides evidence that only single occupancy of hydrogen in the small cavity can be expected in this system, at least up to 70 MPa [HES 06]. Neutron scattering is especially suited for the determination of the position and quantity of hydrogen (deuterium) and for clarifying the difference in the thermal vibrations of the same kind of atoms at crystallographically distinct sites and, if full structural refinement is performed, locates the hydrogen atom positions with high precision [HES 06]. In addition, in the case of single crystals, it also allows the refinement of multiple cage occupancy and orientation ordering in the cages (with e.g. pure H_2 hydrates [LOK 04], binary H_2/THF hydrates [HES 06]). Further quantitative NP analysis shows that, instead of the observation of "tuning" of the THF concentration in favor of hydrogen uptake in the large cages, reducing the THF content of this clathrate below its stoichiometric 1:17 ratio leads to phase separation of the pure clathrate and liquid water, which becomes ice at lower temperatures [MUL 08]. Thus, the stabilization of the THF-depleted clathrate phase by D_2 in the vacant large cages compensating for the lower THF concentration is not observed. This investigation also measured a hydrogen occupancy in the small cages of the clathrate (at pressures of up to 95 bar and $T = 264$ or 274 K) significantly smaller than expected, reaching only 30% at 274 K and 90.5 bar, which is equivalent to 0.27 wt% H_2 storage. Any improvement in the hydrogen content requires further knowledge about the mechanisms of hydrogen loading in clathrate hydrate. In addition, such an approach may help to design a system that overcomes the typical limitations of kinetics, mass and heat transfer for the gas storage (natural gas or H_2).

1.4.3.3. *Mechanism of hydrogen loading in THF clathrate hydrate*

The hydrogen loading mechanism in THF clathrate assumes hydrogen adsorption onto the clathrate particle surface and the subsequent diffusion of hydrogen into the clathrate hydrate particle. This diffusion is the process-limiting step and is characterized by a high activation energy of 78.7 kJ/mol [NAG 09]. The thermally assisted rapid diffusion of hydrogen has been observed during its loading in the mixed THF hydrogen clathrate [MUL 08] by means of *in situ* powder NP (Figure 1.21). At 264 and 274 K, the clathrate hydrate takes up hydrogen rapidly at pressures between 60 and 90 bar as D_2 molecules rapidly diffuse through the THF clathrate, with a diffusion coefficient estimated of at least 10^{-11} cm^2 s^{-1} at 264 K. The diffusion slows down below 220 K, leading to the stagnation of further uptake at lower temperatures. The maximum hydrogen occupancy in the small cage of about 0.43 achieved at around 50 K is reduced by increasing temperature up to 220 K and progressively drops to zero when approaching 282 K. Subsequent application of high hydrogen pressure at 282 K reproduces a D_2 cage occupation of approximately 0.3. Diffusion of hydrogen through the crystalline clathrate is thus found to be fast enough to enable (un-)loading from the gas phase at temperatures down to about 220 K. Above this temperature, the clathrate behaves as a nanoporous material.

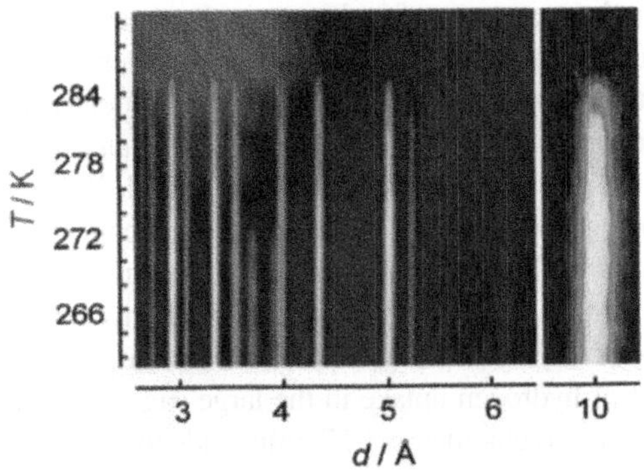

Figure 1.21. *Diffraction patterns as a function of the temperature for deuterated THF·17 D_2O + 0.07 D_2O. Ice Ih Bragg peaks are observed at 3.7 and 3.9 Å (below ca. 273 K) next to the TDF·17 D_2O clathrate peaks (TDF holds for deuterated THF). At the applied pressure (90 bar), the clathrate melting point is approximately 284 K, which is consistent with that given in [FLO 04]. Above the melting points of the crystalline phases, broad diffraction features corresponding to the liquids are visible (i.e. liquid D_2O between 284 and 274 K and a liquid mixture above 284 K) (from [MUL 08]). For a color version of this figure, see www.iste.co.uk/broseta/hydrates1.zip*

Considering the size of the pentagonal windows connecting the small cages (4.24 Å at 264 K) and taking van der Waals radii into account, the approximate window openings is ca.1.2 Å. Such a value is significantly narrower than the kinetic diameter of a hydrogen molecule (2.89 Å). The potential energy barrier associated with H2 guest molecules migration through hexagonal face of the large cage and smaller pentagonal faces of the small cages of the s-II clathrate hydrate are, respectively, estimated about 23 kJ/mol and between 104.7 and 121.4 kJ/mol (depending on the orientation of the H2 molecule) [ALA 07]. As for CO2 diffusion during ice to CO2 hydrate conversion, hydrogen molecular transport is thus facilitated by (transient) defects in the structure, leading to a temporary opening of the cage window. Diffusion of D^+ ions in THF clathrate lattice results from the combination of deuteron translational diffusion (deuteron exchange between two O atoms involving OD^- and/or D_3O^+) with a correlation time around 264 K of about 10^{-3} s, and Bjerrum defects mobility (reorientation of an O–D bond) up to two orders of magnitude faster [KIR 03]. Assuming that the timescale of the molecular transfer of D_2 from a small cage to another is related to the passage of defects, its diffusion coefficient will thus be between 10^{-12} and 10^{-10} cm²/s, depending on which type of defects dominates. The magnitude of the above-estimated experimental lower limit for the D_2 diffusion constant is inbetween these two extremes [MUL 08] Besides, the hypothesis assuming that Bjerrum and ionic defects facilitate D_2 diffusion could also explain the observed slowdown of hydrogen uptake and release at lower temperatures by the strong temperature dependence of the occurrence of such defects (as reported in [KIR 03]). In this respect, the clathrate behaves very differently from other surface-adsorption materials, because the diffusion pathways between the available adsorption sites are progressively obstructed upon reducing the temperature. Furthermore, as the transport of hydrogen molecules between the cages may only proceed in the neighborhood of a cage vacant guest site, the hydrogen diffusion coefficient value also gradually decreases with increasing pressure or increasing hydrogen concentration in the hydrate [STR 06].

1.4.3.4. *Hydrogen dynamics in mixed THF/H_2 clathrate hydrates*

At room temperature, free hydrogen molecules consist of two different species (namely ortho and para) separated by an unperturbed energy difference of 14.7 meV. The ortho population is almost three times that of para at room temperature but, upon cooling, the lower energy para-population increases because of an ortho- to para-transition induced by the strong magnetic field generated by a neighboring H_2 molecule. When a hydrogen molecule is incorporated into a hydrate cavity, its chemical environment adds a potential barrier to the rotational motion of the molecule, which usually results in a decrease of the ortho- to para-transition energy. IINS is especially sensitive to this transition (Figure 1.22). IINS experiments have led to a detailed analysis of rattling modes of the hydrogen molecule inside the small cage of mixed THF/H_2 [ULI 07, XU 08, CHO 10, XU 11, COL 13]. At 2 K, a single peak at about 13.6 meV is observed and, increasing the temperature above 10 K, a second peak appears at 14.7 meV in addition. The small shift between the peaks suggests that, even if the trapped hydrogen molecules are still relatively free from their host at 2 K, a rotational barrier certainly exists [CHO 10]. Finally, at 20 K, a threefold splitting of both the para- to ortho-transition attributed to the anisotropic structure of the cage was reported at 13.64, 14.44 and 15.14 meV [ULI 08]. Both H_2 roton and vibron bands are at lower frequencies in clathrate cage structure than in the free gaseous phase [FLO 04, ULI 07, TAI 07, ULI 08, CHO 10], which is representative of the attractive interaction existing between the H_2 molecule and the aqueous cage.

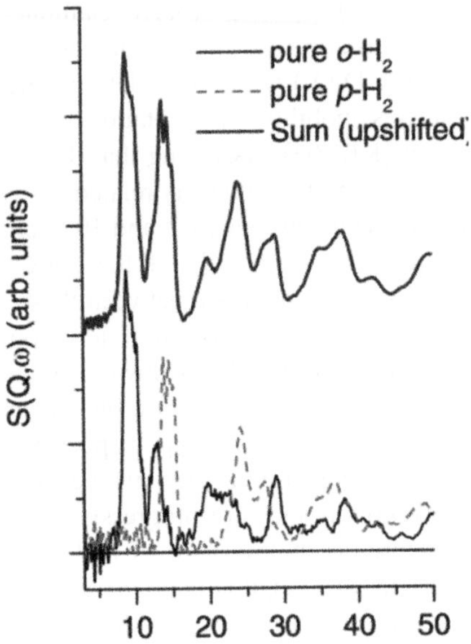

Figure 1.22. *The clathrate IINS spectrum (thick blue line) decomposed into the sum of the ortho-H_2 (black, solid) and para-H_2 (red, dashed) contributions. The para-H_2 spectrum exhibits the strong $J = 0 \rightarrow 1$ rotational band (a triplet at E ~14 meV), and the ortho-H_2 spectrum exhibits the splitted band around 10 meV, due to the quantum rattling motion of the molecule in the cage. Adapted from [ULI 07]. For a color version of this figure, see www.iste.co.uk/broseta/hydrates1.zip*

In general, the molecules enclosed within the cages of mixed hydrogen clathrate hydrates are trapped because the small hexagonal and pentagonal windows make transport between the cages impossible. However, from structural point of view, hydrogen molecules could be assumed to occasionally jump from one small cage to an unoccupied neighboring small cage, as observed during the microscopic hydrogen loading or unloading process in THF clathrate. At low temperatures, hydrogen molecules are not energetic enough to penetrate the potential barrier between the small cages but at high enough temperatures the occurrence of such a process could be observed in neutron scattering spectra. While only a slight increase in the signal of IINS spectra of mixed THF/H_2 clathrate hydrate is observed up to 60 K, this signal shows a sudden jump between 60 and 65 K, which persists at higher temperatures [CHO 10]. As this strong signal jump is attributed neither to macroscopic unloading of hydrogen at 120 K nor to any structural change, it strongly suggests hydrogen molecular diffusion in the crystalline lattice. QENS spectra on a picosecond timescale above 100 K confirm the observation of diffusive motions of the hydrogen molecules (Figure 1.23) [PEF 12]. However, these H_2

molecular translations, characterized by a low activation energy of 1.59 ± 0.06 kJ/mol, are found to occur within localized spherical area of the small s-II cage: the dynamical diameter of volume visited by the H_2 molecules varies from 2.08 Å at 250 K to 1.64 Å at 100 K and the diffusion constant ranges from 0.16 ± 0.03 rad ps^{-1} at 100 K to 0.49 ± 0.03 rad ps^{-1} at 250 K [PEF 12]. No diffusion between the cages is observed in the picosecond time scale [PEF 12, CHO 10, TAI 07]. Regardless of temperature and pressure, the atomic displacement factor for the D atom of the D_2 molecule in the small cages is about 0.17 Å² according to NP. Such a value is close to what may be expected for a tumbling D_2 molecule that cannot leave the small cage on the picosecond timescale [MUL 08], in agreement with the QENS analysis [PEF 12].

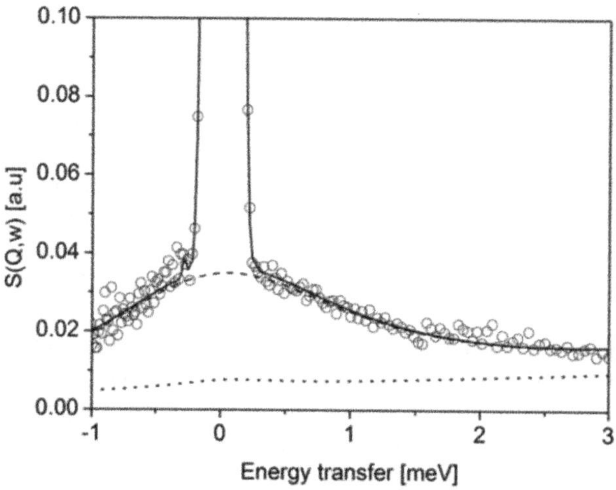

Figure 1.23. *Quasi-elastic neutron scattering (QENS) spectrum of the H_2-TDF-17D_2O (TDF stays for deuterated THF) at Q = 1.8 $Å^{-1}$. The dotted line represents the QENS signal of H_2 confined into the small cage of the s-II clathrate hydrate. The dotted line represents the QENS contribution arising from the deuterated storing clathrate hydrate TDF-17D_2O. The elastic line is a direct signature of the non-existence of intercage H_2 diffusion. NEAT @HZB (Berlin, Germany), $\Delta E \approx 100$ µeV, $\lambda_0 = 5$ Å. Adapted from [PEF 12].*

1.4.4. *Ionic clathrate hydrates and semiclathrates*

Another research area focuses on the specific properties met in the case of ionic clathrate hydrates. In such systems, the ionicity originates from the incorporation of ionic species mainly as guest molecules. The inclusion of ionic guest molecules (such as bases, strong acids or quaternary ammonium salts) leads to the formation of a water substructure containing ionic defects, through the incorporation of counter-ions within the water framework. Such chemical defects – leading to the violation of the ice rule – induce modification of the intrinsic properties of the clathrate hydrates. Their structural properties exhibit a large variability ranging from a semiclathrate structure of quaternary ammonium salts [SLO 08] to a rich structural phase diagram depending on the hydration number as in the case of strong acids [BOD 55, MOO 87, CHA 08]. The thermodynamics of gas hydrates, such as their heat capacities, is modified by the addition of ionic species such as KOH [YAM 88] or ammonium salts [SAN 14]. Among the fascinating properties of ionic clathrate hydrates made with alkyl ammonium hydroxide [BOR 96, OPA 95, CAP 94] or strong acids [CHA 08, HUA 88, ASC 94], their superprotonic conductivity is due to a supermobility of protons in the host substructure [DES 04, DES 13, BED 14]. The modification of proton dynamics promoted by the addition of ionic species in the host substructure may also switch the thermal conductivity from an abnormal "glass-like" behavior to a standard crystalline behavior [KRI 08]. Recently, it has been shown that the substitution of few THF molecules by strong acid molecules within the s-II THF clathrate hydrate modifies the thermodynamics and the cage flexibility of the clathrate hydrate [DES 15]. Finally, unusual properties of ionic clathrate hydrates also include their superoxide ion formation [CHA 10], superexchange-like interaction [SHI 11a] or abnormal positioning of host molecules [SHI 14].

Finally, due to the ionic interactions between host and guest molecules, ionic clathrate hydrates exhibit many original physicochemical properties. These unique features make them particularly promising in various practical application fields such as solid electrolytes, gas sensors and gas storage. This section will address two specific properties linked to ionic clathrate hydrates and for which neutron scattering provides invaluable information: the potentiality for storing gaseous species as mainly investigated by means of NP and the microscopic mechanism of superprotonic conductivity as revealed by means of QENS.

1.4.4.1. *Small molecule storage in semiclathrates*

The development of fast and reversible methods for storing gas is of utmost importance (e.g. hydrogen storage [VEL 14], carbon dioxide capture and separation [YAN 08]). New alternatives propose improved gas encapsulation capacities in clathrate hydrate by storing the gaseous molecules both inside the clathrate cavities and as molecular species participating to the host lattice. Semiclathrates formed with quaternary ammonium salts are good candidates for that purpose. Semiclathrates are named from the fact that, while the cationic part of these typically ionic compounds occupies the cages of hydrate structure like a guest molecule, its anionic part is involved in the lattice formation along with water molecules. These compounds have been extensively investigated as promoters for gas storage because of their improved stability and ease of hydrate formation close to room temperature and pressure compared to s-I and s-II hydrates (Figure 1.12).

In the particular case of hydrogen storage, semiclathrate hydrates such as tetrabutyl ammonium borohydride (TBABh) semiclathrate has been reported as a very promising hybrid material [SHI 09]. The improved storage capacities also arise from both guest-based and host-based atomic hydrogen generation [KOH 12]. The phase equilibrium data from the literature of mixed hydrogen clathrate with tetrabutyl phosphonium bromine, trimethyl amine (TMA) or tetrabutyl ammonium bromide (TBAB), fluoride (TBAF), chloride (TBAC) and nitrate (TBAN) are reported in the Figure 1.24 [VEL 14]. As revealed by this figure, the hydrogen/TBAF mixed hydrate appears to be the most stable hydrate formed at room temperature. However, as for the other mixed hydrogen clathrate formed with semiclathrate promoters, the storage capacity of hydrogen is less than that of hydrogen/THF mixed hydrates [VEL 14]: the maximum hydrogen storage capacity reported so far for the semiclathrate is in TBABh (0.5 wt%).

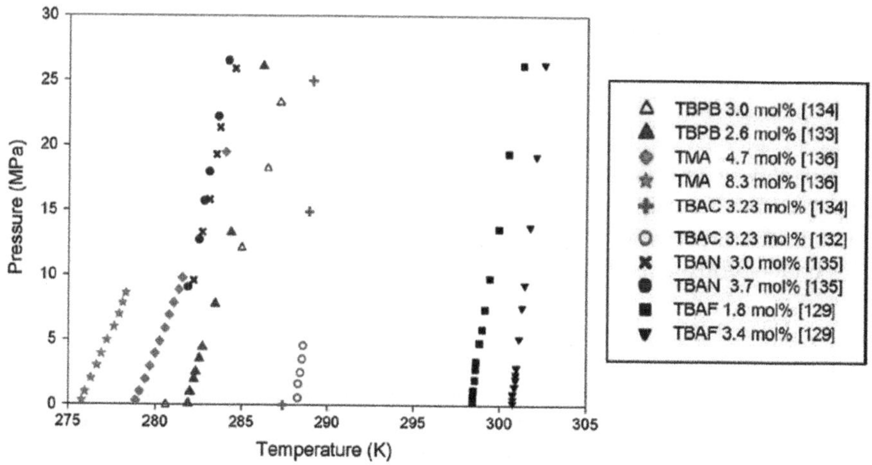

Figure 1.24. *Phase equilibrium data for mixed hydrogen hydrates formed in the case of various semiclathrate promoters at different concentration (from [VEL 14]). For a color version of this figure, see www.iste.co.uk/broseta/hydrates1.zip*

The inclusion of small molecules such as H_2, D_2, N_2, O_2 or CH_4 as secondary guests has been investigated in tetramethylammonium hydroxide (TMAO) and perchloric acid clathrates hydrates by means of powder NP [SHI 11b]. TMAOD (deuterated TMAOH) can form several clathrate structures but the inclusion of small non-ionic guest molecules in TMAOD–$16D_2O$ induces its transformation in the s-II structure in which small gas molecules occupy the small cages and TMA cations are inserted inside large cages with the corresponding hydroxide group incorporated in the host lattice [SHI 08, CHA 10]. $DClO_4$–$5.5D_2O$ usually crystallizes within the s-I structure [MOO 87]: the acidic excess protons are delocalized on the host lattice and the $DClO_4^-$ anions occupy both cavities. Moreover, small non-ionic guest molecules can be included in the small cages over the anions under favorable host–guest interactions. Neither the basic s-II clathrate nor the acid s-I clathrate is profoundly affected by the size and mass of the secondary guest [SHI 11b]. The lattice expansion of binary s-II TMAOD hydrates clearly seems dominantly affected by the charge interaction between the lattice hydroxide ions and the secondary guest, especially in the presence of electronegative O_2 guests. Furthermore, the thermal expansivity of the mixed perchloric acid clathrate hydrates is considerably higher than that measured for non-ionic s-I hydrates. This may be explained by partial incorporation of ClO_4^- oxygen atoms in the host framework (through the occupation of host water vacancies [MOO 87, SHI 11a]), leading to an extraordinary host–guest interaction.

1.4.4.2. *Superprotonic conductors*

The superconductivity, the relatively soft formation conditions and the high melting temperature of ionic clathrate hydrates render these compounds promising as solid proton conductor materials. In acidic clathrate hydrates, the high conductivity is associated with the delocalization of the acidic proton in the cage structure, thus formed by hydronium ions in addition to the water molecules. Such ionic defects induce the violation of the ice rule and are dynamically delocalized. This proton diffusion is observed within the host lattice framework.

As detailed previously in sections 1.3.2 and 1.4.3.4, QENS is the appropriate technique to access displacements of hydrogen nuclei with spatial and time resolution. Using the time-filtering property of the instrumental resolution, QENS have thus enabled the disentangling of the host molecule dynamics in the $HClO_4$ and HPF_6 acid clathrate hydrates. Such investigations provide a comprehensive model for proton diffusion in ionic clathrate [DES 04, DES 13, BED 14].

While perchloric acid clathrate adopts the s-I structure [MOO 87], the hexafluorophosphoric acid clathrate hydrate crystallizes either in the s-I structure or in the type VII cubic structure (denoted s-VII), depending on the hydration number (typically varying between 4 and 7.67 H_2O molecules per HPF_6 acidic molecules).

The s-VII unit cell (space group *Im3m* with ≈7.7Å) contains 24 water molecules forming one single $4^6 6^8$ type of cages, all occupied by PF_6^- ions, leading to a ratio of six water molecules per HPF_6 molecules [BOD 55]. A structural phase transition between s-I and s-VII structures is observed at ca. 230 K for hydration numbers greater than 6 [CHA 08, BED 14].

Proton diffusion in s-I $HClO_4$–$5.5H_2O$ clathrate has been characterized by a diffusion coefficient reaching 3.5×10^{-8} cm²/s at 220 K and an activation energy of about 0.3 eV [DES 04]. In the s-I phase of the HPF_6 clathrate hydrate (i.e. at temperatures lower than 230 K for the studied sample), the diffusion coefficient of protons was found smaller than ca. 10^{-7} cm²/s [BED 14], which is in agreement with results in the perchloric acid clathrate hydrate adopting the same structure. In contrast, in the s-VII phase of HPF_6 clathrate hydrate (i.e. between ca. 230 K and room temperature), the long-range proton diffusion coefficient was measured up to 3.85×10^{-6} $cm^2 \cdot s^{-1}$ at 275 K, with an activation energy of about 0.19 eV. To our knowledge, such diffusion coefficients appear to be one of the most important ones measured so far in ionic clathrate hydrates.

For both clathrate hydrates, excellent agreement between data from proton pulse field gradient NMR (^{1}H PFG NMR) and QENS experiments (which characterize proton diffusion at long distance and at a molecular scale, respectively) lead the representation of long-range proton diffusion microscopic mechanism with the help of the Chudley–Elliot jump diffusion model [DES 04, BED 14], indicating that the proton jump occurs between host oxygen sites. On a molecular scale, the proton diffusion coefficient has indeed been characterized by its mean residence time and its mean jump distance, because of the momentum transfer sensitivity of QENS technique. The mean jump distance was measured as 2.8 and 2.79 Å for $HClO_4$ and HPF_6 acid clathrate, respectively. Such values match the O–O distance of water cages. The mean residence time of protons on each crystallographic site was determined as 3.7 ns at 220 K for the perchloric acid case. In the HPF_6 case, the higher proton conductivity is related to significantly shorter mean residence times: 0.2, 0.052 and 0.03 ns at 230, 255 and 280 K, respectively.

From these results, the superconductivity of acidic clathrate hydrates, attributed to proton diffusion along the clathrate aqueous framework, appears to result from a series of protons jumps from one to another oxygen, which is in good agreement with the Grotthuss mechanism. Grotthuss mechanism assumes that protons are transferred along hydrogen bonds from one water molecule of the host framework to another with the assistance of the reorganization of this diffusing proton environment, including lattice molecule reorientation. According to this mechanism, several localized dynamical processes occurring on a broad timescale are thus involved. In the HPF_6 acid clathrate, the mean time associated with these localized dynamical processes has been measured as 2.5 ps at 255 K with an activation energy of 0.24 ± 0.04 eV [BED 14]. Localized diffusive motions of the proton surrounding the oxygen sites thus occur more frequently than the proton jumps between oxygen sites over a long-range. However, despite this timescale difference, the energy barrier associated with the localized diffusive motions appears slightly higher than that of long-range diffusion (about 0.19 eV). The limiting step for the proton conductivity thus occurs on local spatial scale, which confirms the key role played by the surrounding of the acidic proton in the long-range proton diffusion.

QENS experiments performed by tuning the observation time to the timescale of the probed dynamical process met in the perchloric acid clathrate hydrate at 220 K have led to disentangle three different localized dynamical processes occurring on this clathrate lattice. First, water molecules reorientations are reproduced well by a model describing proton jumps between two sites separated by 1.45 Å on which the mean residence time is about 0.7 ns at 220 K [DES 04]. These motions are characterized by an activation energy of about 0.18 eV and can be considered as the limiting elementary step in the mechanism of the proton conductivity met in the perchloric acid clathrate hydrate. Then, two dynamic elementary processes are identified for the hydronium ions: hydronium reorientations coupled to intermolecular proton transfer [DES 13]. Hydronium ions of the clathrate lattice are found to undergo reorientations over four different orientations tetrahedrally distributed with a jump distance of about 1.31 Å and a characteristic time of about 42 ps at 220 K. The transfer of a hydronium proton within the hydrogen bridge formed between a "hydronium" oxygen site and a neighboring "water" oxygen site is described by means of a jump model between two proton sites separated by about 0.96 Å and occurs with a characteristic time of about 1.4 ps at 220 K.

While this jump distance is in full agreement with the structure determination, the values obtained for both H_2O and H_3O^+ molecular reorientations appear shorter than the 1.6 Å value expected assuming a "standard" configuration of these molecules in the clathrate lattice [MOO 87]. This suggests that the oxygen atoms do not coincide with the center of rotation of the reorientation motions. Such an assumption is supported by results from X-ray diffraction analysis, which show large isotropic thermal parameters of cage atoms (values between 2.05 and 2.95 Å^2 for oxygen atoms and of 4 Å^2 for the protons) and indicate that the distribution of proton site around the oxygen atom is slightly distorted from perfect tetrahedra [MOO 87]. As shown in the examples reported in Figure 1.25, the complete model including components due to long range proton diffusion, to H_2O and H_3O^+ reorientational motions and to proton transfers through hydrogen bonds lead to an excellent reproduction of the perchloric acid clathrate QENS spectra whatever the observation times (energy resolution) and the momentum transfers are [DES 13].

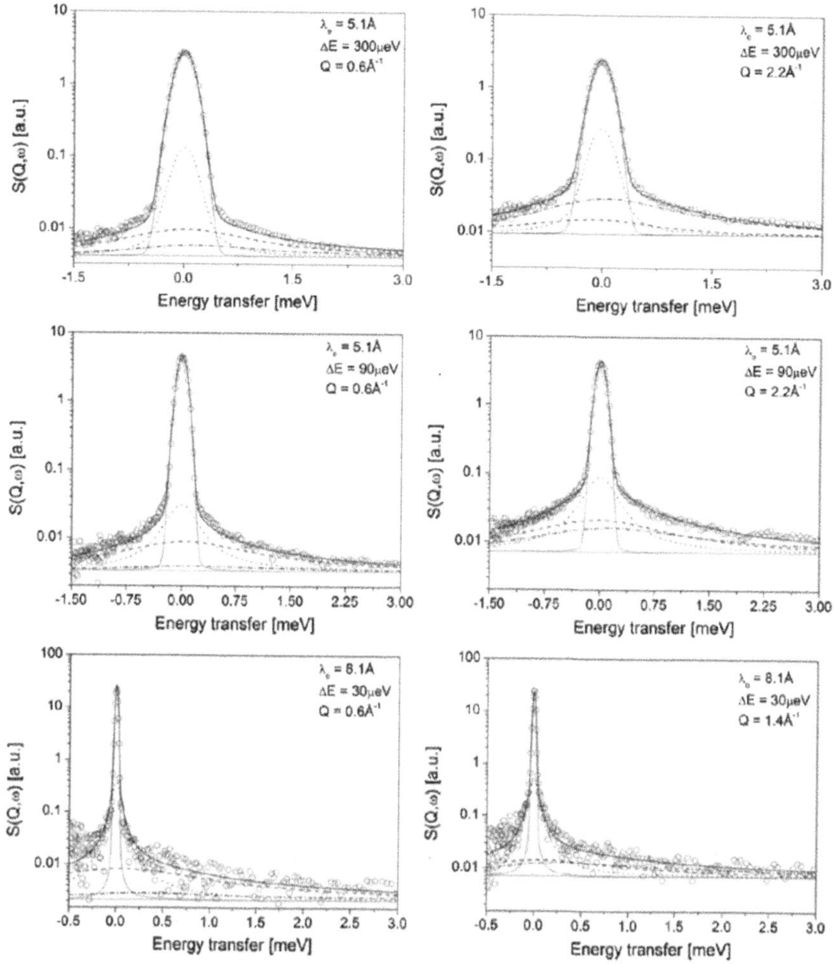

Figure 1.25. *Experimental (symbols) and fitted (thick continuous line) QENS spectra of perchloric acid clathrate at 220 K. QENS contributions due to water molecule reorientations and long-range proton diffusion are shown as thin continuous lines. The dotted line represents the contribution related to the lattice hydronium ions reorientation, the dashed line represents the contribution due to proton transfer within hydrogen bond and the dashed-dotted line stands for the contribution related to the coupling of these two phenomena (from [DES 13])*

1.5. Concluding remarks

The array of investigations carried out on clathrate hydrates is probably a result of the peculiar implications in the broad field of energy, technology and environment. Whatever the hydrate research area, a fundamental understanding of the microscopic mechanisms governing the properties specific to their cage structure constitutes the necessary basis for further developments. This chapter has reviewed the contribution of neutron scattering techniques to fundamental and applied issues. In addition to the description of the basics of neutron scattering, the invaluable information revealed by these techniques has been reviewed not only for structural aspects, but also for spectroscopic aspects, providing information ranging from Brownian relaxations to intramolecular vibrations. Detailed examples are treated, when possible, in conjunction with applications of gas hydrates, e.g. in the fields of energy recovery and hydrogen storage. The main information includes thermodynamic aspects (i.e. knowledge of phase boundaries and equilibrium guest compositions) and the kinetic features associated with the formation of clathrate hydrates, which play a role in the determination of limitations and of viable approaches for gas replacement (e.g. CO_2–CH_4) and in the design of systems overcoming typical limitations of kinetics, mass and heat transfer for gas storage. NP is particularly appropriate to structural investigations, but also to follow the structural changes that occur as a sample changes under "extreme" conditions (high pressure, low temperature). This technique has enabled some recent successes in studying the clathrate hydrate guest occupancy, kinetics of formation/ decomposition, guest replacement reactions (especially when accompanied by a structural transition) and also provided useful information on the time dependency of hydrate transformations. Furthermore, inelastic neutron scattering provides information about excitations met in clathrate hydrates with the advantage of no application of selection rules unlike in optical spectroscopies (e.g. IR, Raman). It has allowed for a detailed description of gas adsorption at the cage surfaces and gas molecule dynamics in the clathrate hydrate cages. Finally, QENS provides access to relaxation processes; it has enabled us to disentangle the dynamics of clathrate hydrates occurring on a broad timescale and has for instance, provided direct experimental evidence about the fundamental aspect of inter- or intracage diffusion. In addition to "standard" properties accessed by means of neutron scattering techniques, such studies provide spectroscopic signatures of clathrate hydrates that might find numerous applications in astrophysics (hydrates might be involved in numerous models of planets and comets to geology (e.g. for *in situ* mapping of gas hydrates in deep ocean) and in various technological area (e.g. for improving the gas storage capacity in clathrate hydrates).

As detailed in this chapter, neutron spectroscopy techniques are very useful tools to serve various issues of "hydrate science" because of the striking features of the neutrons: they are neutral particles (matter is transparent to neutron), they interact with the nuclei rather than with the electronic cloud (neutron is sensitive to light element like hydrogen), they have (de Broglie) wavelength in the range of interatomic distances, their mass are in the atomic mass range and they carry similar energies and momentum than those of clathrate hydrates. From a methodological point of view, the contribution of neutron scattering is significant and complementary to results obtained with other techniques such as X-ray scattering, Raman scattering or NMR. A powerful advantage of neutron scattering resides in the basic observable reached experimentally (via Fourier transforms): the time evolution of atomic positions. Such observable can be "easily" accessed with the help of numerical modeling by means of *ab initio* or classical simulations. The possibility of comparing experimental and theoretical results, by using such a "simple" common observable, gives an extraordinary rich opportunity for future research in clathrate science and other fields. The limitation of neutron scattering for the investigation of clathrate hydrates (and for many other systems) is probably linked to the difficulty of preparing a large single crystal. Nevertheless, this limitation will probably be overcome in the near future with the opportunity of using small samples because of the construction of high-brilliance neutron source like the European Spallation Source at Lund (Sweden) and to the development of highly sensitive neutron instruments on existing high-brilliance reactor-based sources and on future small/medium-brilliance sources. Such technological evolution of the neutron sources and instruments will finally be a real breakthrough for *in situ* or *operando* experiments on clathrate hydrates.

1.6. Bibliography

[ALA 07] ALAVI S., RIPMEESTER J.A., "Hydrogen gas migration through clathrate hydrate cages", *Angewandte Chemie International Edition*, vol. 46, pp. 6102–6105, 2007.

[ALA 09] ALAVI S., SUSILO R., RIPMEESTER J.A., "Linking microscopic guest properties to macroscopic observables in clathrate hydrates: guest-host hydrogen bonding", *Journal of Chemical Physics*, vol. 130, p: 174501, 2009.

[AND 07] ANDERSON R., CHAPOY A., TOHIDI B., "Phase relations and binary clathrate hydrate formation in the system H_2-THF-H_2O", *Langmuir*, vol. 23, pp. 3440–3444, 2007.

[ASC 94] ASCHRAFI-MAHABADI S., CAPPADONIA M., STIMMING U., "Proton transport in solid electrolytes with clathrate structure", *Solid State Ionics*, vol. 70–71, pp. 311−315, 1994.

[BAC 01] BACH-VERGES M., KITCHIN S.J., HARRIS K.D.M. *et al.*, "Dynamic properties of the tetrahydrofuran clathrate hydrate, investigated by solid state 2H NMR spectroscopy", *Journal of Physical Chemistry*, vol. 105, pp. 2699–2706, 2001.

[BAU 03] BAUMERT J., GUTT C., SHPAKOV V.P. *et al.*, "Lattice dynamics of methane and xenon hydrate: observation of symmetry-avoided crossing by experiment and theory", *Physical Review B*, vol. 68, p. 174301, 2003.

[BED 91] BEDNAREK J., ERICKSON R., LUND A. *et al.*, "The return of the trapped electron in x-irradiated clathrate hydrates. En ESR investigation", *Journal of the American Chemical Society*, vol. 113, pp. 8990–8991, 1991.

[BED 96] BEDNAREK J., LUND A., SCHLICK S., "Unstable intermediates in x-irradiated hydrates: ESR and ENDOR of tetramethylammonium hydroxide pentahydrate (TMNOH)", *Journal of Physical Chemistry*, vol. 100, pp. 3910–3916, 1996.

[BED 14] BEDOURET L., JUDEINSTEIN P., OLLIVIER J. *et al.*, "Proton diffusion in the hexafluorophosphoric acid clathrate hydrate", *Journal of Physical Chemistry B*, vol. 118, pp. 13357–13364, 2014.

[BEE 86] BEE M., *Quasielastic Neutron Scattering*, Adam Hilger, 1986.

[BOD 55] BODE V., TEUFER G., "Die kristallstruktur der hexafluorophos-phorsaüre", *Acta Crystallographica*, vol. 8, pp. 611–614, 1955.

[BOR 96] BORKOWSKA Z., TYMOSIAK A., OPALLO M., "Conductivity of stoichiometric $(CH_3)_4NOH$ clathrate hydrates", *Journal of Electroanalytical Chemistry*, vol. 406, pp. 109–117, 1996.

[BOS 09] BOSWELL R., "Is gas hydrate energy within reach?", *Science*, vol. 325, p. 957, 2009.

[BUC 09] BUCH V., DEVLIN J.P., MONREAL I.A. *et al.*, "Clathrate hydrates with hydrogen-bonding guests", *Physical Chemistry Chemical Physics*, vol. 11, pp. 10245–10265, 2009.

[BRE 99] BREWER P.G., FRIEDERICH C., PELTZER E.T. *et al.*, "Direct experiments on the ocean disposal of fossil fuel CO_2", Science, vol. 284, pp. 943–945, 1999.

[CAP 94] CAPPADONIA M., KORNYSHEV A.A., KRAUSE S. *et al.*, "Low temperature proton transport in clathrates", *Journal of Chemical Physics*, vol. 101, pp. 7672–7682, 1994.

[CHA 05] CHATTI I., DELAHAYE A., FOURNAISON L. *et al.*, "Benefits and drawbacks of clathrate hydrates: a review of their areas of interest", *Energy Conversion and Management*, vol. 46, pp. 1333–1343, 2005.

[CHA 07] CHAPOY A., ANDERSON R., TOHIDI B., "Low-pressure molecular hydrogen storage in semi-clathrate hydrates of quaternary ammonium compounds", *Journal of the American Chemical Society,* vol. 129, pp. 746–747, 2007.

[CHA 08a] CHA J.H., SHIN K., CHOI S. *et al.*, "Maximized proton conductivity of the HPF_6 clathrate hydrate by structural transformation", *Journal of Physical Chemistry C*, vol. 112, pp. 13332–13335, 2008.

[CHA 08b] CHA J.H., SHIN K., CHOI S. *et al.*, "Ionic conductivity enhancement due to coguest inclusion in the pure ionic clathrate hydrates", *Journal of Physical Chemistry C*, vol. 112, pp. 10573–10578, 2008.

[CHA 09a] CHA J.H., LEE W., LEE H., "Thermal stability and ionic conductivity of the ionic clathrate hydrates incorporated with potassium hydroxide", *Journal of Materials Chemistry*, vol. 19, pp. 6542–6547, 2009.

[CHA 09b] CHA J.H., LEE W., LEE H., "Hydrogen gas sensor based on proton-conducting clathrate hydrate", *Angewandte Chemie International Edition,* vol. 48, pp. 8687–8690, 2009.

[CHA 10] CHA M., SHIN K., KWON M. *et al.*, "Superoxide ions entrapped in water cages of ionic clathrate hydrates", *Journal of the American Chemical Society*, vol. 132, pp. 3694–3696, 2010.

[CHO 10] CHOI Y.N., SUNGIL PARK J.M., STRASSLE T., *et al.*, "Dynamics of hydrogen molecules in the channels of binary THF-H_2 clathrate hydrate and its physicochemical significance on hydrogen storage", *International Journal of Hydrogen Energy*, vol. 35, pp. 13068–13072, 2010.

[CLA 01] CLARK M., BISHNOI P.R., "Determination of the activation energy and intrinsic rate constant of methane gas hydrate decomposition", *Canadian Journal of Chemical Engineering*, vol. 79, pp. 143–147, 2001.

[COL 02] COLLETT T.S., "Energy resource potential of natural gas hydrates", *AAPG Bulletin*, vol. 86, pp. 1971–1992, 2002.

[COL 13] COLOGNESI D., CELLI M., ULIVI L. *et al.*, "Neutron scattering measurements and computation of the quantum dynamics of hydrogen molecules trapped in the small and large cages of clathrate hydrates", *Journal of Physical Chemistry A*, vol. 117, pp. 7314–7326, 2013.

[CRA 09] CRANGANU C., "In-situ thermal stimulation of gas hydrates", *Journal of Petroleum Science and Engineering*, vol. 65, pp. 76–80, 2009.

[DAV 73] DAVIDSON D.W., RIPMEESTER J.A., "NMR, NQR and dielectric properties of clathrates", in ATWOOD J.L., DAVIES J.E.D., MACNICOL D.D., *Inclusion Compounds*, vol. 3, Academic Press, London, 1973.

[DEM 10] DEMIRBAS A., "Methane hydrates as potential energy resource: part 2-methane production processes from gas hydrates", *Energy Conversion and Management*, vol. 51, pp. 1562–1571, 2010.

[DES 04] DESMEDT A., STALLMACH F., LECHNER R.E. *et al.*, "Proton dynamics in the perchloric acid clathrate hydrate $HClO_4$-5.5H_2O", *Journal of Chemical Physics*, vol. 121, pp. 11916–11926, 2004.

[DES 11] DESMEDT A., SOETENS J.C., PRAGER M. *et al.*, "Dynamics of methyl iodide clathrate hydrate, investigated by MD simulations and QUENS experiments", *Journal of Physical Chemistry C*, vol. 115, pp. 12689–12701, 2011.

[DES 12] DESMEDT A., BEDOURET L., PEFOUTE E. *et al.*, "Energy landscape of clathrate hydrates", *European Physical Journal Special Topics*, vol. 213, pp. 103–127, 2012.

[DES 13] DESMEDT A., LECHNER R.E., LASSEGUES J.C. *et al.*, "Hydronium dynamics in the perchloric acid clathrate hydrate", *Solid State Ionics*, vol. 252, pp. 19–25, 2013.

[DES 15] DESMEDT A., MARTIN-GONDRE L., NGUYEN T.T. *et al.*, "Modifying the flexibility of water cages by co-including acidic species within clathrate hydrate", *Journal of Physical Chemistry C*, vol. 119, pp. 8904–8911, 2015.

[DU 11] DU J.W., LIANG D.Q., DAI X.X. *et al.*, "Hydrate phase equilibrium for the (hydrogen + tertbutylamine+water) system", *Journal of Chemical Thermodynamics*, vol. 43, pp. 617–621, 2011.

[DYA 99] DYADIN Y.A., LARIONOV E.G., ALADKO E.Y. *et al.*, "Clathrate formation in water-noble gas (hydrogen) systems at high pressures", *Journal of Structural Chemistry*, vol. 40, pp. 790–795, 1999.

[DYA 99] DYADIN Y.A., LARIONOV E.G., MANAKOV A.Y. *et al.*, "Clathrate hydrates of hydrogen and neon", *Mendeleev Communications*, vol. 9, pp. 209–210, 1999.

[ENG 03] ENGLISH N.J., MACELROY J.M.D, "Structural and dynamical proeprties of methane clathrate hydrates", *Journal of Computational Chemistry*, vol. 24, pp. 1569–1581, 2003.

[ENG 09] ENGLISH N.J., TSE J.S., CAREY D., "Mechanism for thermal conduction in various polymorphs of methane hydrate", *Journal of Physical Review* B, vol. 80, pp. 134306, 2009.

[FAL 11] FALENTY A., GENOV G.Y., HANSEN T.C. *et al.*, "Kinetics of CO_2 hydrate formation from water frost at low temperatures: experimental results and theoretical model", *Journal of Physical Chemistry C*, vol. 115, pp. 4022–4032, 2011.

[FAL 13] FALENTY A., SALAMATIN A.N., KUHS W.F., "Kinetics of CO_2-hydrate formation from ice powders: data summary and modelling extended to low temperatures", *Journal of Physical Chemistry C*, vol. 117, pp. 8443–8457, 2013.

[FLO 04] FLORUSSE L.J., PETERS C.J., SCHOONMAN J. *et al.*, "Stable low-pressure hydrogen clusters stored in a binary clathrate hydrate", *Science*, vol. 306, pp. 469–471, 2004.

[FOU 04] FOURNAISON L., DELAHAYE A., CHATTI I. *et al.*, "CO_2 hydrates in refrigeration processes", *Industrial & Engineering Chemistry Research,* vol. 43, pp. 6521–6526, 2004.

[GEN 04] GENOV G., KUHS W.F., STAYKOVA D.K. *et al.*, "Experimental studies on the formation of porous gas hydrates", *American Mineralogist*, vol. 89, pp. 1228–1239, 2004.

[GOE 06] GOEL N., "In situ methane hydrate dissociation with carbon dioxide sequestration: current knowledge and issues", *Journal of Petroleum Science and Engineering*, vol. 51, pp. 169–184, 2006.

[GRA 08] GRAUE A., KVAMME B., BALDWIN B.A. *et al.*, "MRI visualization of spontaneous methane production from hydrates in sandstone core plugs when exposed to CO_2", *SPE Journal*, vol. 13, pp. 146–152, 2008.

[GRI 12] GRIM R.G., KERKAR P.B., SHEBOWICH M. *et al.*, "Synthesis and characterization of sI clathrate hydrates containing hydrogen", *Journal of Physical Chemistry C*, vol. 116, pp. 18557–18563, 2012.

[GRI 14] GRIM R.G., BARNES B.C., LAFOND P.G. *et al.*, "Observation of interstitial moleculahydrogen in clathrate hydrates", *Angewandte Chemie International Edition*, vol. 53, pp. 10710–10713, 2014.

[GUT 01] GUTT C., PRESS W., HULLER A. *et al.*, "The isotope effect and orientational potentials of methane molecules in gas hydrates", *Journal of Chemical Physics*, vol. 114, pp. 4160–4170, 2001.

[HAA 09] HAACKE R.R., HYNDMAN R.D., PARK K.P. *et al.*, "Migration and venting of deep gases into the odean through hydrate-choked chimmeys offshore Korea", *Geology*, vol. 37, pp. 531–534, 2009.

[HAL 01] HALPERN Y., THIEU V., HENNING R.W. *et al.*, "Time-resolved in situ neutron diffraction studies of gas hydrate: transformation of structure II (sII) to structure I (sI)", *Journal of the American Chemical Society*, vol. 123, pp. 12826–12831, 2001.

[HAM 34] HAMMERSCHMIDT E.G., "Formation of gas hydrates in natural gas transmission lines", *Industrial & Engineering Chemistry Research,* vol. 26, pp. 851–855, 1934.

[HAN 87] HANDA Y.P., COOK J.G., "Thermal conductivity of xenon hydrate", *Journal of Physical Chemistry*, vol. 91, pp. 6321–6328, 1987.

[HAN 06] HANSEN J.P., MCDONALD I.R., *Theory of Simple Liquids*, 3rd ed., Elsevier, 2006.

[HAQ 99] HAQ B.U., "Methane in the deep blue sea", *Science*, vol. 285, pp. 543–544, 1999.

[HAS 07] HASHIMOTO S., SUGAHARA T., SATO H. *et al.*, "Thermodynamic stability of H_2 + tetrahydrofuran mixed gas hydrate in nonstoichiometric aqueous solutions", *Journal of Chemical and Engineering Data*, vol. 52, pp. 517–520, 2007.

[HEN 00] HENNING R.W., SCHULTZ A.J., THIEU V. *et al.*, "Neutron diffraction studies of CO_2 clathrate hydrate: formation from deuterated ice", *Journal of Physical Chemistry A*, vol. 104, pp. 5066–5071, 2000.

[HER 91] HERZOG H., GOLOMB D., ZEMBA S., "Feasibility, modeling and economics of sequestering power plant CO_2 emissions in the deep ocean", *Environmental Progress*, vol. 10, pp. 64–74, 1991.

[HES 06] HESTER K.C, STROBEL T.A., SLOAN E.D. *et al.*, "Molecular hydrogen occupancy in binary THF-H2 clathrate hydrates by high resolution neutron diffraction", *Journal of Physical Chemistry B*, vol. 110, pp. 14024–14027, 2006.

[HOU 96] HOUGHTON J.T., MEIRA FILHO L.G., CALLANDER B.A. *et al.*, *Climate Change 1995: The Science of Climate Change*, Cambridge University Press, New York, 1996.

[HOV 54] HOVE L., "Correlations in space and time and born approximation scattering in systems of interacting particles", *Physical Review*, vol. 95, p. 249, 1954.

[HUA 88] HUANG T.-H., DAVIS R.A., FRESE U. *et al.*, "Proton mobility in liquid and frozen $HClO_4$-$5.5H_2O$: NMR and conductivity measurements", *Journal of Physical Chemistry,* vol. 92, pp. 6874–6876, 1988.

[HWA 90] HWANG M.L., WRIGHT D.A., KAPUR A. *et al.*, "An experimental study of crystallization and crystal growth of methane hydrates from melting ice", *Journal of Inclusion Phenomena and Macrocyclic Chemistry*, vol. 8, pp. 103–116, 1990.

[IKE 99] IKEDA T., YAMAMURO O., MATSUO T. *et al.*, "Neutron diffraction study of carbon dioxide clathrate hydrate", *Journal of Physics and Chemistry of Solids*, vol. 60, pp. 1527–1529, 1999.

[ITA 67] ITAGAKI K., "Self-diffusion in single crystal ice", *Journal of Physical Society of Japan*, vol. 22, pp. 427–431, 1967.

[JAN 27] JANDER W.Z., "Reaktionen im festen Zustande bei höheren Temperaturen. Reaktionsgeschwindigkeiten endotherm verlaufender Umsetzungen", *Zeitschrift für anorganische und allgemeine Chemie*, vol. 163, pp. 1–30, 1927.

[JEF 96] JEFFREY G.A., "Hydrate inclusion compounds", in ATWOOD J.L., DAVIES J.E.D., MAC-NICOL D.D., *et al.* (eds), *Comprehensive Supramolecular Chemistry*, vol. 6, Pergamon, Oxford, 1996.

[KAS 08] KASTNER M., CLAYPOOL G., ROBERTSON G., "Geochemical constraints on the origin of the pore fluids and gas hydrate distribution at Atwater Valley and Keathley Canyon, northern Gulf of Mexico", *Mar. Petrol. Geol.*, vol. 25, pp. 860–872, 2008.

[KID 09] KIDA M., SUZUKI K., KAWAMURA T. *et al.*, "Characteristics of natural gas hydrates occurring in pore spaces of marine sediments collected from the eastern Nankai Trough, off Japan", *Energy & Fuels*, vol. 23, pp. 5580–5586, 2009.

[KIR 03] KIRSCHGEN T.M., ZEIDLER M.D., GEIL B. *et al.*, "A deuteron NMR study of the tetrahydrofuran clathrate hydrate. Part I: relaxation pathways", *Physical Chemistry Chemical Physics,* vol. 5, pp. 5243–5246, 2003.

[KOH 07] KOH C.A., SLOAN E.D., "Natural gas hydrates: recent advances and challenges in energy and environmental applications", *AIChE Journal*, vol. 53, pp. 1636–1643, 2007.

[KOH 12] KOH D.Y., KANG H., PARK J. *et al.*, "Atomic hydrogen production from semi-clathrate hydrates", *Journal of the American Chemical Society*, vol. 134, pp. 5560–5562, 2012.

[KOM 00] KOMAI T., YAMAMO Y., OHGA K., "Dynamics of reformation and replacement of CO_2 and CH_4 gas hydrates", *Ann. N. Y. Acad. Sci.*, vol. 912, pp. 272–280, 2000.

[KOM 02] KOMAI T., KAWAMURA T., KANG S. *et al.*, "In situ observation of gas hydrate behavior under high pressure by Raman spectroscopy", *J. Phys.: Condens. Matter*, vol. 14, pp. 11395–11400, 2002.

[KOM 13] KOMATSU H., OTA M., SMITH R.L. *et al.*, "Review of CO_2–CH_4 clathrate hydrate replacement reaction laboratory studies – properties and kinetics", *Journal of the Taiwan Institute of Chemical Engineering*, vol. 44, pp. 517–537, 2013.

[KRI 08] KRIVCHIKOV A.I., ROMANTSOVA O.O., KOROLYUK O.A., "The effect of proton ordering on the thermal conductivity of clathrate tetrahydrofuran hydrate", *Low Temp. Phys.*, vol. 34, pp. 648–654, 2008.

[KUH 06a] KUHS W.F., HANSEN T.C., "Time-resolved neutron diffraction studies with emphasis on water ices and gas hydrates", *Rev. Mineral. Geochem.*, vol. 63, pp. 171–204, 2006.

[KUH 06b] KUHS W.F., STAYKOVA D.K., SALAMATIN A.N., "Formation of methane hydrate from polydisperse ice powders", *Journal of Physical Chemistry B*, vol. 110, pp. 13283–13295, 2006.

[KUH 14] FALENTY A., HANSEN T.C., KUHS W.F., "Formation and properties of ice XVI obtained by emptying a type sII clathrate hydrate", *Nature*, vol. 516, pp. 231–233, 2014.

[KUM 09] KUMAR R., ENGLEZOS P., MOUDRAKOVSKI I. *et al.*, "Structure and composition of CO_2/H_2 and $CO_2/H_2/C_3H_8$ hydrate in relation to simultaneous CO_2 capture and H_2 production", *AIChE Journal*, vol. 55, pp. 1584–1594, 2009.

[LEE 03] LEE H., SEO Y., SEO Y.T. *et al.*, "Recovering methane from solid methane hydrate with carbon dioxide", *Angewandte Chemie International Edition,* vol. 42, pp. 5048–5051, 2003.

[LEE 04] LEE H., SEO Y., SEO Y.T. *et al.*, "Replacement of methane hydrate by carbon dioxide: C-13 NMR study for studying a limit to the degree of substitution", *Stud. Surf. Sci. Catal.*, vol. 153, pp. 495–500, 2004.

[LEE 05] LEE H., LEE J.-W., KIM Y.D. *et al.*, "Tuning clathrate hydrates for hydrogen storage", *Nature*, vol. 434, pp. 743–746, 2005.

[LIU 09] LIU Y., STRUMENDO M., ARASTOOPOUR H., "Simulation of methane production from hydrates by depressurization and thermal stimulation", *Industrial & Engineering Chemistry Research*, vol. 48, pp. 2451–2464, 2009.

[LIV 97] LIVINGSTON F.E., WHIPPLE G.C., GEORGE S.M., "Diffusion of HDO into single-crystal $H2^{16}O$ ice multilayers: comparison with $H2^{18}O$", *Journal of Physical Chemistry B*, vol. 101, pp. 6127–6131, 1997.

[LOK 04] LOKSHIN K.A., ZHAO Y., HE D. *et al.*, "Structure and dynamics of hydrogen molecules in the novel clathrate hydrate by high pressure neutron diffraction", *Phys. Rev. Lett.*, vol. 93, p. 125503, 2004.

[LOK 06] LOKSHIN K.A., ZHAO Y., "Fast synthesis method and phase diagram of hydrogen clathrate hydrate", *Appl. Phys. Lett.*, vol. 88, p. 13109, 2006.

[LOV 84] LOVESEY S.W., *Theory of Neutron Scattering from Condensed Matter*, Oxford University Press, 1984.

[MAT 11] MATSUMOTO R., RYU B.J., LEE S.R. *et al.*, "Occurrence and exploration of gas hydrate in the marginal seas and continental margin of the Asia and Oceania region", *Mar. Petrol. Geol.*, vol. 28, pp. 1751–1767, 2011.

[MAO 02] MAO W.L., MAO H.-K., GONCHAROV A.F. *et al.*, "Hydrogen clusters in clathrate hydrate", *Science*, vol. 297, pp. 2247–2249, 2002.

[MAO 04] MAO W.L., MAO H.-K., "Hydrogen storage in molecular compounds", *Proc. Natl. Acad. Sci. U.S.A.*, vol. 101, pp. 708–710, 2004.

[MAX 00] MAX M.D., *Natural Gas Hydrate: In Oceanic and Permafrost Environments*, Kluwer Academic Publishers, Boston, 2000.

[MIZ 87] MIZUNO Y., HANAFUSA N., "Studies of surface properties of ice using nuclear magnetic resonance", *J. Phys. Colloq.*, vol. 48, pp. 511–517, 1987.

[MOO 87] MOOTZ D., OELLERS E.-J., WIEBKE M., "First examples of type I clathrate hydrates of strong acids: polyhydrates of hexafluorophosphoric, tetrafluoroboric, and perchloric acid", *Journal of the American Chemical Society*, vol. 109, pp. 1200–1202, 1987.

[MOO 94] MOOTZ D., STABEL D., "Hydrates of weak and strong bases. 10. $Cs(Me_4N)_2(OH)_3$.cntdot.$14H_2O$: a metal ion in a clathrae hydrate polyhedral cage", *Journal of the American Chemical Society*, vol. 116, pp. 4141–4142, 1994.

[MUL 08] MULDER F.M., WAGEMAKER M., VAN EIJCK L. *et al.*, "Hydrogen in porous tetrahydrofuran clathrate hydrate", *Chem. Phys. Chem.*, vol. 9, pp. 1331–1337, 2008.

[MUR 09] MURSHED M.M., KUHS W.F., "Kinetic studies of methane-ethane mixed gas hydrates by neutron diffraction and raman spectroscopy", *Journal of Physical Chemistry B*, vol. 13, pp. 5172–5180, 2009.

[MUR 10] MURSHED M.M., SCHMIDT B.C., KUHS W.F., "Kinetics of methane-ethane gas replacement in clathrate hydrates studies by time-resolved neutron diffraction and raman spectroscopy", *Journal of Physical Chemistry A*, vol. 114, pp. 247–255, 2010.

[NAG 08] NAGAI Y., YOSHIOKA H., OTA M. *et al.*, "Binary hydrogen-tetrahydrofuran clathrate hydrate formation kinetics and models", *AIChE Journal*, vol. 54, pp. 3007–3016, 2008.

[NIS 92] NISHIKAWA N., MORISHITA M., UCHIYAMA M. *et al.*, "CO_2 clathrate formation and its properties in the simulated deep ocean", *Energy Conversion Management*, vol. 33, pp. 651–657, 1992.

[NOG 08] NOGAMI T., WATANABE S., "Development of natural gas supply chain by means of Natural Gas Hydrate (NGH)" *Proceedings of International Petroleum Technology Conference (IPTC 12880)*, Kuala Lumpur, Malaysia, December 3–5, 2008.

[OGA 06] OGAWA T., ITO T., WATANABE K. *et al.*, "Development of a novel hydrate-based refrigeration system: a preliminary overview", *Appl. Therm. Eng.*, vol. 26, pp. 2157–2167, 2006.

[OHG 96] OHGAKI K., TAKANO K., SANGAWA H. *et al.*, "Methane exploitation by carbon dioxide from gas hydrates-phase equilibria for CO_2-CH_4 mixed hydrate system", *J. Chem. Eng. Jpn*, vol. 29, pp. 478–483, 1996.

[OPA 95] OPALLO M., TYMOSIAK A., BORKOWSKA Z., "Conductivity of tetramethylammonium fluoride tetrahydrate", *Journal of Electroanalytical Chemistry*, vol. 387, pp. 47–52, 1995.

[OTA 05] OTA M., ABE Y., WATANABE M. *et al.*, "Methane recovery from mathane hydrate using pressurized CO_2", *Fluid Phase Equilibria*, vol. 228–229, pp. 553–559, 2005.

[OTA 07] OTA M., SAITO T., AIDA T. *et al.*, "Macro and microscopic CH_4-CO_2 replacement in CH_4 hydrate under pressurized CO_2", *AIChE Journal*, vol. 53, pp. 2715–2721, 2007.

[PAR 06] PARK Y., KIM D.Y., LEE J.W. *et al.*, "Sequestering carbon dioxide into complex structures of naturally occurring gas hydrates", *Proc. Natl. Acad. Sci.*, vol. 103, pp. 12690–12694, 2006.

[PAT 03] PATCHKOVSKII S., TSE J.S., "Thermodynamic stability of hydrogen clathrates", *Proc. Natl. Acad. Sci. USA*, vol. 100, pp. 14645–14650, 2003.

[PEF 10] PEFOUTE E., Vers une meilleure compréhension du stockage de l'hydrogène dans les clathrate hydrates, PhD Thesis, University of Bordeaux 1, 2010.

[PEF 12] PEFOUTE E., KEMNER E., SOETENS J.C. *et al.*, "Diffusive motion of molecular hydrogen confined in THS clathrate hydrate", *Journal of Physical Chemistry C*, vol. 116, pp. 16823–16829, 2012.

[PEF 16] PEFOUTE E., PRAGER M., RUSSINA M. *et al.*, "Quasi-elastic neutron scattering investigation of the guest molecule dynamics in the bromomethane clathrate hydrate", *Fluid Phase Equilibria*, vol. 413, pp. 116–122, 2016.

[POH 09] POHLMAN J.W., KANEKO M., HEUER V.B. *et al.*, "Methane sources and production in the northern Cascadia margin gas hydrate system", *Earth Planet Sci. Lett.*, vol. 287 pp. 504–512, 2009.

[PRA 04a] PRAGER M., PIEPER J., BUCHSTEINER A. *et al.*, "Methyl rotational potentials as a probe of the cage potential surface in methyl iodide clathrate", *Physica B*, vol. 350, pp. E399–E402, 2004.

[PRA 04b] PRAGER M., PIEPER J., BUCHSTEINER A. *et al.*, "Probing adsorption sites in a cubic II water clathrate cage by methyl group rotation of CH3I guest molecules", *J. Phys.: Condens. Matter*, vol. 16, pp. 7045–7061, 2004.

[PRA 05] PRAGER M., BAUMERT J., PRESS W. *et al.*, "Adsorption sites and rotational tunneling of methyl groups in cubic I methyl fluoride water clathrate", *Physical Chemistry Chemical Physics*, vol. 7, pp. 1228–1234, 2005.

[PRA 07] PRAGER M., DESMEDT A., ALLGAIER J. *et al.*, "Methyl group rotation and whole molecule dynamics in methyl bromide hydrate", *Phase Transitions*, vol. 80, pp. 473–488, 2007.

[PRA 08] PRAGER M., DESMEDT A., UNRUH T. *et al.*, "Dynamics and adsorption sites for guest molecules in methyl chloride hydrate", *J. Phys.: Condens. Matter*, vol. 20, p. 125219, 2008.

[REK 95] REKOSKE J.E., BARTEAU M.A., "In situ studies of carbonyl coupling: comparisons of liquid-solid and gas-solid reactions with reduced titanium reagents", *Industrial & Engineering Chemistry Research*, vol. 34, pp. 2931–2939, 1995.

[RIP 88] RIPMEESTER J.A., RATCLIFFE C.I., TSE J.S., "The nuclear magnteic resonance of 129Xe trapped in clathrates and some other solids", J. *Chem. Soc., Faraday Trans.*, vol. 84, pp. 3731–3745, 1988.

[ROS 81] ROSS R.G., ANDERSSON P., BACKSTROM G., "Unusual PT dependence of thermal conductivity for a clathrate hydrate", *Nature*, vol. 290, pp. 322–323, 1981.

[RYD 07] RYDZY M.B., SCHICKS J.M., NAUMANN R. *et al.*, "Dissociation enthalpies of synthesized multicomponent gas hydrates with respect to guest composition and cage occupancy", *Journal of Physical Chemistry B*, vol. 111, pp. 9539–9545, 2007.

[RYU 09] RYU B.J., RIEDEL M., KIM J.H. *et al.*, "Gas hydrates in the western deep-water Ulleung Basin, East Sea of Korea", *Mar. Petrol. Geol.*, vol. 26, pp. 1483–1498, 2009.

[SAJ 92] SAJI A., YOSHIDA H., SAKAI M. *et al.*, "Fixation of carbon dioxide by clathrate-hydrate", *Energy Conversion Management*, vol. 33, pp. 643–649, 1992.

[SAN 14] SANGWAI J.S., OELLRICH L., "Phase equilibrium of semiclathrate hydrates of methane in aqueous solutions of tetra-n-butyl ammonium bromide (TBAB) and TBAB–NaCl", *Fluid Phase Equilibria*, vol. 367, pp. 95–102, 2014.

[SCH 01] SCHLAPBACH L., ZSTTEL A., "Hydrogen-storage materials for mobile applications", *Nature*, vol. 414, pp. 353–358, 2001.

[SCH 03] SCHOBER H., ITOH H., KLAPPROTH A. *et al.*, "Guest-host coupling and anharmonicity in clathrate hydrates", *Eur. Phys. J. E*, vol. 12, pp. 41–49, 2003.

[SCH 05] SCHUTH F., "Technology-Hydrogen and hydrates", *Nature*, vol. 43, pp. 712–713, 2005.

[SEA 92] SEARS V.V.F., "Neutron scattering lengths and cross sections", *Neutron News*, vol. 3, pp. 26–37, 1992.

[SHI 08] SHIN K., CHOI S., CHA J.H. *et al.*, "Structural transformation due to co-host inclusion in ionic clathrate hydrates", *Journal of the American Chemical Society*, vol. 130, pp. 7180–7181, 2008.

[SHI 09] SHIN K., KIM Y., STROBEL T.A. *et al.*, "Tetra-*n*-butylammonium borohydride semiclathrate: a hybrid material for hydrogen storage", *Journal of Physical Chemistry A*, vol. 113, pp. 6415–6418, 2009.

[SHI 10] SHIN K., CHA J.O., SEO Y. *et al.*, "Physico-chemical properties of ionic clathrate hydrates", *Chem. Asian J.*, vol. 5, pp. 22–34, 2010.

[SHI 11a] SHIN K., CHA M., LEE W. *et al.*, "Superexchange-like interaction of encaged molecular oxygen in nitrogen-doped water cages of clathrate hydrates", *Journal of the American Chemical Society*, vol. 133, pp. 20399–20404, 2011.

[SHI 11b] SHIN K., LEE W., CHA M. *et al.*, "Thermal expansivity of ionic clathrate hydrates including gaseous guest molecules", *Journal of Physical Chemistry B*, vol. 115, pp. 958–963, 2011.

[SHI 14] SHIN K., CHA M., LEE W. *et al.*, "Abnormal proton positioning of water framework in the presence of paramagnetic guest within ion-doped clathrate hydrate host", *Journal of Physical Chemistry C*, vol. 118, pp. 15193–15199, 2014.

[SKO 72] SKOELD K., ROWE J.M., OSTROWSKI G.E. *et al.*, "Coherent and incoherent scattering laws of liquid argon", *Phys. Rev. A - At. Mol. Opt. Phys.*, vol. 6, pp. 1107–1131, 1972.

[SLO 03] SLOAN E.D., "Fundamental principles and applications of natural gas hydrates", *Nature*, vol. 426, pp. 353–363, 2003.

[SLO 08] SLOAN E.D., KOH C.A., *Clathrate Hydrates of Natural Gases*, 3rd ed., Taylor & Francis-CRC Press, Boca Raton, 2008.

[SLO 10] SLOAN E.D., KOH C.A., SUM A.K., *Natural Gas Hydrates in Flow Assurance*, Gulf Professional Publishing, 2010.

[SQU 12] SQUIRES G.L., *Introduction to the Theory of Thermal Neutron Scattering*, Cambridge University Press, Cambridge, 2012.

[STA 03] STAYKOVA D.K., KUHS W.F., SALAMATIN A.N. *et al.*, "Formation of porous gas hydrates from ice powders: Diffraction experiments and multi-stage model", *Journal of Physical Chemistry B*, vol. 107, pp. 10299–10311, 2003.

[STE 98a] STERN L.A., KIRBY S.H., DURHAM W.B., "Polycrystalline methane hydrate: synthesis from superheated ice, and-temperature mechanical properties", *Energy Fuels*, vol. 12, pp. 201–211, 1998.

[STE 98b] STERN L.A., HOGENBOOM D.L., DURHAM W.B. *et al.*, "Optical-cell evidence for superheated ice under gas-hydrate-forming conditions", *Journal of Physical Chemistry B*, vol. 102, pp. 2627–2632, 1998.

[STE 01] STERN L.A., KIRBY S.H., DURHAM W.B., "Comment to the Moudrakovski et al. letter "hydrate layers on ice particles and superheated ice: a H-1 NMR microimaging study", *Journal of Physical Chemistry A*, vol. 105, pp. 1223–1224, 2001.

[STO 79] STOLL R.D., BRYAN G.M., "Physical properties of sediments containing gas hydrates", *J. Geophys. Res.*, vol. 84, pp. 1629–1634, 1979.

[STR 06] STROBEL T.A., TAYLOR C.J., HESTER K.C. *et al.*, "Molecular hydrogen storage in binary THF-H2 clathrate hydrates", *Journal of Physical Chemistry B*, vol. 110, pp. 17121–17125, 2006.

[STR 09] STROBEL T.A., SLOAN E.D., KOH C.A. "Raman spectroscopic studies of hydrogen clathrate hydrates", *Journal of Chemical Physics*, vol. 130, p. 014506, 2009.

[SUG 05] SUGAHARA T., MURAYAMA S., HASHIMOTO S. *et al.*, "Phase equilibria for H2+CO2+H2O system containing gas hydrates", *Fluid Phase Equilibria*, vol. 223, pp. 190–193, 2005.

[SUG 10] SUGAHARA T., HAAG J.C., WARNTJES A.A. *et al.*, "Large-cage occupancies of hydrogen in binary clathrate hydrates dependent on pressures and guest concentrations", *Journal of Physical Chemistry C*, vol. 114, pp. 15218–15222, 2010.

[SUG 09] SUGAHARA T., HAAG J.C., PRASAD P.S.R. *et al.*, "Increasing hydrogen storage capacity using tetrahydrofuran", *Journal of the American Chemical Society*, vol. 131, pp. 14616–14617, 2009.

[SUM 97] SUM A.K., BURRUSS R.C., SLOEN E.D. Jr., "Measurement of clathrate hydrates via raman spectroscopy", *Journal of Physical Chemistry B*, vol. 101, pp. 7371–7377, 1997.

[SUS 08] SUSILO R., ALAVI S., RIPMEESTER J.A. *et al.*, "Molecular dynamics study of structure H clathrate hydrates of methane and large guest molecules", *Journal of Chemical Physics*, vol. 128, p. 194505, 2008.

[TAI 07] TAIT K.T., TROUW F., ZHAO T. *et al.*, "Inelastic neutron scattering study of hydrogen in d8-THF/D2O ice clathrate", *Journal of Chemical Physics*, vol. 127, p. 134505, 2007.

[TAK 00] TAKEYA S., HONDOH T., UCHIDA T., "In situ observation of CO2 hydrate by x-ray diffraction", *Gas Hydrates Challenges for the Future*, vol. 912, pp. 973–982, 2000.

[TSE 88] TSE J.S., WHITE M.A., "Origin of glassy crystalline behavior in the thermal properties of clathrate hydrates: a thermal conductivity study of tetrahydrofuran hydrate", *Journal of Physical Chemistry*, vol. 92, pp. 5006–5011, 1988.

[TSE 97] TSE J.S., RATCLIFFE C.I., POWELL B.M. *et al.*, "Rotational and translational motions of trapped methane. Incoherent inelastic neutron scattering of methane hydrate", *Journal of Physical Chemistry*, vol. 101, pp. 4491–4495, 1997.

[TSE 01] TSE J.S., SHPAKOV V.P., BELOSLUDOV V.R., *et al.*, "Coupling of localized guest vibrations with the lattice modes in clathrate hydrates", *Europhysics Letters*, vol. 54, pp. 354–360, 2001.

[TSE 05] TSE J.S., KLUG D.D., ZHAO J.Y. *et al.*, "Anharmonic motions of Kr in the clathrate hydrate", *Nature Materials*, vol. 4, pp. 917–921, 2005.

[UDA 01] UDACHIN K.A., RATCLIFFE C.I., RIPMEESTER J.A., "Structure, composition, and thermal expansion of CO_2 hydrate from single crystal X-ray diffraction measurements", *Journal of Chemical Physics B*, vol. 105, pp. 4200–4204, 2001.

[ULI 07] ULIVI L., CELLI M., GIANNASI A. *et al.*, "Quantum rattling of molecular hydrogen in clathrate hydrate nanocavities", *Physical Review B*, vol. 76, pp. 161401, 2007.

[VEL 14] VELUSWAMY G.P., KUMAR R., LINGA P., "Hydrogen storage in clathrate hydrates: current state of the art and future directions", *Applied Energy*, vol. 122, pp. 112–132, 2014.

[WAL 09] WALSH M.R., KOH C.A., SLOAN E.D. *et al.*, "Microsecond simulations of spontaneous methane hydrate nucleation and growth", *Science*, vol. 326, pp. 1095–1098, 2009.

[WAN 02] WANG X., SCHULTZ A.J., HALPERN Y., "Kinetics of methane hydrate formation from polycrystalline deuterated ice", *Journal of Physical Chemistry A*, vol. 106, pp. 7304–7309, 2002.

[XU 08] XU M., SEBASTIANELLI F., BACIC Z., "Quantum dynamics of H2, D2, and Hd in the small dodecahedral cage of clathrate hydrate: evaluating H_2-water nanocage interaction potentials by comparison of theory with inelastic neutron scattering experiments", *Journal of Chemical Physics*, vol. 128, p. 244715, 2008.

[XU 11] XU M., ULIVI L., CELLI M. *et al.*, "Quantum calculation of inelastic neutron scattering spectra of a hydrogen molecule inside a nanoscale cavity based on rigorous treatment of the coupled translation-rotation dynamics", *Physical Review B*, vol. 83, p. 241403, 2011.

[YAM 88] YAMAMURO O., OGUNI M., MATSUO T. *et al.*, "Calorimetric study of pure and KOH-doped tetrahydrofuran clathrate hydrate", *J. Phys. Chem. Solids*, vol. 49, pp. 425–434, 1988.

[YAN 08] YANG H.Q., XU Z.H., FAN M.H. *et al.*, "Progress in carbon dioxide separation and capture: a review", *Journal of Environmental Sciences*, vol. 20, pp. 14–27, 2008.

[YOO 04] YOON J.H., KAXAMURA T., YAMAMOTO Y. *et al.*, "Transformation of methane hydrate to carbon dioxide hydrate: in situ raman spectroscopic observations", *Journal of Physical Chemistry A*, vol. 108, pp. 5057–5059, 2004.

[YOU 14] YOUN Y., SEOL J., CHA M. *et al.*, "Structural transition induced by CH_4 enclathration and cage expansion with large guest molecules occurring in amine hydrate systems", *J. Chem. Eng. Data*, vol. 59, pp. 2004–2012, 2014.

[ZHU 14] ZHU J., DU S., YU X. *et al.*, "Encapsulation kinetics and dynamics of carbon monoxide in clathrate hydrate", *Nat. Commun.*, vol. 5, p. 4128, 2014.

2

Spectroscopy of Gas Hydrates: From Fundamental Aspects to Chemical Engineering, Geophysical and Astrophysical Applications

2.1. Introduction

Structural analysis provides a time-averaged description of gas hydrates. However, their "time-dependent" behavior at a molecular level is required for a complete understanding of their fundamental properties as well as for further improvements in the development of their potential applications. In this regard, the knowledge of their fluctuations is a prerequisite. These fluctuations may be accessed by analyzing the linear response of a gas hydrate to an external perturbation (in the framework of linear response theory). Such properties may be investigated by means of various spectroscopic techniques such as nuclear magnetic resonance (NMR), infrared (IR) absorption/reflection, Raman scattering or inelastic neutron/X-ray scattering. With such experimental approaches, the basic information related to these fluctuations concerns the molecular excitations and relaxations occurring in the gas hydrates. The contribution of neutron scattering has been detailed in Chapter 1 and NMR allows relaxation processes to be probed, as described in various reviews or textbooks (e.g. [RIP 91, RIP 99]). In this chapter, we will focus on optical spectroscopies (Raman, IR and X-ray) for investigating gas hydrate excitations (intra- and intermolecular modes) and their implications from fundamental aspects to chemical engineering, geophysical and astrophysical contexts.

Chapter written by Bertrand CHAZALLON, Jennifer A. NOBLE and Arnaud DESMEDT.

The most widely used optical spectroscopic techniques for investigating vibrational properties in condensed matter are probably IR absorption and Raman scattering. Moreover, a complementary technique of growing interest in the physical–chemistry community is inelastic X-ray scattering (IXS). The basic concepts of these spectroscopic techniques will not be described in this chapter (as many textbooks are already available); however, a few basic and complementary aspects of these techniques will be summarized.

Raman and IR spectroscopy employ near-visible light, yielding vibrational modes to be probed at nearly zero momentum transfer (i.e. providing access only to non-dispersive optical modes of intra- and intermolecular origins). The interaction of a dielectric material with an electric field involves fundamental differences between the mechanism of absorption (resonance) for IR spectroscopy and scattering (oscillating dipole) for Raman spectroscopy. This fundamental difference leads to the investigation of complementary vibrational modes resulting from the probed molecular properties (i.e. polarizability variation for Raman, dipolar moment variation for IR); vibrational modes are then usually assigned as Raman-active, IR-active or non-active depending on their symmetry. IR was applied successfully over the last decade to obtain insights into hydrate properties at the molecular level, especially for kinetic experiments. In parallel, the short acquisition times as well as the coupling with a confocal microscope make Raman spectroscopy particularly suited for *in situ* investigations at a micrometer sample scale. Both techniques allow the extraction of fine spectral signatures because of the very good energy resolution on the order of 0.1 meV (~1 cm^{-1}). With IXS spectroscopy, meV (~10 cm^{-1}) energy resolution can be achieved with keV X-rays (to be compared with IR and Raman incident photon energy ranging from meV to eV). With the improvement in energy resolution over recent decades, IXS can perform light inelastic scattering of thermal excitations with a wavelength of the order of interatomic spacing in solids. Thus, it is possible to access the momentum transfer dependence of the vibrational modes, and thus to perform phonon dispersion measurements in crystals (as with inelastic neutron scattering) to access acoustic phonon modes of hydrates, for instance.

This chapter will review the recent advances made in understanding fundamental and applied properties of gas hydrates thanks to optical spectroscopic investigations. To address this issue, section 2.2 is dedicated to a general overview of the information contained in the vibrational spectra of gas hydrates. The spectral signatures of the encapsulated guest molecules and the water cages are detailed. The main features associated with phonons are also described in terms of guest rattling and acoustic cage modes. In section 2.3, one of the most challenging issues regarding hydrates is addressed: their formation mechanism and in particular, the nucleation step, i.e. the process by which small clusters of water and gas molecules aggregate until reaching the critical size above which continuous growth occurs. Such a microscopic phenomenon is extensively investigated by means of computational methods (e.g. MD simulations [RIP 16]), but remains difficult to capture experimentally. On this topic, the recent progress made by means of Raman and IR spectroscopy is reviewed. Following this progress, an interesting outcome from hydrate nucleation and growth studies is described in the context of chemical engineering: the kinetics of gas replacement (gas swapping process) in which methane hydrates are intended to be converted into stable CO_2 hydrates. Section 2.3 focuses on the problematic of natural gas hydrates (NGHs). The existence of large deposits of methane hydrate in the deep ocean makes the understanding of the occurrence and equilibrium behavior of NGH in nature crucial for estimating gas hydrate reserves. In addition to the sampling of NGH directly at or close to the seafloor, useful physical parameters such as structure, composition, distribution of guests in hydrate cages, levels of water saturation in the sediment and degree of water conversion to hydrate can be obtained with a characterization of recovered samples by a suite of laboratory techniques, including Raman measurements. Finally, section 2.4 deals with the fascinating scientific questions related to hydrates in an astrophysical context. Gas hydrates have long been proposed as an ice phase present in various astrophysical environments including comets and planets. To our knowledge, no unambiguous spectral evidence has been identified as observational proof of existence of hydrates in star-forming regions and planetary systems; their potential formation is based on hypotheses derived from gas and solid phase observations, theoretical models and laboratory studies. In this regard, the last section reviews IR spectroscopy of astrophysical ices, including interstellar and solar system observations, as well as information obtained from laboratory spectroscopy of gas hydrates.

2.2. Vibrational spectrum

The vibrational properties of gas hydrates will be described by decomposing a typical spectrum into two parts: the intramolecular modes (guest and host molecule vibrations) and the intermolecular modes (phonon). Section 2.2.1 will be dedicated to the description of intramolecular modes of the hydrates, allowing the identification of spectral signatures of the encapsulated molecules and the water molecules. Section 2.2.2 will illustrate the phonon modes, a characteristic of the hydrate structure. Both types of mode will be discussed on the basis of experimental results obtained by means of Raman scattering and IXS. Contributions of neutron scattering techniques to intermolecular mode characterization are provided in Chapter 1.

2.2.1. *Intramolecular modes*

Raman scattering is an optical spectroscopic technique giving access to detailed information about intramolecular vibration (e.g. [GUI 08]). The short acquisition times as well as the coupling with a microscope make this technique particularly appropriate for *in situ* investigations. In this section, the typical Raman signature will be described in terms of intramolecular modes, i.e. internal vibrational modes (stretching and bending) due to water molecules and guest molecules. The O-H stretching bands of water molecules give rise to a broad feature observed between 2,900 and 3,800 cm^{-1} (Figure 2.1). With respect to hexagonal ice, the main band corresponding to O-H stretching is observed at higher frequencies in the clathrate hydrate phase, as shown in the case of the xenon clathrate hydrate in Figure 2.2 [KLO 13] and in [SUG 05]. In the harmonic approximation, these changes are considered a direct signature of the strengthening of the O–H bond of the water molecule in the clathrate compared to the ice.

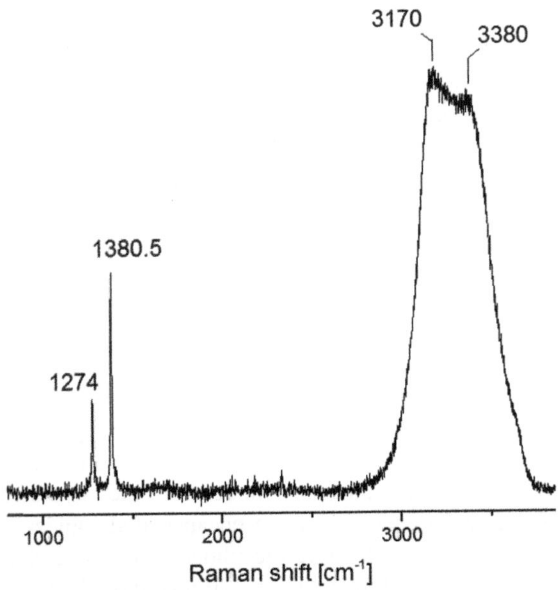

Figure 2.1. *Experimental Raman spectrum of a CO_2 -clathrate hydrate formed at 269 K and 21 bar. The guest molecule bands are observed at 1,274 cm^{-1} and 1,380.5 cm^{-1}. The water molecule stretching bands are represented by a broad band that spreads over 2900 and 3800 cm^{-1}, with two main features at 3,170 cm^{-1} and 3,380 cm^{-1}. Note that the water contribution resembles that of ice (or supercooled water) at the corresponding temperature*

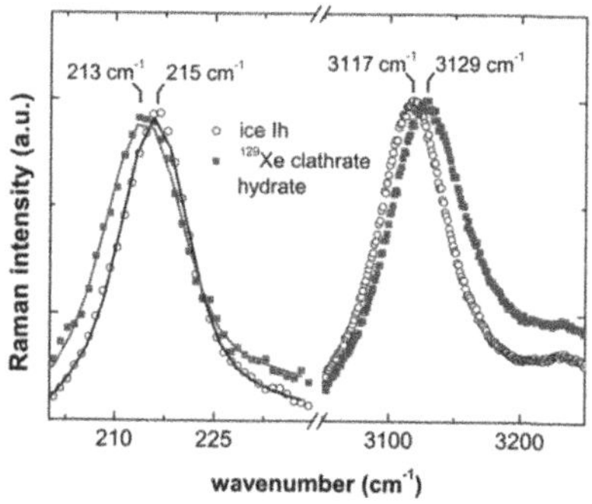

Figure 2.2. *Translational lattice modes (left) and O-H stretching modes (right) of the Xe clathrate hydrate and ice Ih at 150 K. Adapted from [KLO 13]. For a color version of this figure, see www.iste.co.uk/broseta/hydrates1.zip*

Figure 2.3 shows the Raman signal of the CH stretching modes of the methane hydrate in the type I and type II structures. In the gas phase, a single band is observed at 2,917 cm^{-1} while the gas hydrate gives rise to two bands associated with the confinement of the methane molecule in two different types of cages. In addition, the relative intensity of these bands is a direct signature of the guest distribution in the cages. In the case of the type I structure, knowing that the intensity at 2,904 cm^{-1} is three times more than that at 2,915 cm^{-1} and that there are three large $5^{12}6^2$ cages for every small 5^{12} cage, the band at 2,904 cm^{-1} is attributed to CH stretching of methane molecules located within the large cages (LCs), and the band at 2,915 cm^{-1} to methane molecules within the small cages (SCs). In the case of the type II structure, the band at 2,905 cm^{-1} is due to methane molecules encapsulated within the large $5^{12}6^4$ cages and that at 2,910 cm^{-1} to methane molecule within the small 5^{12} cages.

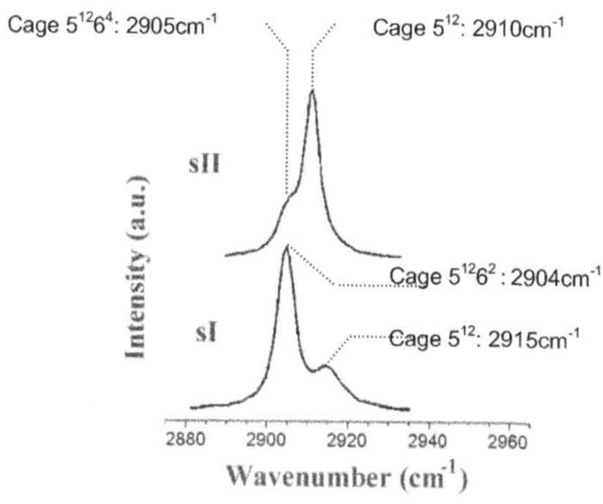

Figure 2.3. *Raman spectra of the stretching region v(CH) for methane in the type I and II structures (T = 298 K and P = 1,250 bars). Adapted from [CHO 00]*

The hydrate phase gives rise to some rather intriguing behavior in the stretching frequencies as compared to the gas phase. The C–H stretching frequency of gaseous methane (denoted V_{gas}) is higher than that of methane encapsulated in the cages of a clathrate hydrate. On the other hand, this stretching vibration is observed at lower frequency in the LCs (denoted V_{LC}) than in the SCs (denoted as V_{SC}), whatever (type I or type II) the clathrate structure. These observations are contradictory. Indeed, by considering the gas cage as a "confinement in a cage of infinite size", one would expect $V_{gas} > V_{LC} > V_{SC}$ instead of the relative values experimentally determined, i.e. $V_{gas} > V_{SC} > V_{LC}$. This behavior can be explained with the help of the "loose-cage tight-cage" model developed by Buckingham *et al.* ([SCH 03] and references therein). On the basis of the first and second derivatives of the interaction potential between a simple diatomic molecule and its environment, this model allows the investigation of the confinement effect on vibrational frequency. By applying this model to the case of the clathrate hydrate and by considering an attractive methane-cage interaction, one obtains a "gas" frequency higher than the "clathrate" frequency. Moreover, $V_{SC} > V_{LC}$ because of a shorter guest-cage distance in the SC than in the LC together with a shortening of the C–H bond in the SC [TSE 02], as confirmed by *ab initio* MD simulations [HIR 12].

2.2.2. *Intermolecular modes*

The low frequency modes of gas hydrates have been investigated by means of various spectroscopic techniques. Among them, inelastic neutron scattering and Raman scattering are probably the most widely used, while IXS is increasingly applied. The contribution of neutron scattering has been described in Chapter 1. This section will thus be focused on the description of the contributions of optical spectroscopy.

Analyses of neutron-derived vibrational density of state revealed that the confined guest molecules undergo rattling modes at very low frequencies, typically of the order of 1 meV (8 cm^{-1}), i.e. at energy matching that of neutron scattering. The observation of such modes with optical spectroscopy requires very high-energy resolution, which can hardly be reached with Raman spectrometers. The use of IXS is then required. An illustration of IXS spectrum recorded on Xe hydrate is shown in Figure 2.4. The rattling modes of Xe atoms encapsulated in LCs and in SCs of the type I hydrate structure are observed at ca. 2.9 meV (averaged value of the two LC contributions at 2.45 meV and at 3.35 meV) and at 4.35 meV, respectively.

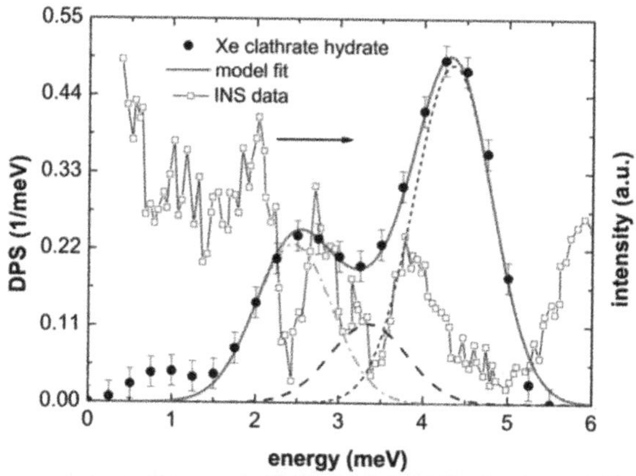

Figure 2.4. *IXS spectrum of the Xe hydrate at 17K (type I structure with large and small cages filled with guest molecules) compared with INS spectrum recorded at 50 K [GUT 02]. The bold (red) solid line represents the fit of the Xe contribution in LC and SC. The Xe encapsulated in LC give rise to two inelastic bands represented with a dashed-dotted line and dashed line. The Xe in SC corresponds to the fitted band represented with a dotted line. The specific guest atom signature is obtained by means of nuclear resonance inelastic X-ray scattering technique onto ^{129}Xe enriched hydrate. From [KLO 13]. For a color version of this figure, see www.iste.co.uk/broseta/hydrates1.zip*

Characterizing intermolecular vibrations in hydrates is useful for understanding many properties of gas hydrates. As an example, heat transport properties in gas hydrates can be better understood if thermal conductivity, which is proportional to its heat capacity (see the following chapters dedicated to the thermodynamics of gas hydrates), and to the mean free path of phonon in the cage structure, is described with the knowledge of guest–host intermolecular vibrations and specific interactions. Understanding the elementary mechanism involved in the thermal transport properties of gas hydrates is relevant not only for its applied impact (e.g. pipeline blocking), but also because of its unusual fundamental feature [STO 79, ROS 81]. Indeed, thermal conductivity of gas hydrates follows an abnormal behavior. Despite the crystalline structure of gas hydrates, their thermal behavior possesses a "glassy" character: the thermal conductivity is very low (below that of ice) and almost temperature independent. This observation has been made on numerous hydrates (see examples cited in [BAU 03]). The explanation for such a trend has been associated with phonon scattering involving coupling of low-frequency cage modes with localized guest modes [TSE 88], as revealed by IXS in the case of methane hydrate [BAU 03]. With an energy resolution of ca. 1.5 meV and a momentum transfer ranging from 0.15 to 1.1 $Å^{-1}$, IXS allowed the analysis of dispersion of phonons in the first Brillouin zone on a powder sample of gas hydrate (the type I unit cell corresponds to a reciprocal parameter $2\pi/a$

~0.5 Å^{-1}). As shown in Figure 2.5, the methane rattling mode (no dispersion) is observed at ca. 5 meV and crosses the acoustic mode dispersing in the IXS accessible energy range. From lattice dynamic calculations, the simulated IXS spectra can be computed by considering a powder average. These simulation-derived dispersions reproduce the experimental results quite well (Figure 2.5). In particular, intensity transfer is observed from the acoustic mode to the rattling mode after the crossing of these two branches at ca. 0.2 Å^{-1}. This result confirms the coupling between localized guest mode and cage phonon (longitudinal acoustic mode). Such resonant diffusive mechanism is likely to contribute to the abnormal thermal conductivity of gas hydrates.

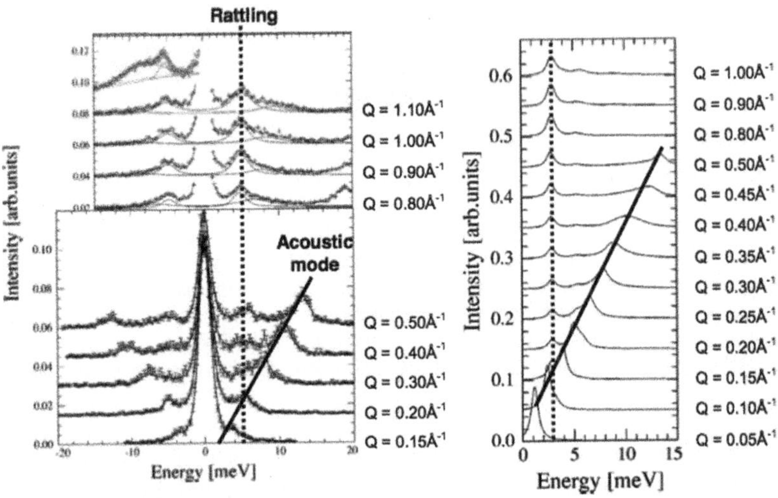

Figure 2.5. *Left: IXS spectra (bottom curve) recorded on a powder sample of methane hydrate (CH_4–H_2O) and inelastic neutron scattering DOS (top curve) recorded on partially deuterated sample (CH_4–D_2O). Rattling of methane molecule located in the large cages of the type I structure is observed at ca. ~5 meV (dotted line). The continuous line indicates the longitudinal acoustic mode. Right: Simulated inelastic X-ray scattering spectra by considering a powder average. Adapted from [BAU 03]. For a color version of this figure, see www.iste.co.uk/broseta/hydrates1.zip*

An additional and remarkable signature of hydrate resides in the phonon mode involving the water cage. This intermolecular mode appears in the 150–350 cm^{-1} spectral region, as illustrated in Figure 2.2 (left). It corresponds to a translational lattice mode involving the intermolecular O–O stretching of water molecules. It is observed at 213 cm^{-1} (see Figure 2.2) in the xenon hydrate [KLO 13, SUG 05]. This value is lower than that observed for ice Ih at ~215 cm^{-1} (Figure 2.2). In the harmonic approximation, this difference can be associated with the O–O–O angle difference in the ice phase and in the type I hydrate phase: the tetrahedral angles between H-bonded water molecules in the ice are distorted by 3.7° in the hydrate [SLO 08]. Moreover, this angle modification implies an increase in the O–H stretching frequency of the xenon clathrate hydrate with respect to that of ice Ih, as previously described (Figure 2.2).

Recently, it has been shown that the translational lattice mode is influenced by the ionicity of the water cages. Type II hydrate can be formed by mixing tetrahydrofuran (THF) and perchloric acid guest molecules, leading to the insertion of perchlorate anion in the LC of the cationic host substructure of the THF type II clathrate hydrate (Figure 2.6) [DES 15]. Shrinkage of the type II unit cell is measured together with a softening of the host lattice mode. The intermolecular O–O stretching of water molecules forming the clathrate cage is shown in Figure 2.6 at 230 K. As expected, this lattice mode is observed at lower frequency in the THF clathrate hydrate (214 cm^{-1}) [PRA 07] with respect to the ice (216.5 cm^{-1}) at 230 K. In the case of the mixed clathrate hydrate, this mode is further lowered to 212.5 cm^{-1} at 230 K. As shown in Figure 2.6, this behavior is systematically observed in the studied temperature range. This trend represents a signature of the softening of the lattice mode by adding acidic molecules within the THF clathrate hydrate. In other words, the addition of acidic protons within the host H-bond network modifies the flexibility of the cages.

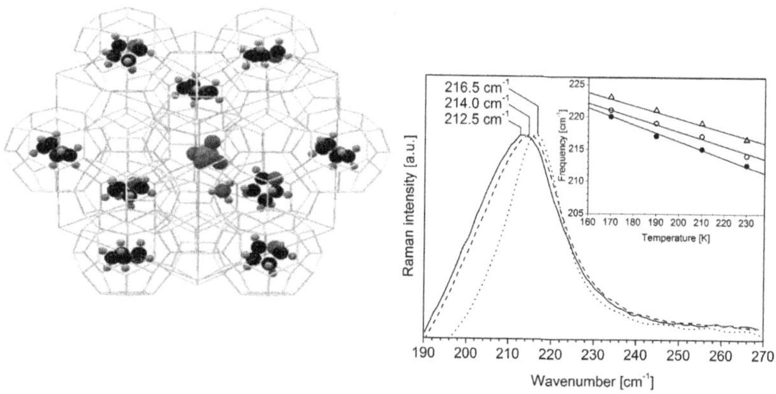

Figure 2.6. *Left: Structural representation of the type II 7THF-1$HClO_4$-136H_2O hydrates (a = 17.185 Å). While all small 5^{12} cages are empty, the large $5^{12}6^4$ cages are filled with either perchlorate anion (middle; maximum of one anion per unit cell) or THF molecules. The hydronium cation participating to the cage substructure is also represented (water molecules are not represented for clarity). Right: Translational lattice modes of the mixed THF-$HClO_4$ hydrate (continuous line), the THF hydrate (dashed line) and the ice Ih (dotted line) at 230 K. The inset shows the thermal evolution of the lattice frequencies for the mixed hydrate (filled circles), the THF hydrate (open circles) and the ice (triangles). From [DES 15]. For a color version of this figure, see www.iste.co.uk/broseta/hydrates1.zip*

2.3. Applications to the investigation of formation mechanism

The thermodynamics of hydrate formation/dissociation is relatively well established for pure gas systems, whereas a lot of questions regarding hydrate nucleation and growth kinetics are still open. They represent the most important issues to be understood for developing gas hydrate derived applications. Accurate knowledge of how and when hydrate starts to form will help to develop sustainable hydrate management strategies in energy storage and recovery (from natural hydrates), gas separation and CO_2 capture, seawater desalination, refrigeration and flow assurance. This part briefly reviews the added value brought by the experimental spectroscopic point of view.

2.3.1. *Formation mechanism: nucleation and growth*

One of the most challenging issues regarding hydrates is their formation mechanism: it starts with the nucleation step, i.e. the process by which small clusters of water and gas molecules aggregate until reaching the critical size above which continuous growth occurs. This is a microscopic phenomenon difficult to capture experimentally. So far, different formation models based on different hypotheses on the molecular scale have been extensively investigated by molecular dynamics (MD) simulations. Most of the advances come in support or against these hypotheses and cannot be corroborated by experimental determination due to the fact that the clathrate nucleus is generally too small to be characterized. The labile cluster model proposed by Christiansen and Sloan [CHR 94] and Sloan and Fleyfel [SLO 91] deals with the bulk liquid and is based on cluster nucleation theory. It suggests that precursor water cages (labile ring structures of pentamers and hexamers) already exist in water without gas. When the gas is solubilized, labile clusters (of 20 or 24 coordination shell) form spontaneously and associate with each other to assemble the hydrate nucleus. The association is made under many different configurations of which only a few will lead to stable hydrate structures. In another model dealing with the vapor–liquid interface and introduced by Rodger [ROD 90], and further extended by Kvamme [KVA 00], a surface-driven model is developed where gas molecules adsorb on the water surface and are trapped on partially completed water cages (the so-called "interfacial cluster hypothesis"). The nucleation is then observed to arise on the vapor side of the interface, with a large number of clusters forming at every instant, whose size progresses with time. However, isolated empty or guest-filled clathrate cages in solution are extremely rare events according to MD simulations [GUO 04]. So the viability of these models has been questioned as the energy barrier for these clusters to agglomerate and promote growth is much higher than that required for these entities to disintegrate [RAD 02]. Therefore, a new conceptual model based on local structural ordering (the so-called "local structure hypothesis", LSH) has been proposed by Radhakrishnan and Trout [RAD 02], who investigated the water–liquid CO_2 interface. In this guest-rich interface represented by an aqueous solution, a group of guest molecules become arranged in a configuration similar to that in the hydrate phase (Figure 2.7).

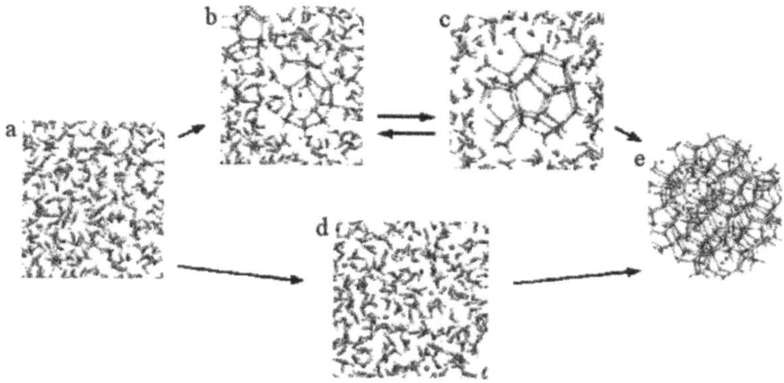

Figure 2.7. *Comparison of the labile cluster model (a-b-c-e) and the local structuring hypothesis (a-d-e). a) Water without dissolved gas molecules (initial condition). b) Cluster precursors form immediately after dissolution of gas molecules. c) Cluster precursors agglomerate by sharing faces. These agglomerated clusters may be unstable (step back to (b) is possible). d) No cluster formation after dissolution of gas molecules. e) Hydrate nucleation. Adapted with permission from [LEH 09]. For a color version of this figure, see www.iste.co.uk/broseta/hydrates1.zip*

This local ordering involves thermal fluctuations in guest-water solution and induces ordering of the host molecules at the nearest and next-to-nearest-neighboring shells. The perturbed water molecules surrounding the guest-cluster become locally ordered around the guests until the number of guest molecules locally arranged exceed a critical size and results in a continuous hydrate growth. In this model, no hydrate precursor is needed.

More recently, Rodger and co-workers [MOO 07] and later Walsh *et al.* [WAL 09] were able to produce spontaneous hydrate nucleation in unconstrained atomistic MD simulations of water methane systems under conditions of high driving force. In agreement with the LSH model, a large number of guest molecules were found to surround a water cage when it forms. The structure of the resulting nucleus lacked long-range crystalline order, as was also found for the LSH model, but instead presented an assemblage of cages 5^{12} and $5^{12}6^{n}$ with $n = 2$, 3 and 4, that were neither organized nor followed the proportions found in clathrate crystals.

At the same time, an alternative nucleation model based on an amorphous agglomerate (the so-called "blob mechanism") has been proposed [JAC 10a, JAC 10b]. As in the local structuring model, local ordering of guest molecules leads to small water-guest molecular clusters (prior to the amorphous clathrate phase). This assemblage diffuses in the solution as described in the framework of the labile cages model (the difference being that the "diffusing particle" is not a single clathrate cage, but rather an amorphous assembly of water clusters and guest molecules with interstitial water that continually rearranges to form transient clathrate cages – the "blob") (Figure 2.8). Thus, the nucleus arises from a multistep mechanism involving dense "blobs" of solvent-separated guests formed in equilibrium with the dissolved guest in solution.

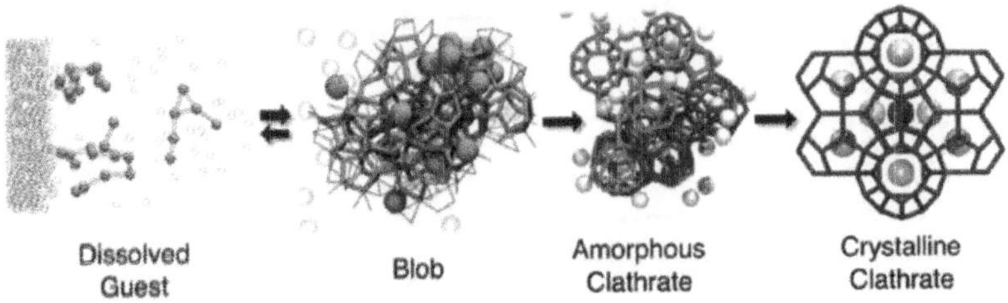

Figure 2.8. *Crystallization pathway of hydrates of hydrophobic guests. First step involves the formation of dense blobs of solvent-separated (SSP) guest molecules in equilibrium with the very dilute solution (SSP guests shown as pink balls; the other guests shown in gray; the blue lines in the blob represent half-cages and the red lines full clathrate cages). In the second step, the water in the blob organizes into clathrate cages, producing an amorphous clathrate nucleus (cages colored according to type as in Figure 1). In the final step, the amorphous nucleus grows a crystalline phase. This last step may be preceded by the maturation of the amorphous nuclei to form crystalline nuclei. At conditions of high supercooling, the amorphous nuclei grow to a metastable amorphous clathrate phase instead of to the stable crystal. Adapted with permission from [JAC 10b]. For a color version of this figure, see www.iste.co.uk/broseta/hydrates1.zip*

These blobs survive in solution and give birth to several clathrate cages, which under conditions of high driving force can grow into a metastable "amorphous" clathrate phase that develops into a crystalline clathrate after annealing [VAT 10].

2.3.2. *The Raman contribution*

At present, there is no direct experimental evidence in favor of the existence of an amorphous intermediate phase or "blobs" during the clathrate formation, probably due to the limited size of the clathrate nuclei [KOH 02] and to the difference of time and space scales probed in experimental and theoretical approaches [WAL 11].

Within most experimental studies, hydrate formation is observed as an interfacial process. However, optical and calorimetric methods probe the hydrate formation process after the initial nucleation has already taken place. In general, the quantity measured in all of these experiments is the induction time t_i for hydrate nucleation (see [KAS 03]). This induction time depends on the kind of guest molecules, the degree of supercooling and the history of the water-gas system [FAN 14]. Thus, so far there are no data regarding the initial nucleation step, which occurs at the nanometer scale, before the macroscopic formation begins. Moudrakovski *et al.* [MOU 01] investigated the formation of Xe hydrates on the surface of ice by NMR spectroscopy. An initial hydrate film of 500 Å was observed to form before the reaction became extremely slow. The cage occupancy ratio θ_L/θ_S was monitored as a function of time and used to follow the nature of the material formed. The ratio changed from a value close to 1 during the early part of the reaction to its equilibrium value of ~3–4 after the nucleation process was finished and rapid growth commenced. The initial low cage occupancy value can thus be seen as an evidence of a precursor hydrate phase with a large number of SCs that is quite different from the equilibrium hydrate. These authors mentioned that this should not necessarily be seen as a cage with a definite geometry, as it is in the sI crystal, but as a hydration sphere of about 20 water molecules that may be quite flexible, with distinct symmetries. Lehmkühler *et al.* [LEH 09] attempted to address nucleation at the CO_2–water and other hydrocarbon–water interfaces on a molecular length scale using X-ray reflectivity (Figure 2.9).

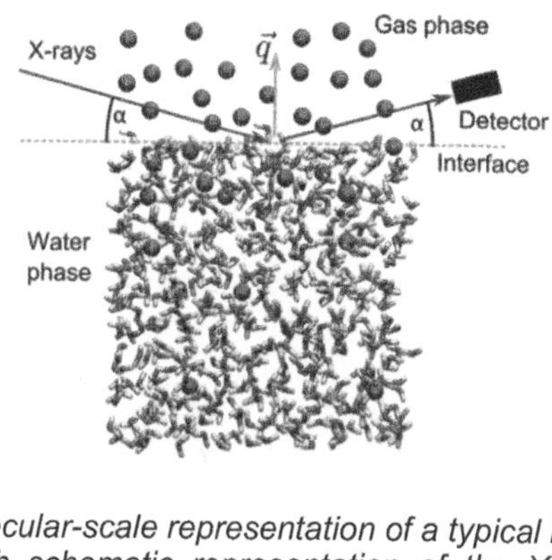

Figure 2.9. *Molecular-scale representation of a typical interface under investigation with schematic representation of the X-ray scattering geometry. Adapted with permission from [BOE 12]. For a color version of this figure, see www.iste.co.uk/broseta/hydrates1.zip*

They observed that the precondition of hydrate formation (as also suggested by [KAS 03]) is linked to the accumulation of guest molecules in nanometer-thick supersaturated layers at the interface between water and guest molecules that serve to trigger hydrate formation. According to these authors, this result supports the local structuring hypothesis.

On the other hand, the probe of micrometer-sized areas on sample surfaces using Raman spectrometers coupled with confocal microscopy appears to be particularly well suited for *in situ* studies of clathrate hydrate properties. For instance, clathrate hydrate formation kinetics has been monitored by micro-Raman spectroscopy [SUB 99, SLO

03]. Figure 2.10 shows the time evolution of the methane stretching mode monitored during the methane clathrate hydrate formation (spectrum at time $t = 12{,}000$ s) from a water-methane solution at a pressure of 317 bar (spectrum at time $t = 0$ s). The band at 2,911 cm^{-1}, characteristic of methane solvated in liquid water, splits into two bands characteristic of the methane encapsulated in the type I clathrate hydrate structure. The deconvolution of the two bands reveals the relative cage occupancies during the clathrate formation [SUB 99]. From the cavity ratio profile (i.e. cavity ratio of LC to SC as a function of time) obtained in the early stage of formation (of ~0.5 at the time $t = 0$ s), it was suggested that the formation (or filling) of large sI cages ($5^{12}6^2$) may be rate limiting in the hydrate formation process. Such experimental approaches [UCH 00, SLO 98] may support the labile cluster hypothesis in general, i.e. labile 5^{12} cages formed initially around guest molecules in solution and then further reconstruction of water clusters to large cavities $5^{12}6^2$ remains a limiting step [SLO 91, LED 96, CHR 94].

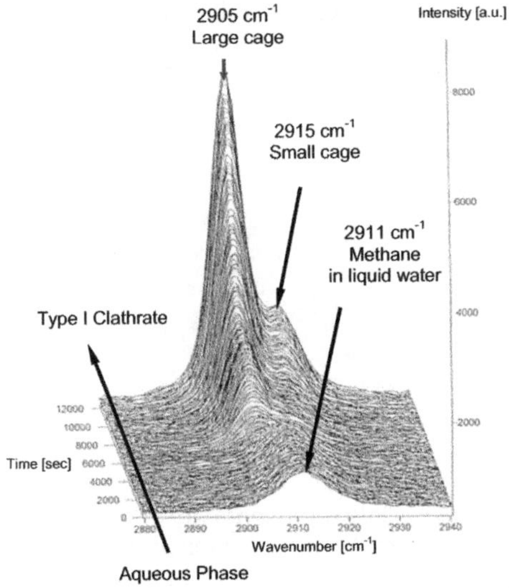

Figure 2.10. *Raman spectra (25 s per spectrum) recorded during the formation of methane hydrate at P = 317 bar. t = 0 s: methane in water at T = 297.15 K. t = 12,000 s: methane hydrate at T = 275.65 K. Cooling rate: 0.1 K·min^{-1}. Adapted with permission from [SUB 99]*

Using the ice seeding method with pure CH_4 gas [SCH 13] Schicks *et al.*, reported induction times for enclathration of gas molecule into a hydrate cavity via Raman spectroscopy. They provided Raman spectra of the formation of dodecahedral 5^{12} cavities during the first 15 min of the formation, whereas the signal of the LC $5^{12}6^2$ starts to develop afterward. This striking difference with the water case described above may arise from the availability of mobile water molecules at the ice surface for reorganization of clusters of 5^{12} cages into $5^{12}6^{2}$ cages that is less pronounced than in liquid water. Thus, nucleation steps of such transition may therefore be better separated in time when starting from ice. It should be noted that some authors have found evidence for different activation energies of hydrate nucleation from ice and water, as well as a substantial decrease in sensitivity of hydrate formation to supercooling in the presence of ice [WAL 09]. From their Raman data Schicks and Luzi [SCH 13] suggested that the guest incorporation into the 5^{12} cages is a first step during the initial stage of hydrate formation and that the hindered reconstruction of water clusters to form large $5^{12}6^{2}$ cages is a second step. They proposed that their experimental study may support the labile cluster hypothesis described above.

Such results may also highlight two distinct routes for hydrate formation and growth: one resulting in CH_4-sI and the second resulting in metastable CH_4-sII. Indeed, a solid phase distinct from that of a stable crystalline clathrate phase is shown to co-exist concomitantly with CH_4-sI clathrate. It is tentatively attributed to a remaining amorphous hydrate phase, which transforms over time into a crystalline hydrate structure CH_4-sII with well-defined cavities. The uncommon $5^{12}6^{3}$ cavities involved in the "blob" model extracted from simulations [JAC 10a] are in this case proposed to be the initial building blocks of an amorphous nucleus, which has to reorganize into a crystalline metastable CH_4-sII nucleus [SCH 13].

2.3.3. *Insights from IR spectroscopy*

IR spectroscopy has been used successfully over the last decade to obtain insights into the kinetics of clathrate hydrate formation from ice and contribute to detailing the nucleation mechanism at the molecular level. Historically, a large fraction of the molecular data for clathrates was derived from NMR measurements. This was due to the typical clathrate preparation methods: high pressure methodology at temperatures close to the melting point of ice. The resulting samples were not suited for *in situ* IR measurements, and most of the time, *ex situ* experiments were carried out on ground clathrate powders formed at high pressure [BER 78]. On this basis, it is often possible to quickly discern the clathrate formation from ice crystals, as well as which cages are occupied by a particular guest and whether the structure is sI, sII or a more unusual structure. However, new approaches developed during the 1980s provided molecular IR spectra of clathrates and information on their formation (and decomposition) over a temperature range generally of interest for space science (see also section 2.5). The methods of preparation of crystalline clathrates consisted of introducing an appropriate water: guest vapor mixture that may be deposited (by direct growth or by epitaxial growth) at $T < 150$ K, then annealed at higher temperature [FLE 91, ZHA 04, CHA 08]. The experimental set-up generally consisted of a cryogenic IR (or Raman) sample cell in which thin films (μm thick) of clathrates are produced by deposition on a cold substrate (which can be CsI [ZHA 04]). The clathrates formed at low temperature under vacuum normally include a proton-acceptor molecule (ether, ketone, aldehyde, etc.). Small non-polar guests (CO_2, C_2H_6, CH_3Cl, etc.) do not form by deposition without the help of guests with proton-acceptor properties. An alternative way based on production of large clusters as described, for example, by Ewing and Sheng [EWI 88], and later adapted for ice or clathrate samples suited for FT-IR spectra of surface features [DEV 95] consists of the formation of nanocrystals (aerosols) after injection (rapid expansion and cooling) of appropriate mixtures in an inert carrier gas into a precooled IR cell [FLE 90]. A variation in this preparation method consists of coating the ice with H-bonding adsorbates (e.g. ether), simultaneously with their formation at 70 K [HER 98]. In this way, Devlin and co-workers produced CO_2-clathrates on doped ice nanocrystals composed of several layers of 20–30 nm thickness [HER 98], and demonstrated a range of unusually rapid processes possible for clathrate formation with H-bond active guests at low temperatures and low pressures [BUC 09]. Although the reported formation rates are exceptionally high, their significance with respect to hydrate formation at high temperature is not established, because the conditions differ greatly from those common to more traditional clathrate experiments; i.e. $T > 220$ K, larger ice particles (and thus large guest transport distance to the fresh ice front), pressures higher by orders of magnitude. Further development of the preparation method consisted of the use of mixed guests from originally single vapor phase. Subsequent pulsing of the warm vapor mixture into a cold chamber can produce aerosols of 100% crystalline gas-hydrate particles on a subsecond timescale. The high rates observed (timescales in the order of the millisecond) include two nucleation and two growth events: the nucleation of nanodroplets (water + ether) and their growth followed by clathrate nucleation and crystallization [DEV 10]. A competition between clathrate nucleation and ice nucleation from ether aqueous solution takes place during cooling at 100 K or 70K. Solidification of the nano-droplets takes place in the range ~240–200 K followed by a subsequent cooling to 100 or 70 K. It is noted that nucleation of pure ice is preferred when the cooling rate is higher. Furthermore, the addition of guest CO_2 results in IR spectra of clathrates that are not consistent with CO_2 in either sI, sII or a combination of the two crystals, suggesting that a transient structure such as that reported previously [JAC 10a, VAT 06] may have formed. With elapsed time (2–18 h), the experimental spectra showed that the structure undergoes transformation leading to the "normally" expected CO_2-guest distribution and relative occupancy of CO_2 (Figure 2.11).

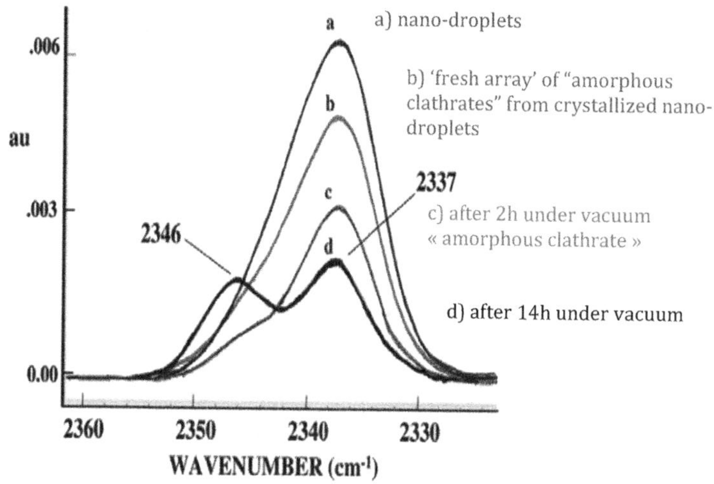

Figure 2.11. *Infrared spectra of CO_2 in the mixed clathrate with TMO (trimethylene oxide) formed from nanodroplets rich in both TMO and CO_2. The spectra show the slow redistribution of CO_2 from the large to the small cages at 100 K: a) fresh aerosol; b) fresh array of mostly CH particles; c) array after 2 h with cell evacuated; d) array after 14 h. Evacuation removes adsorbed and vapor phase CO_2 so that (c) is of clathrate particles, only, but a peak absorbance greater than the O-H stretch band is retained. Adapted with permission from [DEV 10] . For a color version of this figure, see www.iste.co.uk/broseta/hydrates1.zip*

A comparison with IR spectra of typical sII-CO_2-clathrate (mixed with THF) and obtained from a deposited thick film shows minor occupation of cages of opposite sizes (Figure 2.12), i.e. minor occupation of CO_2 in LC of type II (THF case) versus minor occupation of CO_2 in SC of type II or H (TMO case with nanodroplets). Projection of the behavior for larger droplets is not straightforward and the spectra obtained with conditions designed to favor clathrate formation from larger droplets (i.e. equivalent to bulk at higher temperature) are expected to be more complex and change as droplet size increases, requiring the simpler nanodroplet results for their interpretation.

Conversely, the drawback of such an approach may come from the difficulty in controlling the degree of supersaturation (and thus the driving force conditions), which are suggested to control the nucleation rate at least at the microscopic scale. This may lead to a distinct route for hydrate nucleation.

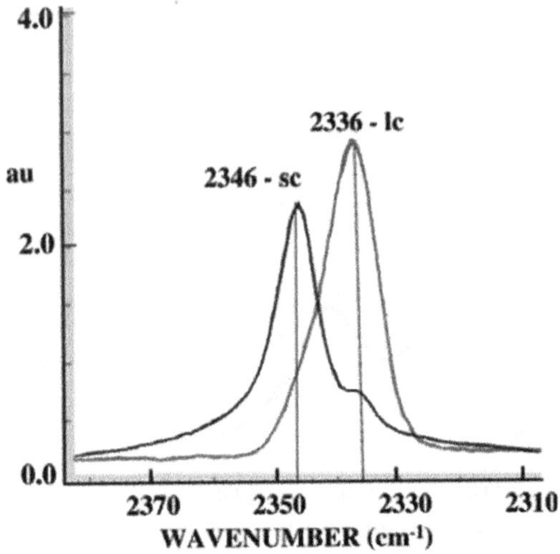

Figure 2.12. *Comparison of peak intensity and band position of SC (small cage) CO_2 in a thick film of s-II THF–CO_2 clathrate with that of primarily LC (large cage) CO_2 of the clathrate of mixed TMO–CO_2 from nanodroplet crystallization. The spectra are normalized to the O–H stretch mode intensity. The au scale is for the thick film only. Adapted with permission from [DEV 10]. For a color version of this figure, see www.iste.co.uk/broseta/hydrates1.zip*

The subsequent approach concerns the kinetics of clathrate formation of proton-acceptor guest molecules with the determination of transport rates and associated activation energies by which ether guest molecules move through the condensed phase during conversion of ice nanocrystals to clathrate nanocrystals. Devlin and co-workers showed time evolution of IR spectra during this conversion and successfully applied the well-established shrinking-core model of particle adsorbate reactions. They observed that the rate-controlling factor is the transport of the guest through the crust of clathrates encasing a reacting particle core. The reaction zones where surface-adsorbed reactant molecules need to move consist of the interface of ice and clathrate. They noticed that the diffusion through the clathrate crust and thus the formation rate at 120 K of mixed ether clathrate with co-guests of small gas molecules (N_2, CO, CH_4, CO_2 and N_2O) resembles that of gas hydrates (CO_2 and CH_4) at 260 K. They evidenced that the rapid conversion rate is derived from defect-facilitated transport of reactants to an interfacial reaction zone, with the defect population enhanced through H-bonding of guest-ether proton acceptor groups with O–H groups of the hydrate cage walls. Molecular simulations support this conclusion and suggest that the unique properties of this family of clathrates are due to the exceptional amount of the host lattice in point defects, caused by defect stabilization by H-bonding of water to the guests [BUC 09]. In contrast, formation kinetics of gas hydrates of simple molecules like CH_4 and CO_2 is not observed at low temperature (< 200 K), except if a high pressure is applied [HEN 00, WAN 02], or with the help of a low percentage of guest molecules with proton-acceptor character, or if epitaxial growth is considered [FLE 91].

2.3.4. *Formation mechanism: chemical engineering applications*

An interesting outcome from hydrate nucleation and growth studies is related to the kinetics of gas replacement (gas swapping process) in which methane hydrates are intended to be converted into stable CO_2 hydrates [LEE 03, PAR 06]. Because vast resources of natural gas in the form of clathrate hydrates have been estimated worldwide, with as much as 2.5×10^6 km^3 to 20×10^6 km^3 of methane being available in ocean floors and permafrost zones across a number of regions in the world [COL 08], considerable efforts have been made to develop efficient methods to extract hydrocarbons from hydrate-bearing sediments. These methods are principally based on promoting dissociation of gas hydrates by depressurization, heating [LIU 09] or injection of inhibitors [DEM 10]. However, the decomposition of NGHs produces sediments, sand and water, and can induce mechanical instability and technical failures. On the other hand, the anthropogenic CO_2 emissions resulting from fossil fuel combustion, which are known to be a major contributor to global warming, have triggered the development of CO_2 capture technologies in order to mitigate and ideally reduce the rise of emissions. The idea of conversion of CH_4- to CO_2-hydrate is described as a two-in-one approach with CO_2 sequestration obtained by replacing methane from the hydrate sediments in ocean, thus being a potentially eco-friendly method for fossil fuel recovery and greenhouse gas control [JAD 06]. Since then, the preliminary results from the field trial at Alaska's North Slope in 2012 showing CH_4 production concomitantly with CO_2 trapping have promoted the interest in gas hydrates from a new source of gas to a potential storage medium for CO_2. Numerous laboratory-scale studies of gas replacement in hydrocarbon hydrates (mainly CH_4) have been conducted over the last two decades using many different techniques (neutron diffraction, NMR spectroscopy, etc.) and injection of CO_2 in various situations (liquid CO_2, emulsion, supercritical CO_2 and gaseous CO_2). Despite these considerable efforts, the work on swapping reactions in gas hydrates embedded in sediments is still in its infancy in comparison to thermodynamics or even kinetics (see [KOM 13] for a review). Moreover, a molecular-level understanding of the swapping mechanism is still far from being achieved. The conversion of a simple CH_4 hydrate or mixed hydrocarbon hydrate to CO_2 containing hydrate is considered as a two-step process: (1) decomposition of the original hydrate structure and (2) rearrangement of molecules and formation of a new hydrate phase. The formation of an ice phase as an intermediate state is possible, depending on the pressure and temperature conditions [SCH 11]. Extensive Raman studies on the kinetics of swapping have been conducted, as the technique could deliver key features on guest replacement process: structure discrimination, gas or hydrate phase composition, guest distribution, etc. As an example, Figure 2.13 shows the time-resolved Raman spectra for guest replacement of methane hydrate by carbon dioxide gas at 4 MPa and 281 K [YOO 04]. The initial replacement rate of methane hydrate was found to be very fast within 200 min, and then became relatively slow. In particular, a sharp increase in the Raman peaks of carbon dioxide hydrate was observed for the initial 100 min, while no further changes in the relative intensity are observed beyond that time. Yoon [YOO 04] suggested that the replacement by carbon dioxide stops when CO_2-hydrate layers of sufficient thickness becomes a strong barrier against the diffusion of carbon dioxide.

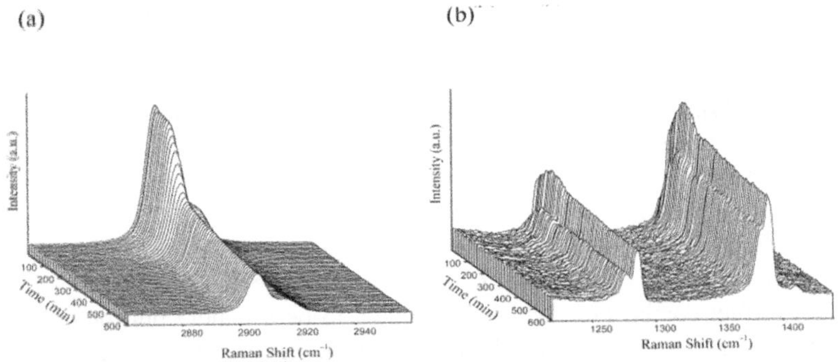

Figure 2.13. *Time-resolved Raman spectra for the C-H stretching mode of CH_4 molecules encaged into hydrate a) and for CO_2 molecules encaged into the large cavity of hydrate b). The experimental conditions for guest replacement are 281 K and 4 MPa (CO_2). Adapted with permission from [YOO 04]*

Furthermore, it is shown that 100% of the methane can be recovered from the hydrate phase in ~150 h, while only 10% of the methane can be recovered when the conditions are not favorable (0.8 MPa and 253 K), i.e. the fluid pressure (gas) is lower than the dissociation pressure of CH_4-hydrate. On the other hand, Park [PAR 06] showed that up to 85% of the methane can be recovered in only 24 h when using a gas mixture $CO_2 + N_2$, a result that is much more effective than when only CO_2 is used on sI-CH_4-clathrate (~64%) [LEE 03]. However, this result depends greatly on the morphology of the original hydrate sample, and disagreement with this result has often been reported [SCH 11]. Furthermore, it is known that the determination of the guest distributions among cages is a quite challenging task and specific physical parameters such as particle size, subtle local dissociation or formation effects, hydrate microstructure, permeation constant, as well as the ongoing change of composition of the vapor phase have to be well characterized [MUR 10, LEE 14]. Schicks [SCH 11] showed the time-dependent behavior of the CH_4 substitution with CO_2 gas as determined with Raman spectroscopy on hydrates formed with ice particles of size < 10 μm (see Figure 2.14) [SCH 11]. Methane recovery was estimated to be 30% in 115 h. In contrast, the quantitative assessment obtained recently by Lee *et al.* [LEE 14] using combined p-T measurements and gas chromatography showed that the recovery is actually around 15–20% in 115 h, with however liquid CO_2 (at 9 MPa) as exchange medium, and size-controlled ice particles of ~80 μm and 125–150 μm. Their data can be explained by a combination of Avrami and shrinking core models [LEE 14, WAN 02]. The former is applicable in the initial reaction stage of the hydrate film covering the ice particles, while the latter is for diffusion through the hydrate film. Further experimental studies are certainly needed to remove the observed inconsistencies concerning the achievable CH_4-recovery and CO_2 storage ratios via a simple CO_2–CH_4 replacement in hydrates.

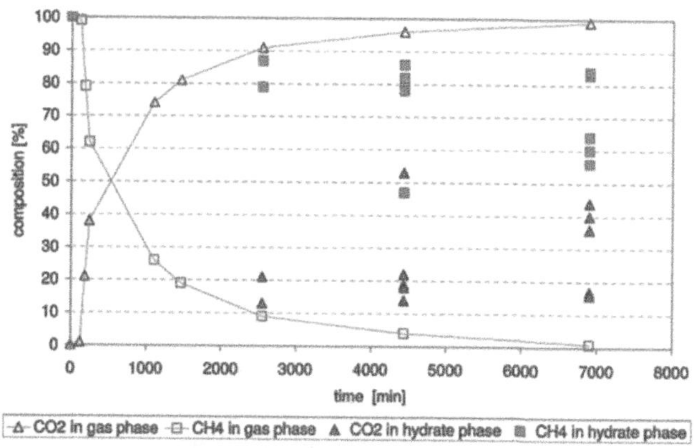

Figure 2.14. *Results of Raman spectroscopic measurements illustrating the exchange process of pure CH_4 hydrate when exposed to CO_2 gas. Open symbols show the change in gas composition from 100% CH_4 to 100% CO_2, filled show triangles and squares show the respective CO_2 and CH_4 concentration in the hydrate. Experimental conditions: p = 3.2 MPa, T = 274 K. Adapted with permission from [SCH 11]. For a color version of this figure, see www.iste.co.uk/broseta/hydrates1.zip*

In particular, the evolution of CO_2 cage occupancies during a swapping process with binary mixtures (flue gas) and/or applied on hydrocarbon mixtures remains a crucial and challenging question. Furthermore, the influence of pressure, temperature and salinity remain issues for the future. A better characterization of the hydrate structure and particles morphology during the replacement experiments is also needed in order to develop a detailed picture of the swapping mechanism for the design of practical processing in sediments and NGHs from oceans. Although mechanically stable [HYO 14] following displacement of CH_4 by CO_2, hydrate-bearing sediments may be affected by reverse reactions in the natural environment, and the CO_2-rich hydrate phase may not be stable if it is exposed to hydrocarbon gases coming from a deeper source [SCH 11].

2.4. NGHs: contribution of spectroscopy

NGHs represent considerable methane hydrate reservoirs which, according to the most recent estimate, can reach ~3,000 trillion cubic meters, i.e. an order of magnitude higher than the conventional gas resource [BOS 11]. They are therefore considered as potential future sources of energy and are very attractive for the gas industry, which has started to provide possible future approaches to their exploitation [CHO 16]. In contrast to this attractive approach, their role has been repeatedly quoted in geological hazards [JIA 16] or global climate change [JAC 01, PHR 12, MIE 12, DEC 12] (see also volume 2 of the present series "Geoscience and Applications"). Hydrocarbon gases that form NGHs originate from mainly two sources: biogenic or thermogenic. However, recent studies showed that the origin of gas-forming hydrates can also come from abiotic sources [JOH 15]. Biogenic natural gases are formed from CH_4 – generating microorganisms either via reduction of CO_2 or via acetate fermentation [WHI 99]. Because methane produced by microbial activity is far more profuse than higher hydrocarbons, structure sI is considered as the most abundant hydrate structure on Earth [KVE 95], despite a limited stability domain of existence compared to other structures (sII or sH). In contrast, thermogenic natural gases are formed from the thermal degradation of kerogen and oil [SCH 88], a common process in marine sediments [BRO 84]. Hydrocarbon hydrates produced this way generally contain methane through butane within the sII structure [BOU 09, KID 06] and methane through pentane within the structure sH [YOU 04, LU 07].

The p–T conditions required to initiate hydrate formation and to stabilize its structure are encountered in marine sediments, mainly in deep waters along continental margins and, to a lesser extent, in polar regions associated with permafrost [KVE 01, LU 11]. Since the first discovery of NGHs in permafrost in the 1960s [MAK 07] and later in oceanic sediments [KVE 80], new gas hydrate sites have been discovered regularly, with more than 230 NGHs deposits discovered worldwide in 79 countries at the present time [SON 14]. Field locations where hydrate expeditions and projects around the world have been undertaken so far can be found elsewhere [CHO 16].

The understanding of the occurrence and equilibrium behavior of gas hydrates in nature represent a key question in estimating gas hydrate reserves. The bottom-simulating reflector (BSR) is a first indicator of the presence of a gas hydrate system, as it is known to closely approximate the base of the gas hydrate stability zone (GHSZ) [HOL 96]. However, the prediction of the vertical extent of the GHSZ and the hydrate distribution within the GHSZ remains a challenging task with this technique. Indeed, major concerns have been raised regarding its unequivocal use in hydrate detection since hydrates have been observed in areas lacking clear and continuous BSR signals. Furthermore, it has been recently recognized that in geological settings dominated by thermogenic gas migration the GHSZ may extend much deeper than previously assumed by BSR measurements, with varying fractionation in hydrate composition as depth decreases and occurrence of both hydrates of structure sI and sII at shallow or deeper depth in the sediment column, respectively [PAG 16]. In this regard, the global carbon budget in hydrates remains difficult to predict [BOS 12].

Such results raise the question of sampling, as most gas hydrates have been directly sampled at or close to the seafloor, i.e. without a complete penetration and direct sampling throughout the GHSZ. Nevertheless, useful physical parameters such as structure, composition, distribution of guests in hydrate cages, levels of water saturation in the sediment and degree of water conversion to hydrate can be obtained with a characterization of recovered samples by a suite of laboratory techniques. Furthermore, initial formation conditions as well as stability fields and decomposition characteristic can be gained from a thorough characterization of recovered samples. Vibrational spectroscopy (Raman) is commonly used to obtain direct information on the features of the hydrates and hydrate-bearing. Charlou *et al.* [CHA 04] reported a detailed physicochemical characterization of a shallow (12 m below sea floor) NGH collected in the Congo-Angola basin. As a part of the ZAI-ROV Leg 2 cruise (December 2000), the deep water remote-controlled system VICTOR 6000 of IFREMER was operated to observe and sample a giant crater formed in the seabed at 3,160 m water depth (at 2.4°C) by the eruption of gas and subsequent seepage of gas and pore fluids. Geochemical tracers were used as a guide for exploration and detection of seepage, and massive gas hydrates were collected in a sediment core (at site KZR-42), representing the first recovery of a NGH sample in France. Intense fluid circulation was reported to transport methane-rich turbid fluids into the overlying bottom waters with elevated concentrations of particles: manganese, iron and methane detected in the water column close to the sediment surface. The recovered specimens occur as small fragments and massive crystal aggregates, mostly disseminated irregularly in the highly disturbed sediment and escaping in the overlying deep seawater, creating methane-rich plumes. The dissociation of solid CH_4 hydrate particles rising in a turbulent flow may explain the high heterogeneity of methane concentration and methane peaks measured in vertical profiles. Major minerals found in the sediments hosting these gas hydrates are quartz, kaolinite and pyrite with traces of pyrrhotite. Granules were shown to consist of an assemblage of magnesian calcite and aragonite. Raman analysis of a massive specimen (granular aspect, Figure 2.15) revealed that the gases contained in the hydrate are mainly 100% methane and form structure I, with H_2S and CO_2 co-clathrated with CH_4 in the cages (Figure 2.16).

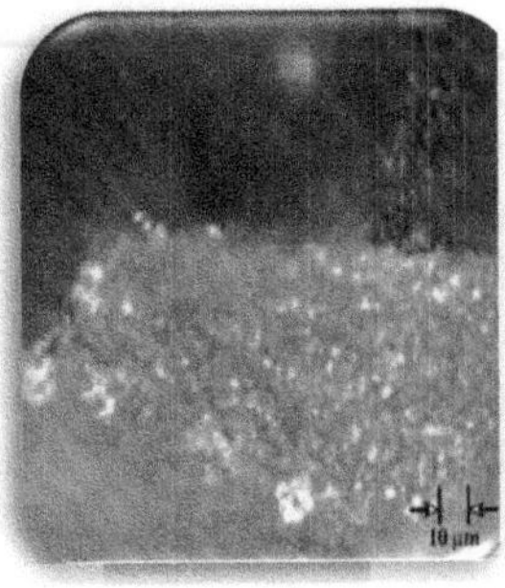

Figure 2.15. *Characteristic microscopic view (objective × 50) of the recovered natural gas hydrate. The opaque zones with a rough and irregular surface correspond to the presence of gas hydrate. Light areas (white marks of ~10 µm) correspond to the presence of ice. Adapted with permission from [CHA 04]. For a color version of this figure, see www.iste.co.uk/broseta/hydrates1.zip*

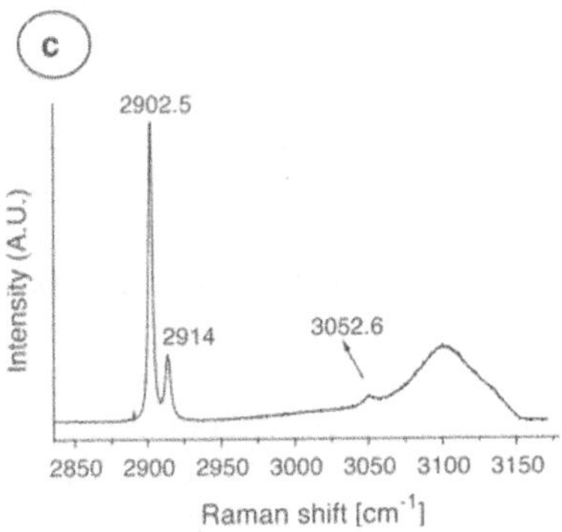

Figure 2.16. *Raman spectra of the C–H and O–H stretching modes of CH4 and H2O in natural gas hydrates collected at 133 K. The arrow highlights a weak contribution of methane in the hydrate [CHA 07a]. Adapted with permission from [CHA 04]*

A detailed structural work using synchrotron X-ray diffraction revealed the temperature-dependent behavior of the lattice constant and the influence of the presence of small amounts of CO_2 and H_2S on the thermal evolution of the lattice expansion of these hydrates [BOU 07].

Furthermore, a comparative Raman analysis of intact NGHs (Figure 2.17) recovered from different geological settings has been performed. Detailed features of the gas distribution within the specimens can be described.

Figure 2.17. *Recovered gas hydrate samples from the Nigerian margin. Adapted with permission from [CHA 07]*

Comparison with Raman signatures of synthetic hydrates of H_2S, CH_4 and CH_4-deuterohydrates allows the unambiguous assignment of weak band overtones of trapped methane and co-clathrated H_2S molecular vibrations [CHA 07a] (Figure 2.18).

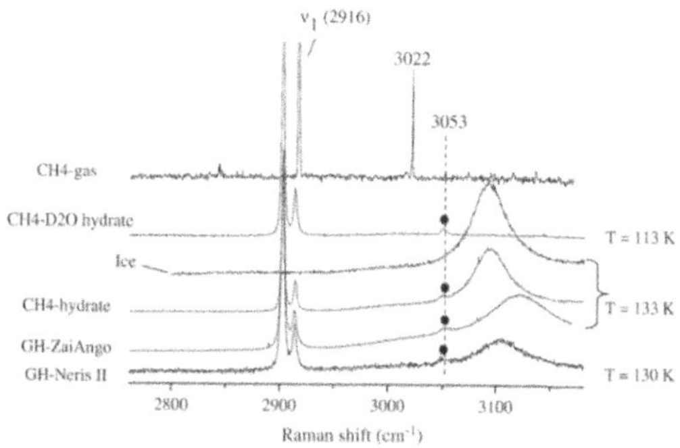

Figure 2.18. *Comparative spectra of the O–H spectral region of water at 3,100 cm^{-1}. The weak additional component at 3,053 cm^{-1} characterizes the presence of hydrates. It is assigned to an overtone of methane. Its presence is confirmed on the spectra of the CH_4–deutero hydrate where the contribution of water molecules (O–D stretching) is absent from the spectral region. Variations in the O–H band position are due to differences in temperature and/or reflect the structural differences between ice and hydrate, which are both present in the variable amounts under the laser spot. The spectra of methane in the free gas phase is shown for comparison. Adapted with permission from [CHA 07a]*

These overtones can now be used to discriminate between the contribution of occluded free methane gas and/or other higher hydrocarbons on *ex situ* recovered samples [SCH 10] or directly with field Raman study on the sea floor [HES 07]. In this latter study, massive seafloor hydrate outcrops were observed at Barkley Canyon with an appearance from yellow, due to oil staining, to white. While the "yellow" hydrate cannot be analyzed by Raman due to significant interference from fluorescence coming from oil residue, detailed analysis of the "white" hydrate was possible. It was shown that the "white" hydrate was mainly sII, but the presence of sI was also detected (Figure 2.19) in these complex gas hydrates containing hydrocarbon gases from C1 to C7.

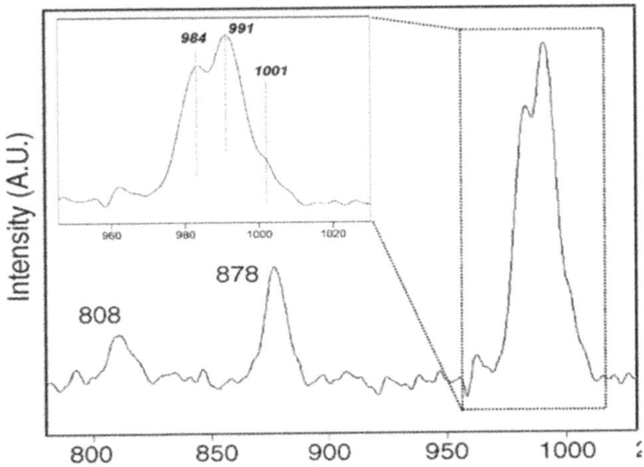

Figure 2.19. *Representative Raman spectrum from natural hydrate at Barkley Canyon (8.5 MPa, 278 K, collection time: 150 s). Peaks labeled at 808 cm^{-1} (i-C_4H_{10} ν7, sII), 878 cm^{-1} (C_3H_8 ν8, sII), 984 cm^{-1} (SO_4^{2-}, dissolved), 991 cm^{-1} (C_2H_6 ν1, sII), 1,001 cm^{-1} (C_2H_6 ν3, sI), 2,904 cm^{-1} (CH_4 ν1, sI–sII large cage) and 2,914 cm^{-1} (CH_4 ν1, sI–sII small cage). Asterisks marked for the C–H stretching modes of C2+ molecules in the hydrate phase. Adapted with permission from [HES 07]*

A deep petroleum reservoir is thought to be supplying this hydrate accumulation [POH 05]. Actually, thermogenic hydrate accumulations can be highly heterogeneous and vary in composition and structure on the subcentimeter scale. This heterogeneity occurs in gas-limited systems as the heavier hydrocarbons are selectively concentrated in the hydrate phase [KID 06, KLA 10].

Thus, a complete characterization of these accumulations becomes challenging, but unlike pure methane systems, gas fractionation provides insights into their growth pathways and system dynamics. For example, compositional variations of hydrocarbons such as ethane and propane in hydrates provide clues about the system dynamic [CHE 03]. In addition, the coexistence of multiple structures in hydrate accumulations is suggested to originate from fractionation or metastability [KID 06], but several other mechanisms may account for these effects [KLA 10]. It should also be noted that sH hydrates have long been suggested to exist in nature and the first direct evidence of NGHs containing sI and small amounts of sII and sH was obtained only recently by [YOU 04] on hydrates from the Gulf of Mexico (Jolliet Field, ~540 m water depth). They were collected off the Louisiana coast of the Gulf and these hydrates are examples of direct association between an oil accumulation and a thermogenic gas hydrate. Oil and gas are suggested to be trapped in Pleistocene–Pliocene reservoir sands at approximately 2–3 km depth. Later on, Lu *et al.* [LU 07] confirmed the occurrence of small quantities of sH mixed with sII in hydrates collected from Barkley Canyon.

Bourry *et al.* [BOU 09] studied NGHs and gas bubbles during the MARNAUT cruise (May–June 2007) in the Sea of Marmara along the Turkey cost. During this campaign, they investigated the relationships between active faults, fluid emission and landslides. The aim was to determine the depth of origin of the fluids and, in particular, whether fluids originate from relatively shallow syn-tectonic sediments or from basement predating the active basin and could therefore be related to the seismogenic zone.

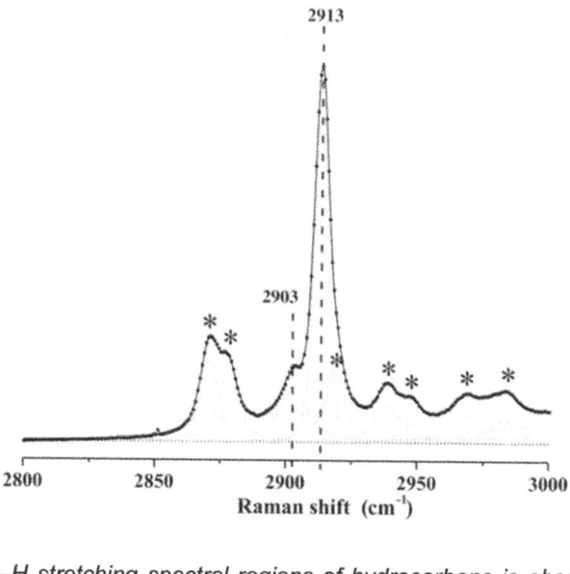

Figure 2.20. *C–H stretching spectral regions of hydrocarbons is observed between 2,800 and 3,000 cm^{-1}. The twin bands at 2,903 and 2,913 cm^{-1} are attributed to CH_4 trapped in the large cages ($5^{12}6^4$) and small cages (5^{12}) of the type II structure. Asterisks marked the bands attributed to C–H stretching modes of C_2H_6, C_3H_8 and i-C_4H_{10}. Adapted with permission from [BOU 09]*

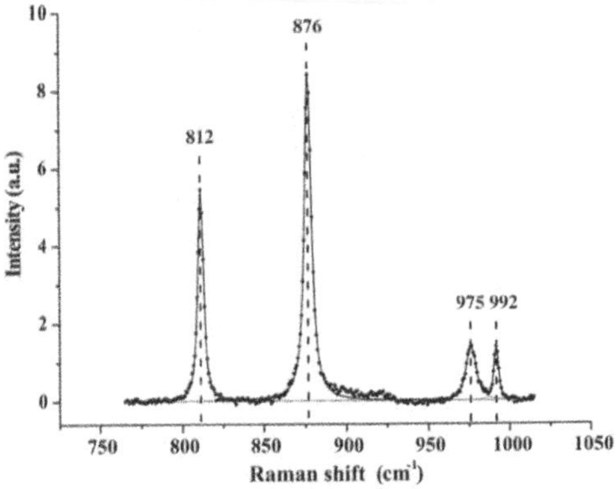

Figure 2.21. *C–C stretching spectral region of hydrocarbons (between 750 and 1,050 cm^{-1}). The bands at ~812, ~876 and ~992 cm^{-1} are attributed to the C–C stretching of i-C_4H_{10}, C_3H_8 and C_2H_6, respectively, in the large cages ($5^{12}6^4$) of type II structure. The band at ~975 cm^{-1} has not yet been assigned. Adapted with permission from [BOU 09]*

Hydrates specimens (Figure 2.22) and gas bubbles from the Western High basin were shown to contain thermogenic methane (~66%) and heavier hydrocarbon gases such as C_2, C_3 and i-C_4 at relatively high concentrations, whereas gas bubbles collected in the vicinity of the gas hydrates were shown to contain more methane (~91%), suggesting that a molecular fractionation occurs during gas hydrate formation with preferential enclathration according to $CH_4 < C_2 < C_3$ and i-C_4. Other parts (Eastern part of Marmara sea) were shown to have primarily microbial origin (~99.6% methane). The yellowish tint of the gas hydrate samples (Figure 2.22) and distinctive smell indicated the presence of traces of petroleum condensate in the pores. As observed by Hester *et al.* [HES 07], "yellow" hydrates could not be analyzed by visible excitation line with Raman due to high background fluorescence. Bourry *et al.* [BOU 09] performed UV-Raman analysis of the "yellow" hydrates from Marmara with excitation line at 266 nm to solve this analytical problem.

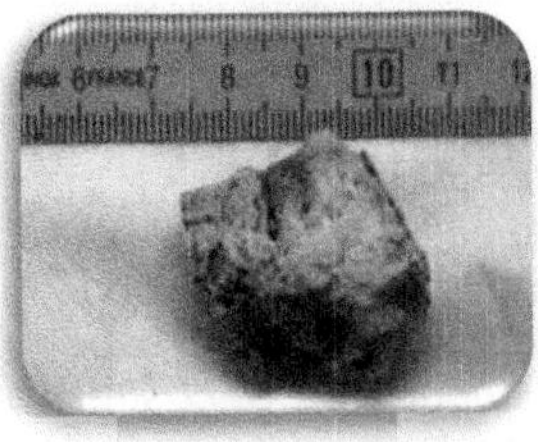

Figure 2.22. *"Yellow" hydrate recovered from Marmara Sea.*
For a color version of this figure, see www.iste.co.uk/broseta/hydrates1.zip

In Figures 2.20 and 2.21, the Raman spectra of the NGHs from the Marmara Sea show the characteristic contribution of methane, ethane, propane and i-butane in the C–H and C–C spectral range. The relative intensity ratio of the methane bands indicates that this hydrate is structure II, with a certain fraction of LCs occupied with C_2H_6, C_3H_8 and i-C_4H_{10}. The bands at ~812 and ~876 cm^{-1} can be, respectively, assigned to the ν7 symmetric C–C stretching vibration of i-C_4 [HES 07, UCH 07] and to the ν8 C–C stretching of C_3 [HES 07, KAW 06] in the large $5^{12}6^4$ cavities of sII structure. The unidentified peak at ~975 cm^{-1} also reported by Uchida *et al.* [UCH 07] has not yet been assigned. Lastly, the C_2 ν3 symmetric C–C stretching in the $5^{12}6^4$ cages can be clearly seen at ~992 cm^{-1} [HES 07, UCH 02], while the corresponding peak for ethane in the LC of sI hydrate at 1,001 cm^{-1} is absent, confirming that sII is the preponderant structure.

We should stress also that rapid progress has been made in the applications of Raman spectroscopy to deep-ocean science. This was made possible by deployment of instrumentation on remotely operated vehicles used for providing power and data flow and for precise positioning on targets of interest. Robust units have been deployed well over 100 times on an expeditionary basis over a very wide range of ocean depths without failure. Real-time access to the spectra obtained in the vehicle control room allows quantification of some of the solutes in seawater or pore waters. The applications include detection of the structure and composition of complex thermogenic gas hydrates both occurring naturally on the sea floor and in controlled sea floor experiments designed to simulate the growth of such natural systems. New developments in the ability to probe the chemistry of sediment pore waters in situ, long thought impossible candidates for Raman study due to fluorescence observed in recovered samples, have occurred. This approach has provided accurate measurement of the abundance of dissolved methane and sulfide in sediment pore waters (see [ZHA 12], for a review).

Besides, commercial exploitation with gas recovery from NGH reservoirs requires a detailed knowledge of the equilibrium thermodynamic and kinetic properties of gas hydrates in sediment matrices. In fact, these properties largely depend on the concentration, the actual fabrics of gas hydrates within the sediments and on the nature of the sediments [UCH 04]. Natural sedimentary matrices contain complex constituents, such as mineral particles, organic debris, pore water and fossils. The role of each constituent on the properties of NGHs is difficult to determine. For example, it is suggested that the characteristics of host-sediments such as grain-size distribution, which determine pore size and permeability have a primary role in the control of gas hydrates morphology and saturation. Therefore, determining the role of a specific factor in the system gives a better understanding of how it may influence equilibrium, formation and decomposition. Furthermore, most previous studies focused on quartz and silica sand as analogues for sediment matrices. For example, decomposition was observed to be completed faster in finer grain sediments (53–106 μm) [LIU 08], which was attributed to lower initial hydrate saturations in finer grain sediments.

It was suggested that the LCs lose methane more rapidly than the SCs in silica sands, unlike what is observed in bulk hydrate (Figure 2.23). In this latter, the decomposition of the unit cell of the hydrate is considered as a single entity [GUP 07]. However, rather complex behavior in the decomposition of NGHs has been identified [LIU 12], with a higher rate in NGH containing fine sediment particles (~63 μm) and a slower rate, with several steps, in sediments containing larger particles (63–500 μm). Unequivocal results that will allow the explanation of these phenomena are not yet available. In particular, the role of the nature of the mineral (morphology, surface chemistry) and how the size of the particles may influence hydrate formation, saturation and decomposition remains a further intringuing aspect that merits specific consideration.

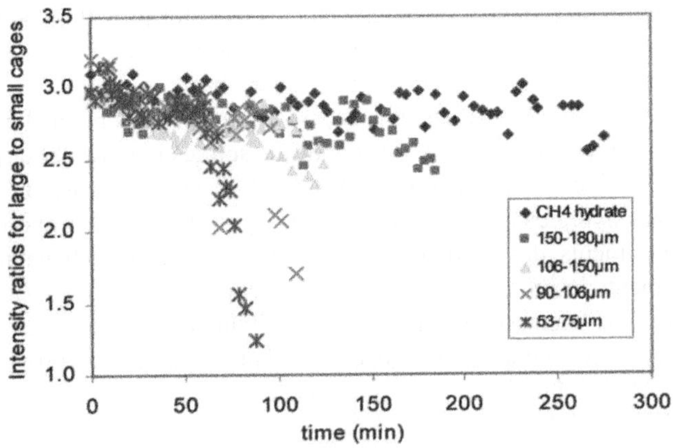

Figure 2.23. *Variations of intensity ratios of large and small cages through hydrate dissociation in silica sands. Adapted with permission from [LIU 08]. For a color version of this figure, see www.iste.co.uk/broseta/hydrates1.zip*

2.5. Clathrate hydrates in astrophysical environments

Water ice is ubiquitous in our solar system and throughout the interstellar medium. Clathrate hydrates have long been proposed as an ice phase present in various astrophysical environments including comets and planets. On Earth, clathrate hydrates are widely observed in ocean floor sediments and ice sheets and, as the thermodynamic conditions necessary for their formation and stability also exist in other extraterrestrial environments, astrophysical clathrates are expected to be observed as well. To date, there is no unequivocal spectral evidence identifying such clathrates, but gas and solid phase observations, theoretical models and laboratory studies all provide insights into the potential formation and evolution of these mixed water ices in star forming regions and planetary systems. This section reviews IR spectroscopy of astrophysical ices, including interstellar and solar system observations, as well as information obtained from laboratory spectroscopy of clathrate hydrates.

2.5.1. *IR spectroscopy of astrophysical ices*

IR imaging and spectroscopy are powerful tools to probe the physics and chemistry of astrophysical environments. Figure 2.24 presents two false-color images of the "Mystic Mountains" of the Carina Nebula taken by the Hubble space telescope. On the left is an image observed at visible wavelengths. Only the dusty exterior of the nebula is seen due to the dominance of light scattered from dust grains at these wavelengths. However, moving into the IR domain (right hand side), the interior of the nebula is probed. This image reveals embedded stars, which were not observed in the visible. These stars can be used to probe the solid phase matter (dust and ice) present in this nebula. Absorption features in the IR spectra of such objects can be attributed to solid phase species along the line of sight toward the star (see, e.g. [BOO 15]). In addition to IR absorption spectroscopy, ices in our solar system can be probed using IR reflectance spectroscopy (see, e.g. [GRU 99]).

Figure 2.24. *False color images of the Carina Nebula observed at visible (left-hand panel) and infrared (right-hand panel) wavelengths. Credit: NASA/ESA/M. Livio & Hubble 20th Anniversary Team (STScI). For a color version of this figure, see www.iste.co.uk/broseta/hydrates1.zip*

IR studies of astrophysical water ice can be performed on ground-based telescopes, for example the NIRSpec instrument on the Keck II telescope that is routinely used to observe comets and solar system bodies and was used to detect ^{13}CO in interstellar ices for the first time [BOO 02]. The advantage of such telescopes is that they provide a much higher spectral resolution than instruments on satellites (typical resolving powers, $R = \lambda/\delta\lambda$, of a few thousands to a few tens of thousands). However, telluric water absorption can make interpretation of the spectra complicated. In particular, the blue wing of the 3 μm OH stretching band is unavailable, making quantification of water ices difficult. There is, therefore, a reliance on airborne or satellite missions to provide near- and mid-IR spectra of astrophysical objects. General purpose missions such as ISO, *Spitzer* and Akari provide a wealth of spectral data, increasing our knowledge of the composition of ices in the solar system and the interstellar medium, while more targeted missions such as New Horizons or Deep Impact afford scientists a more specific, detailed look at a few solar system objects. High-resolution ground-based spectroscopy of particular absorption features is often used to complement lower resolution spectra obtained by instruments on spacecraft.

2.5.2. *Interstellar ices*

IR studies of interstellar ices have informed the development of models describing star formation. In the early stages of the star formation process, dust and atomic gas accumulate in higher density regions of the interstellar medium, forming diffuse clouds. As the density increases, atoms collide with cold dust grains and stick to their surfaces, leading to the formation of molecules, some of which desorb into the gas phase, while others remain on the grain surfaces as part of a developing icy mantle. This ice is amorphous in nature and is composed mainly of water, but also includes relatively high abundances of molecules such as CO, CO_2 and CH_3OH (see Table 2.1) (see [BOO 15] and references therein). The cloud will undergo gravitational collapse, via a dense core, until a protostar is formed. During the accretion phase of the protostar, the matter from the cloud flattens into a disk rotating around the central object, and it is within this disk that collisional grain growth processes occur, leading to the formation of planetesimals and eventually a planetary system. In principal, icy grains that form in the molecular cloud and remain beyond the snowline[1] in the disk midplane during the accretion phase could be delivered to the planet-forming regions of the disk without experiencing irradiation or thermal processing, and thus retaining their amorphous nature [VIS 09]. If this is the case, it is possible that amorphous interstellar ices are conserved in the coldest regions of a planetary system, implying that comets in our

1 Snowline: the distance from the central star where a given molecular species is frozen onto dust grains rather than in the gas phase.

own solar system are composed of pristine amorphous ices from the planetary nebula. However, accretion and grain growth processes are likely to result in heating of the icy grains after each collision, leading to crystallization, melting and segregation processes within their icy mantles. Heating of amorphous mixed ices has been shown to lead to the formation of clathrate hydrates (see section 2.5.4), and thus it is possible that ices present in the planet-forming regions of disks could contain clathrate hydrates.

Molecule	Comets	Dense clouds	Low-mass protostars
CO	0.4–30	9–36	0–100
CO_2	2–30	15–44	2–68
CH_4	0.4–1.6	<3	2–8
CH_3OH	0.2–7	5–12	1–30
H_2CO	0.11–1	–	~6
HCOOH	0.06–0.14	~2	1–9
NH_3	0.2–1.4	<6–9	2–15
$HNCO/OCN^-$	0.02–0.1	<2	<0.9
H_2S	0.12–1.4	<1–4	–
OCS	0.1–0.4	<0.2	–

Table 2.1. *Typical abundances of molecules in cometary and interstellar ices (relative to water). Adapted from Table 3 in [MUM 11]*

Laboratory IR spectra of clathrate hydrates have revealed that the spectral signature of the water ice network in a clathrate is difficult to disentangle from that of crystalline water ice (see section 2.5.4), making direct identification of astrophysical clathrate hydrates via water ice absorption features unlikely. The spectral bands of guest molecules do show characteristic shifts due to encapsulation, and these represent the best chance of directly observing clathrates. One spectral signature widely observed in the spectra of interstellar ices is the band at 3.47 μm in the red wing of the water OH stretching feature. This band is potentially attributable to ammonia hydrates, although other candidates for carrier of the feature include nanodiamonds and aromatic and aliphatic CH stretches on dust grain surfaces [BRO 99].

Recent IR studies do provide some clues as to the interstellar environments where clathrate hydrates may exist. Observations of protoplanetary disks around T Tauri stars using the *Herschel*/PACS spectrometer have revealed that both amorphous and crystalline ices are present in these objects [MCC 15]. Although the observations themselves are not spatially resolved, models suggest that the crystalline ice is generated by dust grain or planetesimal collisions in a proto-Kuiper Belt at > 30 AU. These types of energetic collisions could help to link unprocessed ices in the disk midplane to the crystalline-dominated ices observed in our own outer solar system (e.g. [JEW 04]). It should also be taken into account that processes such as luminosity outbursts could lead to intermittent processing of ices in the midplane, causing previously mixed ices to sublimate and then recondense in segregated layers [TAQ 16]. As small molecules will be mobile at temperatures below their accretion temperature, diffusion into the recondensed water ice layer could lead to the formation of clathrate hydrates.

In addition to the planets in our solar system, which will be discussed below, clathrate hydrates could be present in extrasolar planets. Over the course of the past 20 years, observations have revealed the presence of thousands of planets orbiting stars other than our own sun. The study of the composition of these planets and their atmospheres is in its infancy compared to studies of our solar system, but the serendipitous passage of exoplanets in front of their star has allowed the transmission spectra of around two dozen exoplanetary atmospheres to be recorded (see, e.g. [SEA 10]). A recent study using light reflected off the moon Ganymede to probe Jupiter's atmosphere in transmission from 0.3 to 2.5 μm [MON 15] highlighted the potential power of transit spectra in revealing ices in exoplanet atmospheres. The study, using Jupiter as a proxy for a Jupiter-type exoplanet, was able to attribute spectral absorptions in Jupiter's atmosphere to both water ice and aerosol haze, suggesting the presence of water ice clouds in the stratospheric layer of Jupiter's atmosphere. With the higher sensitivity, spatial and spectral resolution offered by future instruments such as the JWST and SPICA satellites or the E-ELT ground-based telescope, similar detailed studies of the transmission spectra of extrasolar planets may be possible.

2.5.3. *Solar system ices*

In our solar system, bodies that formed beyond the water snowline (i.e. the distance from the sun at which water ice is stable, currently located at ~3.5–4 AU) are rich in water ice. This includes the gas planets and their satellites, Kuiper Belt objects (KBOs), Trans-Neptunian objects and comets. IR spectroscopy can be used to constrain the temperature of solar system bodies by examination of the water ice band profile [GRU 99]. Most ices in the outer solar system have band profiles characteristic of low temperature I*h* [GRU 99]. Even in the spectra of KBOs, objects in the most distant regions of our solar system with temperatures of ~50 K, some signatures of crystallization have been observed, as shown in Figure 2.25, although grain diffraction can cause profile and peak modifications which complicate attribution [CLA 12]. Clathrate hydrates are expected to be widely abundant in solar system bodies [KAR 98, MOU 15], according to observational evidence and models based on the thermodynamic behavior of clathrates. While no direct spectroscopic identification of clathrates has been made, evidence points to them playing a key role in a number of different observed phenomena in solar system ices. A few examples are given here.

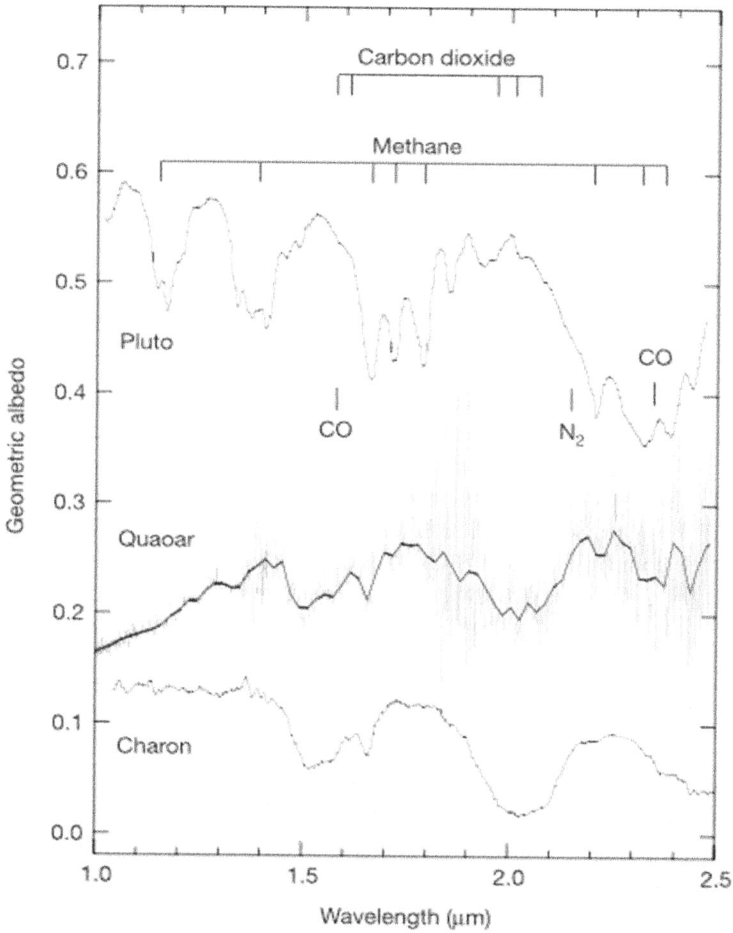

Figure 2.25. *1–2.5 μm IR spectra of three Kuiper-belt objects Pluto, Quaoar and Charon. The wavelengths of absorption bands of solid CO_2, CO, CH_4, and N_2 are labeled. Water ice features are present at 1.5, 1.65 and 2 μm in the Quaoar spectrum, with the 1.65 μm feature being attributed to crystalline ice. The small feature at 2.2 μm is attributed to ammonia hydrate. Reproduced from [JEW 04]*

Cometary ices contain many small volatiles (see Table 2.1) and therefore must be composed of ice able to trap these volatiles at cold temperatures, releasing them into the coma upon heating. There is much debate about whether cometary ices are amorphous or in the form of clathrate hydrates (see, e.g. [CHO 13]). Theoretical studies have argued that clathrates could have formed in ices under early solar system conditions or due to heating during their orbital evolution [MAR 10], and that they explain observations such as the low N_2:CO ratio in comets due to the preferential clathration of CO [IRO 03]. Cometary ices are difficult to directly observe, first because ices are not generally present on the cometary surface (apart from occasional “refrosting” of gases in the comae) and therefore their composition is probed in the gas phase by various techniques, including IR spectroscopy [BRO 03]; second, because clathrate hydrates are more likely to be present only in the inner regions of comets rather than at the surface. The first direct evidence of gaseous H_2O in cometary comae came from observations of comet Halley using an IR spectrometer on the NASA-Kuiper Airborne Observatory [MUM 86]. It was concluded that this water vapor originated from cold water ice due to its derived nuclear spin temperature of 32 K, suggesting that the ice was amorphous at formation.

The EPOXI (Extrasolar Planet Observation and Deep Impact Extended Investigation) mission provided the opportunity to observe the Hartley 2 comet at a distance of 1.1 AU from the sun. Due to the close flyby of the Deep Impact spacecraft, it was possible to image and take spectra of the comet with a high spatial resolution. The spectra presented in Figure 2.26 were observed toward two different regions of the cometary coma: a H_2O vapor outflow and a CO_2 jet. No water ice was detected in the H_2O vapor outflow, which is unsurprising as the H_2O in the spectrum of the CO_2 jet, an absorption band centered at 2.9 μm, is attributable to water ice. The source of this ice is believed to be micrometer-sized particles of pure and mixed water ice carried out of the nucleus by the main volatile component in this region of the coma, CO_2 [AHE 11]. The structural nature of the ice is difficult to determine, both because of the presence of overlapping gaseous water emission in the same spectral region as the ice absorption, and due to the low spectral resolution of the instrument. The recently completed ROSETTA mission studied the composition of comet 67P/Churyumov-Gerasimenko in detail. Evidence from the abundances of O_2, N_2, CO, H_2O and Ar measured in the coma using the ROSINA instrument point to the comet being composed of amorphous ice. It is likely that this formed at temperatures below ~30 K [RUB 15] from pristine Interstellar Medium (ISM) amorphous ice grains [MOU 16]. However, this does not preclude the presence of other comets containing ices composed of clathrates formed during disk cooling.

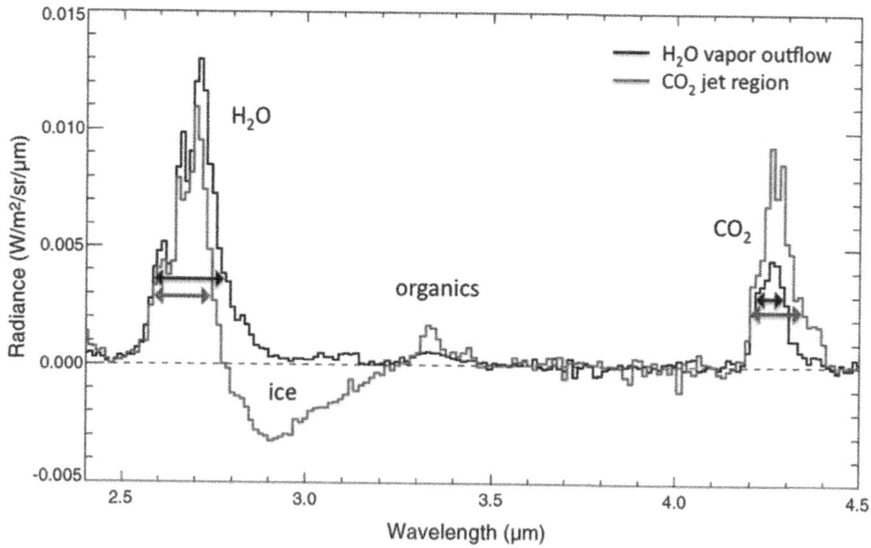

Figure 2.26. *2.4–4.5 μm IR spectra of Hartley 2 comet taken by the Deep Impact (EPOXI) High-Resolution Instrument. Reproduced from [AHE 11]. For a color version of this figure, see www.iste.co.uk/broseta/hydrates1.zip*

One solar system body with a high enough atmospheric pressure to allow stable surface clathrate hydrates is Titan, where methane clathrate hydrates are suspected to exist. Cooling of a body with a mixed N_2:CH_4 atmosphere and a liquid surface could have led to the formation of a 3–5 km thick layer of clathrate, which would explain both the modern day gas phase CH_4 abundance due to continual replenishment due to dissociation of the clathrate [LUN 09, MOU 15] and the low abundance of photochemically produced ethane due to its sequestration in the clathrate layer [MOU 08]. Clathrates in the icy interior could also explain observations performed on Saturn's moon, Enceladus, where the *Cassini* mission discovered plumes of water vapor mixed with various volatiles in the South pole region. These have been attributed to the disruption of mixed clathrate hydrate reserves in the satellite's interior by tectonic activity [KIE 06].

It has long been suggested that the CO_2 rich ices at the Northern and Southern Martian poles could have a clathrate hydrate component [MIL 61] while the permafrost could store CH_4 in clathrate form [CHA 07b], although this is a hotly debated topic. A recent model based on experimentally obtained activation energies for the nucleation and growth of CO_2 hydrates at temperatures of 185–195 K suggests that CO_2 hydrates are unlikely to be a constituent of ices in the Martian polar regions under present day conditions because (1) the formation rate is too slow and (2) summer temperatures are above the stability limit of the hydrate [FAL 11]. However, another model assuming the presence of mixed SO_2:CO_2 clathrates (with sulfur released from volcanic activity and explaining the warm atmospheric temperatures), a higher temperature, and including grain size effects, concludes that formation could occur more quickly than anticipated [CHA 13]. Ultimately, further observations, such as those proposed with the JWST satellite [VIL 16], are required to probe the surface and the atmosphere of Mars to study the nature of the ice in the permafrost, at the poles, and in water ice clouds.

To date, only one observational IR study has thus far tentatively assigned observational spectral features to clathrate hydrates. Cruikshank *et al.* [CRU 10] suggest that shifts in the CO_2 band observed toward Saturn's moons Phoebe, Iapetus and Hyperion could be explained by CO_2 clathrates, but stressed that other mixed ices could also be responsible for the features. There is a need for in-depth laboratory studies of clathrate hydrates in order to understand their potential formation mechanisms in the solar system and to provide IR spectra to aid in the interpretation of observations.

2.5.4. *Insights from laboratory spectroscopy*

Laboratory-based IR and RAMAN spectroscopies are vital tools in interpreting observational spectra and in understanding the thermodynamics of clathrate hydrates under astrophysically relevant conditions. Terrestrial clathrates are formed in high-pressure environments such as ocean floor sediments, but extraterrestrial environments where ices are observed can be much colder and much lower pressure (in the most extreme case, as low as $T > 10$ K with a hydrogen density of $nH \sim 10^4$ cm^{-3} in molecular clouds). IR spectroscopic studies of clathrate hydrates at astrophysically relevant temperatures can be performed by forming a clathrate hydrate at higher temperature from I*h* in contact with the guest molecule at a moderate pressure, before evacuating the residual pressure and cooling the ice. This technique has been utilized, for example, to record Fourier Transform Infrared Spectroscopy (FTIR) transmittance spectra of the CH_4 clathrate at temperatures between 7 and 80 K [DAR 08, DAR 10], the CO_2 clathrate between 6 and 150 K [DAR 09] and the CO clathrate between 5 and 140 K [DAR 11]. These studies provide critical insight into cage and guest molecule dynamics, such as the fact that guest molecules rotate at low temperatures before beginning to rattle and become mobile at higher temperatures [DAR 12]. This behavior gives rise to temperature-dependent frequency shifts and profile modifications of the vibrational bands of the guest molecule, which can be useful in interpreting observational spectra. However, as previously stated, the water ice stretching O–O and O–H bands are difficult to distinguish from I*h* [DAR 08] and therefore astrophysical clathrates will only be positively identified based upon changes in the spectra of caged molecules.

Although IR experiments of clathrates initially formed at high temperature and pressure provide the spectroscopic properties of low temperature clathrate hydrates, they do not address the question of whether, and by what mechanism, clathrates could form under astrophysical conditions. It is known that clathrates of molecules such as CO_2 and CH_4 do not form from diffusion of the guest molecule into crystalline ice at temperatures lower than ~200 K and without high pressures [HEN 00, WAN 02]. However, molecules known to efficiently H-bond with water, such as H_2S or HCN, can form clathrates at temperatures of ~100 K and at pressures of only a few mbar by diffusion into Ih [BUC 09]. In addition to diffusive formation, co-deposition experiments provide evidence of two lower temperature clathrate formation mechanisms. First, mixed amorphous ices produced by co-deposition of water with the guest molecule at low temperature (typically 85–100 K) can undergo a phase transition to a clathrate hydrate upon slow heating. This process has been observed with techniques such as electron microscopy [BLA 01a] and IR spectroscopy [RIC 85]. During the heating, much of the volatile guest molecule present in the water ice will be lost in order to obtain the required H_2O:guest ratio of 17:1 for the type II clathrate hydrate [CHA 08]. This can occur via segregation within the ice, for example to the ice grain boundaries or by desorption. Such a heating mechanism is widely believed to be responsible for outgassing of volatiles from comets as they approach the sun. Second, there is some evidence of direct clathrate hydrate formation upon co-deposition, such as the case of direct vapor deposition experiments with CH_3OH at temperatures between 125 and 135 K, which have produced clathrates under conditions likely present in the early solar nebula [BLA 01b]. Methanol is known to act as a catalyst for clathrate formation by gas phase co-deposition [DEV 14], so due to its high abundances in astrophysical ices (see Table 2.1) it may increase the probability of clathrate formation in such environments.

Overall, laboratory experiments have already provided key insights into the potential formation of clathrate hydrates in astrophysical ices. However, there are not yet measured spectra available for all astrophysically relevant clathrate hydrates in the NIR and MIR spectral regions, covering the full temperature range relevant to ices in the ISM or our solar system (~10–150 K). Further studies, especially on mixed clathrate hydrate systems, are required, with a particular focus on determining under which astrophysical conditions clathrate hydrates can form and remain stable. Many of the low temperature and pressure data required to trace the equilibrium curve are unavailable.

2.6. Concluding remarks

This chapter has illustrated the various breakthroughs offered by optical spectroscopy such as Raman scattering, FTIR absorption/reflection and IXS.

Overall, these laboratory experiments have already provided key insights into the formation/dissociation mechanism of hydrates, the characterization of NGHs and the plausibility of the formation of gas hydrates in astrophysical systems. Yet, there is still a crucial requirement for spectroscopic signatures to access guest partition and distribution in gas hydrates containing pure (or mixed) gases, particularly in contact with solid surfaces to simulate more realistic natural conditions and further investigate the formation and dissociation processes. The role of the nature of the mineral (morphology, surface chemistry) and how the size of the particles may influence hydrate formation, saturation and decomposition remains a further intringuing aspect that merits specific considerations. Moreover, the deployment of "field" spectroscopic experiments with direct field Raman study on the sea floor for geosciences is currently greatly beneficial for the understanding of hydrate formation in real system. Furthermore, the development of satellites equipped with IR spectrometers adapted to the study of extraterrestrial environments are required to better constrain in-depth laboratory studies of clathrate hydrates formation in conditions relevant to astrophysical environments. To date, only one observational IR study has thus far tentatively assigned observational spectral features to clathrate hydrates. Additionally, there are not yet measured spectra available for all astrophysically relevant clathrate hydrates in the NIR and MIR spectral regions, covering the full temperature range relevant to ices in the ISM or our solar system. Further studies, especially on mixed clathrate hydrate systems, are required, with a particular focus on determining under which astrophysical conditions clathrate hydrates can form and remain stable. Thus, continued efforts are needed to develop spectroscopic characterization techniques for clathrate hydrates to investigate fundamental aspects as these are crucial in many energy and environmental related applications.

2.7. Bibliography

[AHE 11] A'HEARN M.F., BELTON M.J.S., DELAMERE W.A. *et al.*, "EPOXI at comet Hartley 2", *Science*, vol. 332 (6036), pp. 1396–1400, 2011.

[BAU 03] BAUMERT J., GUTT C., SHPAKOV V.P. *et al.*, "Lattice dynamics of methane and xenon hydrate: Observation of symmetry-avoided crossing by experiment and theory", *Physical Review B – Condensed Matter and Materials Physics*, vol. 68 (17), art. no. 174301, pp. 1743011–1743017, 2003.

[BER 78] BERTIE J.E., JACOBS S.M., "Infrared spectra from 300 to 10 cm^{-1} of structure II clathrate hydrates at 4.3 K", *Journal of Chemical Physics*, vol. 69, no. 9, p. 4105, 1978.

[BLA 01a] BLAKE D., ALLAMANDOLA L., SANDFORD S. *et al.*, "Clathrate hydrate formation in amorphous cometary ice analogs in vacuo", *Science*, vol. 254 (5031), pp. 548-551, 1991.

[BLA 01b] BLAKE D., ALLAMANDOLA L.J., SANDFORD S. *et al.*, "Clathrate type II hydrate formation in vacuo under astrophysical consitions", *Meteoritics*, vol. 26, p. 319, 1991.

[BOE 12] BOEWER L., NASE J., PAULUS M. *et al.*, "On the spontaneous formation of clathrate hydrates at water-guest interfaces", *Journal of Physical Chemistry C*, vol. 116, no. 15, pp. 8548–8553, 2012.

[BOO 02] BOOGERT A.C.A., BLAKE G.A., TIELENS A.G.G.M., "High-resolution 4.7 Micron Keck/NIRSPEC spectra of protostars. II. Detection of the 13CO isotope in icy grain mantles", *Astrophysical Journal*, vol. 577 (1 I), pp. 271-280, 2002.

[BOO 15] BOOGERT A.C.A., GERAKINES P.A., WHITTET D.C.B., "Observations of the icy universe", *Annual Review of Astronomy and Astrophysics*, vol. 53, p. 541, 2015.

[BOS 11] BOSWELL R., COLLETT T.S., "Current perspectives on gas hydrate resources", *Energy & Environmental Science*, vol. 4, no. 4, p. 1206, 2011.

[BOS 12] BOSWELL R., COLLETT T.S., FRYE M. *et al.*, "Subsurface gas hydrates in the northern Gulf of Mexico", *Marine and Petroleum Geology*, vol. 34, no. 1, pp. 4–30, 2012.

[BOU 07] BOURRY C., CHARLOU J., DONVAL J. *et al.*, "X-ray synchrotron diffraction study of natural gas hydrates from African margin", *Geophysical Research Letters*, vol. 34, pp. 1–5, 2007.

[BOU 09] BOURRY C., CHAZALLON B., LUC J. *et al.*, "Free gas and gas hydrates from the Sea of Marmara, Turkey Chemical and structural characterization", *Chemical Geology*, vol. 264, pp. 197–206, 2009.

[BRO 84] BROOKS J.M., KENNICUTT M.C., FAY R.R. *et al.*, "Thermogenic gas hydrates in the Gulf of Mexico", *Science*, vol. 80, no. 225, pp. 409–411, 1984.

[BRO 99] BROOKE T.Y., SELLGREN K., GEBALLE T.R., "New 3 micron spectra of young stellar objects with H2O ice bands", *Astrophysical Journal*, vol. 517 (2 PART 1), pp. 883-900, 1999.

[BRO 03] BROOKE T.Y., WEAVER H.A., CHIN G. *et al.*, "Spectroscopy of Comet Hale-Bopp in the infrared", *Icarus*, vol. 166, p. 167, 2003.

[BUC 09] BUCH V., DEVLIN J.P., ABRREY MONREAL I. *et al.*, "Clathrate hydrates with hydrogen-bonding guests", *Physical Chemistry Chemical Physics*, vol. 11, p. 10245, 2009.

[CHA 04] CHARLOU J.L., DONVAL J.P., FOUQUET Y. *et al.*, "Physical and chemical characterization of gas hydrates and associated methane plumes in the Congo – Angola Basin", *Chemical Geology*, vol. 205, pp. 405–425, 2004.

[CHA 07a] CHAZALLON B., FOCSA C., CHARLOU J.-L. *et al.*, "A comparative Raman spectroscopic study of natural gas hydrates collected at different geological sites", *Chemical Geology*, vol. 244, nos. 1–2, pp. 175–185, 2007.

[CHA 07b] CHASTAIN B.K., CHEVRIER V., "Methane clathrate hydrates as a potential source for martian atmospheric methane", *Planetary and Space Science*, vol. 55, p. 1246, 2007.

[CHA 13] CHASSEFIÈRERE E., DARTOIS E., HERRI J.-M. *et al.*, "CO2-SO2 clathrate hydrate formation on early Mars", *Icarus*, vol. 223, p. 878, 2013.

[CHA 08] CHAZALLON B., OANCEA A., CAPOEN B. *et al.*, "Ice mixtures formed by simultaneous condensation of formaldehyde and water: An in situ study by micro-Raman scattering", *Physical Chemistry Chemical Physics*, vol. 10, pp. 702–712, 2008.

[CHE 03] CHEN D.F., CATHLES L.M., "A kinetic model for the pattern and amounts of hydrate precipitated from a gas steam: application to the Bush Hill vent site, Green Canyon Block 185, Gulf of Mexico", *Geophysical Research Letters*, vol. 108, pp. 1–14, 2003.

[CHO 13] CHOUKROUN M., KIEFFER S.W., LU X. *et al.*, "Clathrate hydrates: implications for exchange processes in the outer solar system", in M.S. GUDIPATI, J. CASTILLO-ROGEZ, (eds), *The Sciences of Solar System Ices*, Springer, Berlin, 2013.

[CHO 16] CHONG Z.R., YANG S.H.B., BABU P. *et al.*, "Review of natural gas hydrates as an energy resource: Prospects and challenges", *Applied Energy*, vol. 162, pp. 1633–1652, 2016.

[CHO 00] CHOU I.M., SHARMA A., BURRUSS R.C. *et al.*, "Transformations in methane hydrates", *Proceedings of National Academy of Science*, vol. 97, p. 13485, 2000.

[CHR 94] CHRISTIANSEN R.L., SLOAN E.D., "Mechanisms and kinetics of hydrate formation", *Annals of New York Academy of Science*, vol. 715, no. 1, pp. 283–305, 1994.

[CLA 12] CLARK R.N., CRUIKSHANK D.P., JAUMANN R. *et al.*, "The surface composition of Iapetus: Mapping results from Cassini VIMS", *Icarus*, vol. 218, p. 831, 2012.

[COL 08] COLLETT T., *World Energy Outlook 2008*, International Energy Agency, 2008.

[CRU 10] CRUIKSHANK D.P., MEYER A.L., BROWN R.H. *et al.*, "Carbon dioxide on the satellites of Saturn: Results from the Cassini VIMS investigation and revisions to the VIMS wavelength scale", *Icarus*, vol. 206, p. 561, 2010.

[DAR 08] DARTOIS E., DEBOFFLE D., "Methane clathrate hydrate FTIR spectrum implications for its cometary and planetary detection", *Astronomy & Astrophysics*, vol. 490, p. L19, 2008.

[DAR 09] DARTOIS E., SCHMITT B., "Carbon dioxide clathrate hydrate FTIR spectrum", *Astronomy & Astrophysics*, vol. 504, p. 869, 2009.

[DAR 10] DARTOIS E., DEBOFFLE D., BOUZIT M., "Methane clathrate hydrate infrared spectrum - II. Near-infrared overtones, combination modes and cages assignment", *Astronomy & Astrophysics*, vol 514, p. A49, 2010.

[DAR 11] DARTOIS E., "CO clathrate hydrate: Near to mid-IR spectroscopic signatures", *Icarus*, vol. 212, p. 950, 2011.

[DAR 12] DARTOIS E., BOUZIT M., SCHMITT B., "Clathrate hydrates: Ftir spectroscopy for astrophysical remote detection", *EAS Publications Series*, vol. 58, p. 219, 2012.

[DEC 12] DECONTO R.M., GALEOTTI S., PAGANI M. *et al.*, "Past extreme warming events linked to massive carbon release from thawing permafrost", *Nature*, vol. 490, no. 7419, p. 292, 2012.

[DEM 10] DEMIRBAS A., "Methane hydrates as potential energy resource: part 2 – methane production processes from gas hydrates", *Energy Conversion and Management*, vol. 51, no. 7, pp. 1562–1571, 2010.

[DES 15] DESMEDT A., MARTIN-GONDRE L., NGUYEN T.T. *et al.*, "Modifying the flexibility of water cages by co-including acidic species within clathrate hydrate",*Journal of Physical Chemistry C*, vol. 119, pp. 8904–8911, 2015.

[DEV 95] DEVLIN J.P., BUCH V., "Surface of ice as viewed from combined spectroscopic and computer modeling studies", *Journal of Physical Chemistry*, vol. 99, no. 45, pp. 16534–16548, 1995.

[DEV 10] DEVLIN J.P., MONREAL I.A., "Clathrate-hydrate ultrafast nucleation and crystallization from supercooled aqueous nanodroplets", *Chemical Physics Letters*, vol. 492, nos. 1–3, pp. 1–8, 2010.

[DEV 14] DEVLIN J.P., "Catalytic activity of methanol in all-vapor subsecond clathrate-hydrate formation", *Journal of Chemical Physics*, vol. 140, p. 164505, 2014.

[EWI 88] EWING G.E., SHENG D.T., "Infrared spectroscopy of CO_2 ultrafine particles", *Journal of Physical Chemistry*, vol. 92, no. 14, pp. 4063–4066, 1988.

[FAL 11] FALENTY A., GENOV G., HANSEN T.C. *et al.*, "Kinetics of CO2 hydrate formation from water frost at low temperatures: Experimental results and theoretical model",*Journal of Physical Chemistry C*, vol. 115, p. 4022, 2011.

[FAN 14] FANDIÑO O., RUFFINE L., "Methane hydrate nucleation and growth from the bulk phase: further insights into their mechanisms", *Fuel*, vol. 117, pp. 442–449, 2014.

[HER 98] HERNANDEZ J., URAS N., DEVLIN J.P., "Coated ice nanocrystals from water–adsorbate vapor mixtures: formation of ether–CO_2 clathrate hydrate nanocrystals at 120 K", *Journal of Physical Chemistry B*, vol. 102, no. 23, pp. 4526–4535, 1998.

[FLE 90] FLEYFEL F., DEVLIN J.P., "Infrared spectra of large clusters containing small ether molecules: liquid, crystalline, and clathrate-hydrate cluster spectra", *Journal of Chemical Physics*, vol. 92, no. 1, p. 36, 1990.

[FLE 91] FLEYFEL F., DEVLIN J.P., "Carbon dioxide clathrate hydrate epitaxial growth: spectroscopic evidence for formation of the simple type-I1 COP hydrate", *Journal of Physical Chemistry*, vol. 95, no. 12, pp. 3811–3815, 1991.

[GRU 99] GRUNDY W.M., BUIE M.W., STANSBERRY J.A. *et al.*, "Near-Infrared Spectra of Icy Outer Solar System Surfaces: Remote Determination of h2o Ice Temperatures" *Icarus*, vol. 142, p. 536, 1999.

[GUI 08] GUILLAUME F., COUZI M., *Spectroscopie optique: les vibrations et les phonons*, EDP Sciences, 2008.

[GUO 04] GUO G.J., ZHANG Y.G., ZHAO Y.J. *et al.*, "Lifetimes of cagelike water clusters immersed in bulk liquid water: A molecular dynamics study on gas hydrate nucleation mechanisms", *Journal of Chemical Physics*, vol. 121, no. 3, pp. 1542–1547, 2004.

[GUP 07] GUPTA A., DEC S.F., KOH C.A. *et al.*, "NMR investigation of methane hydrate dissociation", *Journal of Physical Chemistry C*, vol. 111, pp. 2341–2346, 2007.

[GUT 02] GUTT C., BAUMERT J., PRESS W. *et al.*, "The vibrational properties of xenon hydrate: An inelastic incoherent neutron scattering study" *Journal of Chemical Physics*, vol. 116, no. 9, p. 3795, 2002.

[HEN 00] HENNING R.W., SCHULTZ A.J., THIEU V. *et al.*, "Neutron diffraction studies of CO_2 clathrate hydrate: formation from deuterated ice", *Journal of Physical Chemistry A*, vol. 104, pp. 5066–5071, 2000.

[HES 07] HESTER K.C., DUNK R.M., WALZ P.M. *et al.*, "Direct measurements of multi-component hydrates on the seafloor: pathways to growth", *Fluid Phase Equilibria*, vol. 261, pp. 396–406, 2007.

[HIR 12] HIRATSUKA M., OHMURA R., SUM A.K. *et al.*, "Molecular vibrations of methane molecules in the structure i clathrate hydrate from ab initio molecular dynamics simulation", *Journal of Chemical Physics*, vol. 136, p. 044508, 2012.

[HOL 96] HOLLBROOK W.S., HOSKINS H., WOOD W.T. *et al.*, "Methane hydrate and free gas on the blake ridge from vertical seismic profiling", *Science*, vol. 273, pp. 1840–1843, 1996.

[HYO 14] HYODO M., LI Y., YONEDA J. *et al.*, "A comparative analysis of the mechanical behavior of carbon dioxide and methane hydrate-bearing sediments", *American Mineralogist*, vol. 99, no. 1, pp. 178–183, 2014.

[IRO 03] IRO N., GAUTIER D., HERSANT F. *et al.*, "An interpretation of the nitrogen deficiency in comets", *Icarus*, vol. 161, p. 511, 2003.

[JAC 01] JACOBSEN S.B., "Gas hydrates and deglaciations", *Nature*, vol. 412, pp. 691–693, 2001.

[JAC 10a] JACOBSON L.C., HUJO W., MOLINERO V., "Amorphous precursors in the nucleation of clathrate hydrates", *Journal of American Chemical Society*, vol. 132 no. 33, pp. 11806–11811, 2010.

[JAC 10b] JACOBSON L.C., HUJO W., MOLINERO V., "Nucleation pathways of clathrate hydrates: effect of guest size and solubility", *Journal of Physical Chemistry B phys,*, vol. 114, pp. 13796–13807, 2010.

[JAD 06] JADHAWAR P., MOHAMMADI A.H., YANG J. *et al.*, "Subsurface carbon dioxide storage through clathrate hydrate formation", in LOMBARDI S., ALTUNINA L.K., BEAUBIEN S.E. *et al.* (eds), *Advances in the Geological Storage of Carbon Dioxide*, Springer, The Netherlands, 2006.

[JEW 04] JEWITT D.C., LUU J., "Crystalline water ice on the Kuiper belt object (50000) Quaoar", *Nature*, vol. 432, p. 731, 2004.

[JIA 16] JIA Y., ZHU C., LIU L. *et al.*, "Marine geohazards: review and future perspective", *Acta Geologica Sinica (English Edition)*, vol. 90, no. 4, pp. 1455–1470, 2016.

[JOH 15] JOHNSON J.E., MIENERT J., PLAZA-FAVEROLA A. *et al.*, "Abiotic methane from ultraslow-spreading ridges can charge Arctic gas hydrates", *Geology*, vol. 43, no. 5, pp. 371–374, 2015.

[KAR 98] KARGEL J.S. LUNINE J.I., "Clathrate hydrates on earth and in the solar system", in DE BERGH C., FESTOU M., SCHMITT B. (eds), *Solar System Ices*, Kluwer Academic Publishers, Boston, 1998.

[KAS 03] KASHCHIEV D., FIROOZABADI A., "Induction time in crystallization of gas hydrates", *Journal of Crystal Growth*, vol. 250, nos. 3–4, pp. 499–515, 2003.

[KAW 06] KAWAMURA T., SAKAMOTO Y., OHTAKE M., "Dissociation behavior of pellet shaped mixed gas hydrate samples that contain propane as a guest", *Energy Conversion and Management*, vol. 47, pp. 2491–2498, 2006.

[KID 06] KIDA M., KHLYSTOV O., ZEMSKAYA T. *et al.*, "Coexistence of structure I and II gas hydrates in Lake Baikal suggesting gas sources from microbial and thermogenic origin", *Geophysical Research Letters*, vol. 33, pp. 10–13, 2006.

[KIE 06] KIEFFER S.W., LU X., BETHKE C.M. *et al.*, "A clathrate reservoir hypothesis for enceladus south polar plume", *Science*, vol. 314, p. 1764, 2006.

[KLA 10] KLAPP S.A., MURSHED M.M., PAPE T. *et al.*, "Mixed gas hydrate structures at the Chapopote Knoll, southern Gulf of Mexico", *Earth Planetary Science Letters*, vol. 299, nos. 1–2, pp. 207–217, 2010.

[KLO 13] KLOBES B., DESMEDT A., SERGUEEV I. *et al.*, "Hydrogen gas migration through clathrate hydrate cages", *EPL*, vol. 103, p. 36001, 2013.

[KOM 13] KOMATSU H., OTA M., SMITH R.L. *et al.*, "Review of CO_2-CH_4 clathrate hydrate replacement reaction laboratory studies – properties and kinetics", *Journal of the Taiwan Institute of Chemical Engineers*, vol. 44, no. 4, pp. 517–537, 2013.

[KVA 00] KVAMME B., "A unified nucleation theory for the kinetics of hydrate formation", *Annals of New York Academy of Science*, vol. 912, pp. 496–501, 2000.

[KVE 80] KVENVOLDEN K., MCMENAMIN M.A., "Hydrates of Natural Gas: A review of their geological occurrence", *Geological Survey Circular,* vol. 825, pp. 1–11, 1980.

[KVE 95] KVENVOLDEN K., "A review of the geochemistry of methane in natural gas hydrate", *Organic Geochemistry*, vol. 23, nos. 11–12, pp. 997–1008, 1995.

[KVE 01] KVENVOLDEN K., LORENSON T.D., "The global occurence of natural gas hydrates", in PAULL C.K., DILLON W.P. (eds), *Natural Gas Hydrates, Occurrence, Distribution, and Detection*, AGU Books Board, Washington, DC, 2001.

[KOH 02] KOH C.A., WESTACOTT R.E., ZHANG W. *et al.*, "Mechanisms of gas hydrate formation and inhibition", *Fluid Phase Equilibria*, vol. 194–197, pp. 143–151, 2002.

[LED 96] LEDERHOS J.P., LONG J.P., SUM A. *et al.*, "Effective kinetic inhibitors for natural gas hydrates", *Chemical Engineering Science*, vol. 51, no. 8, pp. 1221–1229, 1996.

[LEE 03] LEE H., SEO Y., SEO Y.T. *et al.*, "Recovering methane from solid methane hydrate with carbon dioxide", *Angewandte Chemie International Edition*, vol. 42, no. 41, pp. 5048–5051, 2003.

[LEE 14] LEE B.R., KOH C.A., SUM A.K., "Quantitative measurement and mechanisms for CH4 production from hydrates with the injection of liquid CO_2", *Physical Chemistry Chemical Physics*, vol. 16, no. 28, pp. 14922–14927, 2014.

[LEH 09] LEHMKÜHLER F., PAULUS M., STERNEMANN C. *et al.*, "The Carbon dioxide-water interface at conditions of gas hydrate formation", *Journal of American Chemical Society*, vol. 131, no. 15, pp. 585–589, 2009.

[LIU 08] LIU C., LU H., YE Y. *et al.*, "Raman spectroscopic observations on the structural characteristics and dissociation behavior of methane hydrate synthesized in silica sands with various sizes", *Energy & Fuels*, vol. 585, pp. 3986–3988, 2008.

[LIU 09] LIU Y., STRUMENDO M., ARASTOOPOUR H., "Simulation of methane production from hydrates by depressurization and thermal stimulation", *Industrial & Engineering Chemistry Research*, vol. 48, no. 5, pp. 2451–2464, 2009.

[LIU 12] LIU C., YE Y., MENG Q. *et al.*, "The characteristics of gas hydrates recovered from Shenhu Area in the South China Sea", *Marine Geology*, vol. 307–310, pp. 22–27, 2012.

[LU 07] LU H., SEO Y., LEE J. *et al.*, "Complex gas hydrate from the Cascadia margin", *Nature*, vol. 445, pp. 303–306, 2007.

[LU 11] LU Z., ZHU Y., ZHANG Y. *et al.*, "Cold regions science and technology gas hydrate occurrences in the Qilian mountain permafrost, Qinghai Province, China", *Cold Regions Science and Technology*, vol. 66, no. 2–3, pp. 93–104, 2011.

[LUN 09] LUNINE J.I., CHOUKROUN M., STEVENSON D.J. *et al.*, "Origin and evolution of Titan", in BROWN R.H., LEBRETON J.P., WAITE H. (eds), *Titan from Cassini-Huygens*, Springer, New York, 2009.

[MAK 07] MAKOGON Y.F., HOLDITCH S.A., MAKOGON T.Y., "Natural gas-hydrates – a potential energy source for the 21st Century", *Journal of Petroleum Science and Engineering*, vol. 56, pp. 14–31, 2007.

[MAR 10] MARBOEUF U., MOUSIS O., PETIT J.-M. *et al.*, "Clathrate hydrates formation in short-period comets", *Astrophysical Journal*, vol. 708 (1), pp. 812-816, 2010.

[MCC 15] MCCLURE M.K., ESPAILLAT C., CALVET N. *et al.*, "Detections of trans-neptunian ice in protoplanetary disks", *Astrophysical Journal*, vol. 799 (2), art. no. 162, 2015.

[MIE 12] MIENERT J., "Signs of instability", *Nature*, vol. 490, pp. 491–492, 2012.

[MIL 61] MILLER S.L., "A theory of gaseous anesthetics", *Proceedings of the National Academy of Sciences*, vol. 47, p. 1798, 1961.

[MON 15] MONTAÑES-RODRIGUEZ P., GONZALEZ-MERINO B., PALLE E. *et al.*, "Jupiter as an exoplanet: UV to near transmission spectrum reveals hazes, a Na layer, and possibly stratospheric H2O-ice clouds", *Astrophysics Journal Letters*, vol. 801, p. L8, 2015.

[MOO 07] MOON C., HAWTIN R.W., RODGER P.M., "Nucleation and control of clathrate hydrates: insights from simulation", *Faraday Discussion*, vol. 136, pp. 367–382, 2007.

[MOU 01] MOUDRAKOVSKI I.L., SANCHEZ A.A., RATCLIFFE C.I. *et al.*, "Nucleation and growth of hydrates on ice surfaces: new insights from 129Xe NMR experiments with hyperpolarized Xenon", *Journal of Physical Chemistry B*, vol. 105, no. 49, pp. 12338–12347, 2001.

[MOU 08] MOUSIS O., SCHMITT B., "Sequestration of ethane in the cryovolcanic subsurface of titan", *Astrophysics Journal*, vol. 677, p. L67, 2008.

[MOU 15] MOUSIS O., CHASSEFIÈRE E., HOLM N.G. *et al.*, "Methane clathrates in the solar system", *Astrobiology*, vol. 15, p. 308, 2015.

[MOU 16] MOUSIS O., RONNET T., BRUGGER B. *et al.*, "Origin of molecular oxygen in comet 67p/churyumov-gerasimenko", *Astrophysics Journal Letters*, vol. 823, p. L41, 2016.

[MUM 86] MUMMA M.J., WEAVER H.A., LARSON H.P. *et al.*, "Detection of water vapor in Halley's comet", *Science*, vol. 232, p. 1523, 1986.

[MUM 11] MUMMA M.J., CHARNLEY S.B., "The chemical composition of comets - emerging taxonomies and natal heritage", *Annual Review of Astronomy and Astrophysics*, vol. 49, p. 471, 2011.

[MUR 10] MURSHED M.M., SCHMIDT B.C., KUHS W.F., "Kinetics of methane-ethane gas replacement in clathrate-hydrates studied by time-resolved neutron diffraction and Raman spectroscopy", *Journal of Physical Chemistry A*, vol. 114, no. 1, pp. 247–255, 2010.

[PAG 16] PAGANONI M., CARTWRIGHT J.A., FOSCHI M. *et al.*, "Structure II gas hydrates found below the bottom simulating reflector", *Geophysical Research Letters*, vol. 43 no.10.1002/2016GL069452, pp. 5696–5706, 2016.

[PAR 06] PARK Y., KIM D.-Y., LEE J.-W. *et al.*, "Sequestering carbon dioxide into complex structures of naturally occurring gas hydrates", *Proceedings of National Academy of Science USA*, vol. 103, no. 34, pp. 12690–12694, 2006.

[PHR 12] PHRAMPUS B.J., HORNBACH M.J., "Recent changes to the Gulf Stream causing widespread gas hydrate destabilization", *Nature*, vol. 490, no. 7421, pp. 527–530, 2012.

[POH 05] POHLMAN J.W., CANUEL E.A., CHAPMAN N.R. *et al.*, "The origin of thermogenic gas hydrates on the northern Cascadia Margin as inferred from isotopic (13 C/12 C and D/H) and molecular composition of hydrate and vent gas", *Organic Geochemistry*, vol. 36, pp. 703–716, 2005.

[PRA 07] PRASAD P.S.R., SHIVA PRASAD K., THAKUR N.K., "Laser Raman spectroscopy of THF clathrate hydrate in the temperature range 90–300 K", *Spectrochimica Acta Part A*, vol. 68, pp. 1096–1100, 2007.

[RAD 02] RADHAKRISHNAN R., TROUT B.L., "A new approach for studying nucleation phenomena using molecular simulations: application to CO_2 hydrate clathrates", *Journal of Chemical Physics*, vol. 117, no. 4, pp. 1786–1796, 2002.

[RIC 85] RICHARDSON H.H., WOOLDRIDGE P.J., DEVLIN J.P., "FT-IR spectra of vacuum deposited clathrate hydrates of oxirane H 2S, THF, and ethane", *Journal of Chemical Physics*, vol. 83, p. 4387, 1985.

[RIP 91] RIPMEESTER J.A., RATCLIFFE C.I., "Solid state NMR studies of inclusion compounds", in J.L. ATWOOD, J.E.D. DAVIES, D.D. MACNICOL (eds), *Inclusion Compounds: Inorganic and Physical Aspects of Inclusion*, Academic Press, 1991.

[RIP 16] RIPMEESTER J.A., ALAVI S., "Some current challenges in clathrate hydrate science: nucleation, decomposition and the memory effect", *Current Opinion in Solid State and Material Science*, vol. 20, no. 6, pp. 344–351, 2016.

[ROD 90] RODGER P.M., "Stability of gas hydrates", *J. Phys. Chem.*, vol. 94, pp. 6080-6090, 1990.

[ROS 81] ROSS R.G., ANDERSSON P., BACKSTROM G., "Unusual PT dependence of thermal conductivity for a clathrate hydrate", *Nature*, vol. 290, p. 322, 1981

[RUB 15] RUBIN M., ALTWEGG K., BALSIGER H. *et al.*, "Molecular nitrogen in comet 67P/Churyumov-Gerasimenko indicates a low formation temperature", *Science*, vol. 348, p. 232, 2015.

[SCH 88] SCHOELL M., "Multiple origins of methane in the earth", *Chemical Geology*, vol. 71, pp. 1–10, 1988.

[SCH 03] SCHOBER H., ITOH H., KLAPPROTH A. *et al.*, "Guest-host coupling and anharmonicity in clathrate hydrates", *European Physical Journal E*, vol. 12, p. 41, 2003.

[SCH 10] SCHICKS J.M., ZIEMANN M.A., LU H. *et al.*, "Spectroscopy Raman spectroscopic investigations on natural samples from the integrated ocean drilling program (IODP) expedition 311: indications for heterogeneous compositions in hydrate crystals", *Spectrochimica Acta Part A Molecular and Biomolecular Spectroscopy*, vol. 77, pp. 973–977, 2010.

[SCH 11] SCHICKS J.M., LUZI M., BEESKOW-STRAUCH B., "The conversion process of hydrocarbon hydrates into CO_2 hydrates and vice versa: thermodynamic considerations", *Journal of Physical Chemistry A*, vol. 115, no. 46, pp. 13324–13331, 2011.

[SCH 13] SCHICKS J.M., LUZI-HELBING M., "Cage occupancy and structural changes during hydrate formation from initial stages to resulting hydrate phase", *Spectrochimica Acta Part A: Molecular and Biomolecular Spectroscopy*, vol. 115, pp. 528–536, 2013.

[SEA 10] SEAGER S., DEMING D., *Annual Review of Astronomy and Astrophysics*, vol. 48, p. 631, 2010.

[SLO 91] SLOAN E.D., FLEYFEL F., "A molecular mechanism for gas hydrate nucleation from ice", *AIChE Journal*, vol. 37, no. 9, pp. 1281–1292, 1991.

[SLO 98] SLOAN E.D., SUBRAMANIAN S., MATTHEWS P.N. *et al.*, "Quantifying hydrate formation and kinetic inhibition", *Industrial & Engineering Chemistry Research*, vol. 37, no. 8, pp. 3124–3132, 1998.

[SLO 03] SLOAN E.D., "Clathrate hydrate measurements: microscopic, mesoscopic, and macroscopic", *Journal of Chemical Thermodynamics*, vol. 35, no. 1, pp. 41–53, 2003.

[SLO 08] SLOAN E.D., KOH C.A., *Clathrate Hydrates of Natural Gases*, CRC Press, Boca Raton, 2008.

[SON 14] SONG Y., YANG L., ZHAO J. *et al.*, "The status of natural gas hydrate research in China: a review", *Renewable & Sustainable Energy Reviews*, vol. 31, pp. 778–791, 2014.

[STO 79] STOLL R.D. BRYAN G.M., "Physical properties of sediments containing gas hydrates", *Journal of Geophysics Research*, vol. 84, p. 1629, 1979.

[SUB 99] SUBRAMANIAN S., SLOAN E., "Molecular measurements of methane hydrate formation", *Fluid Phase Equilibria*, vol. 158–160, pp. 813–820, 1999.

[SUG 05] SUGAHARA K., SUGAHARA T., OHGAKI K., "Thermodynamic and roman spectroscopic studies of Xe and Kr hydrates" "Thermodynamic and roman spectroscopic studies of Xe and Kr hydrates", *Journal of Chemical & Engineering Data*, vol. 50, pp.. 274–277, 2005.

[TAQ 16] TAQUET V., WIRSTRÖM E.S., CHARNLEY S.B., *Astrophysics Journal*, vol. 821, p. 46, 2016.

[TSE 88] TSE J.S., WHITE M.A., "Origin of glassy crystalline behavior in the thermal properties of clathrate hydrates: a thermal conductivity study of tetrahydrofuran hydrate", *Journal of Physical Chemistry*, vol. 92, p. 5006, 1988.

[TSE 02] TSE J.S., KLUG D.D., "Nucleation, decomposition and guest vibrations in clathrate hydrates", *Proceedings of the Fourth International Conference on Gas Hydrates*, Yokohama, p. 669, 2002.

[UCH 00] UCHIDA T., OKABE R., MAE S. *et al.*, "In situ observations of methane hydrate formation mechanisms by Raman spectroscopy", *Annals of New York Academy of Science*, vol. 912, pp. 593–601, 2000.

[UCH 02] UCHIDA T., TAKEYA S., KAMATA Y. *et al.*, "Spectroscopic observations and thermodynamic calculations on clathrate hydrates of mixed gas containing methane and ethane: determination of structure, composition and cage occupancy", *Journal of Physical Chemistry B*, vol. 106, pp. 12426–12431, 2002.

[UCH 04] UCHIDA T., TAKEYA S., CHUVILIN E.M. *et al.*, "Decomposition of methane hydrates in sand, sandstone, clays, and glass beads", *Journal of Geophysics Research*, vol. 109, no. B05206, pp. 1–12, 2004.

[UCH 07] UCHIDA T., TAKEYA S., KAMATA Y. *et al.*, "Spectroscopic measurements on binary, ternary, and quaternary mixed-gas molecules in clathrate structures", *Industrial & Engineering Chemistry Research*, pp. 5080–5087, 2007.

[VAT 06] VATAMANU J., KUSALIK P.G., "Unusual crystalline and polycrystalline structures in methane hydrates", *Journal of American Chemical Society*, vol. 128, no. 49, pp. 15588–15589, 2006.

[VAT 10] VATAMANU J., KUSALIK P.G., "Observation of two-step nucleation in methane hydrates", *Physical Chemistry Chemical Physics*, vol. 12, no. 45, p. 15065, 2010.

[VIL 16] VILLANUEVA G.L., ALTIERI F., CLANCY R.T. *et al.*, "Unique Spectroscopy and Imaging of Mars with the James Webb Space Telescope", *Publications of the Astronomical Society of the Pacific*, vol. 128, p. 018004, 2016.

[VIS 09] VISSER R., VAN DISHOECK E.F., DOTY S.D. *et al.*, "The chemical history of molecules in circumstellar disks I. Ices", *Astronomy & Astrophysics*, vol. 495, p. 881, 2009.

[YOO 04] YOON J., KAWAMURA T., YAMAMOTO Y. *et al.*, "Transformation of methane hydrate to carbon dioxide hydrate: in situ Raman spectroscopic observations", *Journal of Physical Chemistry A*, vol. 108, pp. 5057–5059, 2004.

[YOU 04] YOUSUF M., QADRI S.B., KNIES D.L. *et al.*, "Novel results on structural investigations of natural minerals of clathrate hydrates", *Applied Physics A*, vol. 78, pp. 925–939, 2004.

[WAL 09] WALSH M.R., KOH C.A., SLOAN E.D. *et al.*, "Microsecond simulations of spontaneous methane hydrate nucleation and growth", *Science*, vol. 326, no. 5956, pp. 1095–1098, 2009.

[WAL 11] WALSH M.R., BECKHAM G.T., KOH C.A. *et al.*, "Methane hydrate nucleation rates from molecular dynamics simulations: effects of aqueous methane concentration, interfacial curvature, and system size", *Journal of Physical Chemistry C*, vol. 115, no. 43, pp. 21241–21248, 2011.

[WAN 02] WANG X., SCHULTZ A.J., HALPERN Y., "Kinetics of methane hydrate formation from polycrystalline deuterated ice", *Journal of Physical Chemistry A*, vol. 106, pp. 7304–7309, 2002.

[WHI 99] WHITICAR M.J., "Carbon and hydrogen isotope systematics of bacterial formation and oxidation of methane", *Chemical Geology*, vol. 161, nos. 1–3, pp. 291–314, 1999.

[ZHA 04] ZHANG Z., EWING G.E., "Infrared studies of the SO_2 clathrate hydrate", *Journal of Physical Chemistry A*, vol. 108, pp. 1681–1686, 2004.

[ZHA 12] ZHANG X., KIRKWOOD W.J., WALZ P.M. *et al.*, "A review of advances in deep-ocean Raman spectroscopy", *Applied Spectroscopy*, vol. 66, no. 3, pp. 237–249, 2012.

3

High-Resolution Optical Microscopy of Gas Hydrates

3.1. Introduction

Observational studies using macrophotography or optical microscopy have allowed for considerable progress in the characterization and understanding of a number of gas hydrate properties, including crystal morphologies and hydrate growth and dissociation processes. In these studies, especially when using the optical microscope, the main difficulty is the measurement cell. This cell, typically built around two parallel glass or sapphire windows, and its temperature and pressure control system, should be thin enough to fit between the condenser and the microscope objective. The objective has to be within a few millimeters of one of the windows. Only a few research groups have constructed and used this type of cell for optical microscopy studies of gas hydrates, starting more than two decades ago with Smelik and King [SME 97] and, very recently, Beltrán and co-workers [DUQ 16, ESM 16]. Micromodels, i.e. arrays of pores and channels in glass or silicon mimicking a two-dimensional porous medium, are another type of measurement cell employed in optical microscopy studies of gas hydrates [TOH 01, KAT 06, HAU 16].

To the best of our knowledge, these studies have all made use of conventional bright field imaging, with optical microscopes used either in the transmission or reflection modes. Bright field reflection targets opaque systems, e.g. silicon micromodels [HAU 16].

Chapter written by Nelly HOBEIKA, Maria Lourdes MARTINEZ DE BAÑOS, Patrick BOURIAT, Daniel BROSETA and Ross BROWN.

Seeing "bigger" is just one advantage of using a microscope. The main advantage is the wide variety of complementary contrast modes available on even modest research microscopes. This chapter illustrates the potential and complementarity of various imaging modes including differential interference contrast (DIC), dark field, fluorescence and (confocal) reflection. These modes are readily available in modern, infinity-corrected optical microscopes. Yet the bulk of investigations by other groups are essentially bright field transmission studies, and the statement made in the pioneering work by Smelik and King [SME 97], that "optical microscopy has not been fully exploited in hydrate research", still holds true today. This chapter also illustrates the potential of Brewster angle microscopy for precise imaging of lateral heterogeneities on a planar surface. Polarization microscopy is not discussed, as it is less useful on the two most common hydrate crystal structures, I and II, which are cubic and hence not birefringent.

These less-conventional modes of optical microscopy are presented in section 3.2, which is followed by three sections showing how they can be used, either alone or in combination, to gain insights into:

1) how gas hydrates propagate on a substrate as a "halo" under the guest phase;

2) how gas hydrates crystallize from/in an emulsion of guest droplets in water;

3) the adsorption properties of some polymer molecules known to inhibit gas hydrate formation and dissociation.

The model gas hydrate considered in these examples is cyclopentane (CP) hydrate, which is a structure II hydrate, stable at atmospheric pressure and temperatures below 7°C. These mild conditions strongly ease design problems in the sense that only temperature needs to be controlled. The observed physical behavior is expected to be representative of that of natural gas hydrates, also of structure II and with low molecular weight hydrocarbon guests (methane, ethane, etc.) having solubility in water comparable to that of CP. It is hoped that the optical microscopy observations that are reported below will be carried out soon on more realistic gas hydrate systems using measurement cells as described above.

3.2. Optical methods

3.2.1. *Beyond bright-field modes in optical microscopy*

Gas hydrates are intrinsically transparent and often form very thin crystals, whose refractive index is furthermore very close to that of water (1.34–1.35 versus 1.33, see [SLO 08]). Therefore, bright field imaging, based on absorption, refraction or reflection of light, usually yields images with poor contrast, e.g. hydrate crystal edges may be hard to see in an aqueous medium. Contrast modes based on principles other than light adsorption or refraction are therefore particularly well suited to gas hydrates.

Schematically, and as applied to hydrates, we may write the contrast of bright-field modes as a small change, ε, on a large signal, S, with low contrast, $(S \pm \varepsilon)/S \sim 1$. A better policy is to arrange, at least in principle, for the signal to vanish everywhere where the sample is homogeneous and uninteresting, and for a small signal to arise where there are features of interest. This small signal then gives rise to images with high contrast, $\varepsilon/0 \gg 1$. Such methods are called vanishing or zero background techniques. The case of Brewster angle microscopy is singled out and deferred to another (section 3.2.2).

3.2.1.1. *Differential interference contrast (DIC)*

The intensity of the light transmitted or reflected by transparent objects like gas hydrates yields images with low contrast. An obvious alternative is to use the phase of the light waves. Both phase contrast and DIC achieve this end but DIC is perhaps more readily compatible with simultaneous fluorescence imaging, particularly from the point of view of the choice of objectives compatible with both modes. DIC is more difficult to interpret for birefringent materials, but that is not a concern for gas hydrates with structures I and II.

DIC can be performed both in transmission and reflection. The complete theory is subtle [ALL 69]. In a simplified description of the more common transmission mode, it uses crossed linear polarizers, one on the condenser (light source), the polarizer, P, the other behind the objective, the analyzer, A (see Figure 3.1). The field for a homogeneous, non-polarizing sample would thus be completely dark. But additional Wollaston prisms are placed between the polarizer and the condenser, $Wollaston_1$, and between the objective and the analyzer, $Wollaston_2$. The first prism is aligned so that it splits an incoming ray into two components whose mutually perpendicular polarizations are at ±45° to the polarizer. The birefringent crystal of the Wollaston prism is furthermore cut so that the two rays converge slightly, passing through points in the sample separated by a small distance, of the order of the lateral resolution of the objective. Wollaston prism $Wollaston_2$ recombines the rays. If the optical path lengths of the rays through the sample are identical, $Wollaston_2$ reconstructs the input polarization and the field is dark. If the path lengths differ, due either to a variation of the thickness or a lateral variation of the refractive index, the outgoing ray is in general elliptically polarized, hence has a small component on the analyzer. DIC thus reveals regions of the sample where there is a lateral gradient of refractive index or sample thickness, and thus highlights borders and thicknesses of the specimen, which has a three-dimensional appearance. In practice, the orientations of the prisms are slightly offset so that the recombination yields an elliptical polarization even for a homogeneous sample.

Variations in the optical path then give rise to characteristic images in varying shades of gray, with a cast that resembles grazing illumination of a rough surface.

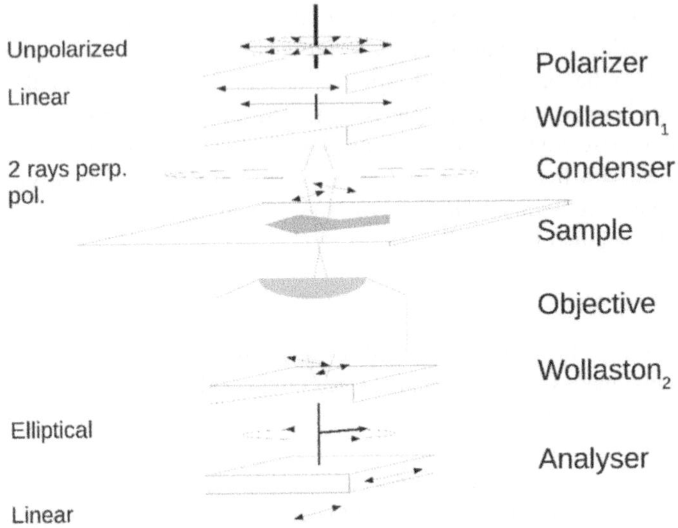

Figure 3.1. *Schematic illustration of DIC imaging. See the main text for details*

Figure 3.2 shows DIC images of a CP hydrate crystal growing in the presence of a non-ionic surfactant: a hollow pyramidal crystal is clearly apparent, similar to that observed recently with the same fluid system by other researchers [KAR 12, MIT 15].

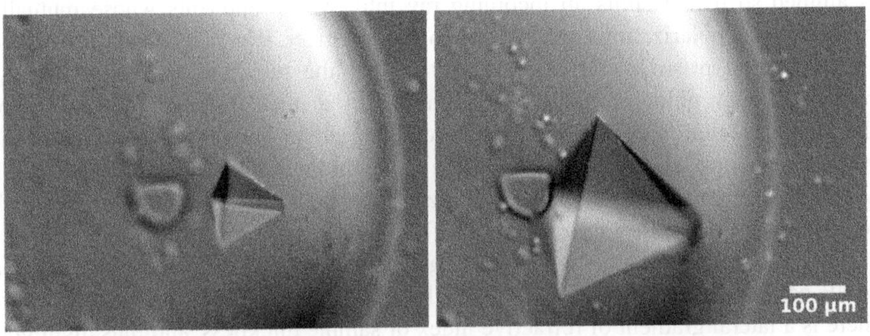

Figure 3.2. *DIC images taken 50 s apart of a hollow pyramidal crystal growing at –1 °C in a drop of water (edge visible on the right) in cyclopentane containing 100 ppm of the surfactant Span 80*

3.2.1.2. *Dark field imaging*

Dark field imaging is another example of a vanishing-background technique. Weak scattering from mostly homogeneous and transparent samples is hard to see against the background of transmitted light. In dark field imaging, therefore, a special, annular stop in the condenser limits illumination to a cone of rays that are so oblique with respect to the optical axis that they cannot be focused directly into the image. The field of view is thus dark, except where scattering in the sample deviates a small fraction of light into directions closer to the axis. Figure 3.3 shows an example.

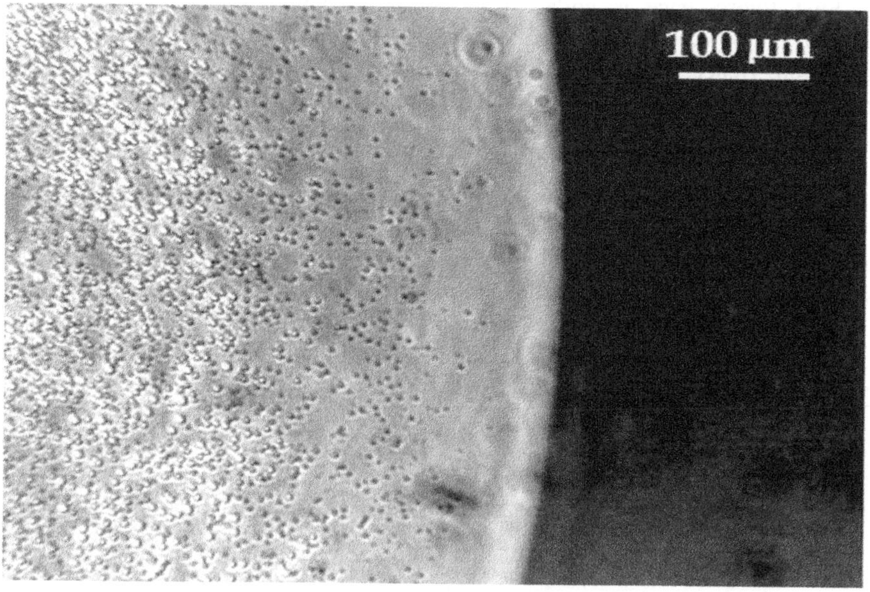

Figure 3.3. *Dark field image of a two-dimensional cyclopentane-in-water emulsion. The edge of the large water drop is visible at center right*

3.2.1.3. *Fluorescence imaging*

Fluorescent molecules, also called fluorophores, absorb light in their ground state to reach a metastable, excited state, that may decay back to the ground state either non-radiatively (releasing heat) or radiatively, by emission of a photon, fluorescence. In general, fluorophores are used to mark selected parts of a sample that are revealed by providing excitation of the appropriate wavelength, unmarked, non-fluorescent parts of the sample remaining dark. Fluorescence imaging is thus a zero-background technique. Both the absorption and the emission bands are broad in solids and liquids. Excitation is usually tuned to the maximum of the absorption band and a broad range of emission wavelengths are collected to maximize the signal.

In practice, perturbation of the system by the fluorophore is minimized by using only trace amounts of dye (many ionic dyes, for example, are also surfactants). Furthermore, in most experimental conditions, molecules in the ground state far exceed those in the excited state. The emission signal is thus very weak compared to the excitation beam. Therefore, fluorescence is rendered visible against the background of excitation light by two main adaptations.

First fluorescence is collected in "epi-illumination" mode, with excitation through the objective, and the detection on the *same* side of the objective as the excitation. Hence, only backscattered excitation light can reach the detector. The excitation beam generally is introduced into the microscope from the side via a dichroic mirror that is strongly reflecting at the excitation wavelength and strongly transmitting at the emission wavelengths. Thus, most backscattered excitation is reflected out of the optical path. Nonetheless, a second, sharp cutoff filter is placed in the detection path to reject scattered excitation light. Figure 3.4 illustrates the excitation-emission processes of a typical fluorophore. Figure 3.5, left, shows the setup schematically. In common practice, a "fluorescence cube" located just behind the objective carries the dichroic mirror and the emission filter. An excitation filter may also be present to narrow the band of excitation wavelengths when a broadband source is used.

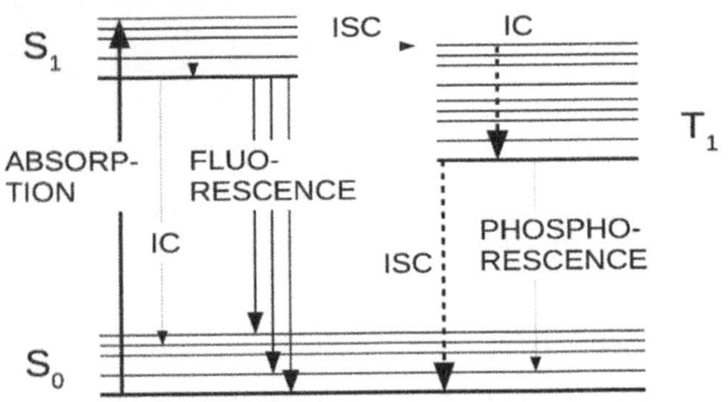

Figure 3.4. *Jablonski or energy level diagram of a typical fluorescent molecule showing the ground (S_0) and excited electronic states (S_1, T_1, heavy horizontal lines) and intramolecular vibrational levels (light horizontal lines). Arrows show photoexcitation (absorption) and emission processes, with thicknesses in proportion to the desired properties of a fluorophore: high fluorescence emission yield, weak intersystem crossing (molecular electronic spin conversion, ISC) to the triplet spin state* T_1 *(weak non-radiative relaxation to the ground state internal conversion, IC)*

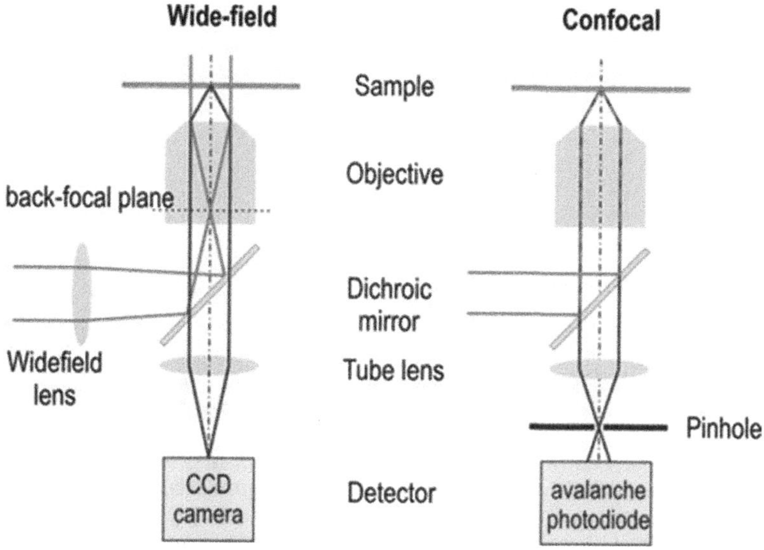

Figure 3.5. *Confocal microscope setup (right) in contrast to the conventional light microscope (left). In the confocal setup, a pinhole is placed in order to block any light outside of the focal plane (the scattered light). For a color version of this figure, see www.iste.co.uk/broseta/hydrates1.zip*

Fluorescent dyes for hydrate research would mostly be chosen to be soluble in the aqueous phase, e.g. rhodamines and other ionic dyes, but could also be dissolved in the guest if the latter is liquid under ambient conditions, e.g. perylene in CP. Another class of dyes is used to highlight interfaces. One water-soluble fluorophore has been proven to be promising in gas hydrate research: DASPI, or trans-4-[4-(dimethylamino)-styryl]-1-methyl-pyridinium iodide. DASPI is a rigidochrome: its fluorescence is enhanced when its geometry is constrained to the planar form, e.g. by confinement within a film or adsorption at an interface. As an example, Figure 3.6 shows DIC and fluorescence images of the same drop of water containing 2×10^{-6} M of DASPI shortly after its deposition on a strongly hydrophilic substrate (glass) under CP. The fluorescent corona seen outside the contact line over a distance of 50 μm (Figure 3.6(b)) corresponds to water seeping out of the drop as a precursor film. The enhanced fluorescence is caused by the confinement of DASPI molecules in this film, which is no thicker than a few nanometers. Interestingly, DASPI also renders CP hydrate and CP-in-water emulsions fluorescent (data not shown), presumably because it adsorbs at the water–hydrate and water–CP interface – the tension of the latter interface decreases slightly when DASPI is added to the water [MAR 16].

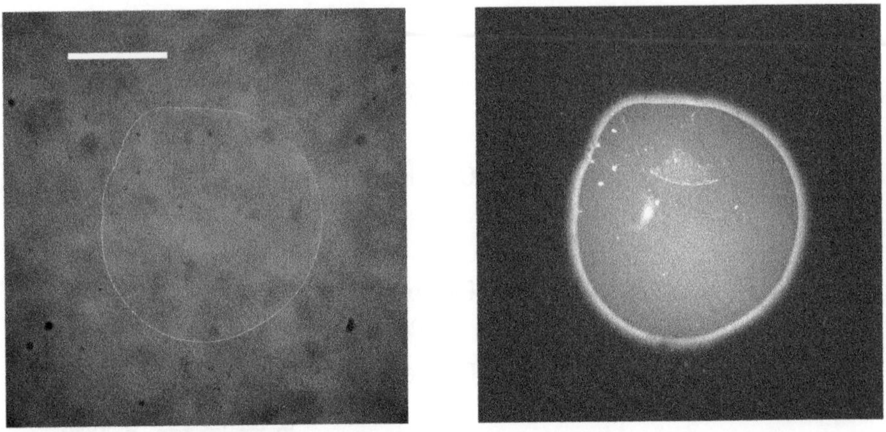

Figure 3.6. *DIC a) and fluorescence b) images of a drop of water deposited on hydrophilic glass under cyclopentane. Scale bar: 500 µm (adapted from [MAR 16])*

Another common fluorophore is rhodamine 6G, which is water soluble as well and can be grafted on latex beads. Observations by fluorescence microscopy of a water drop containing a few ppm of rhodamine-6G-grafted latex beads (27 nm diameter) deposited on glass under the same conditions as those depicted in Figure 3.6 have shown that these nanobeads remain trapped in the drop, confirming that the thickness of the water precursor film does not exceed a few nanometers. These nanobeads can be used as tracers of water flow [MAR 16].

3.2.1.4. *Reflection microscopy*

Hydrate-forming systems sometimes give rise to a mass of small crystals that strongly scatters light, making the sample opaque. Transmission and transmission DIC (which is also possible in reflectance mode) are then powerless and it is useful to be able to perform reflection microscopy. This mode can be achieved easily by reducing the power of the source of the fluorescence epi-illuminator and using a cube equipped with just a half-silvered mirror instead of the dichroic mirror.

While reflection microscopy is a bright field method with a non-vanishing background, it has another use. When the source of excitation on the epi-illuminator is more or less monochromatic (e.g. lasers or some light emitting diodes), reflection microscopy may show up fringes created by interference between light waves, similar in origin to the well-known phenomenon of Newton's rings. When a light wave is incident on the interface between two transparent media, part of the intensity is transmitted, but a small fraction is reflected. The amplitudes of the transmitted and reflected waves depend on the refractive indices on each side of the interface and on the angle of incidence, as determined by the Fresnel relations (section 3.2.2).

Waves reflected off close interfaces may then give rise to interference fringes. The most common case is interference due to waves reflected off the sample cell and off parts of the sample, for example when a thin film of material spreads over the cell wall. If the thickness of the material is e and its refractive index is n, the second reflection then has a phase lag of $2\pi\lambda/(2ne)$. Variations in thickness may then be determined by counting dark or bright fringes. One interfringe, bright to bright or dark to dark, corresponds to a variation of thickness of $\lambda/(2ne)$, about 200 nm for hydrates viewed with green light [LIM 09].

3.2.1.5. *Widefield versus confocal observation*

Reflection, DIC reflection and epi-fluorescence microscopies may be applied either in a widefield or a confocal mode.

In widefield techniques, a point source of light or a collimated laser beam is focused into the *back* focal plane of the objective, causing illumination of a wide area of the sample by a parallel beam of light. Excitation light backreflected or backscattered or fluorescence from points in the *front* focal plane are collected by the objective and transmitted as collimated beams to the tube lens (see Figure 3.5, left) which forms the final image of the focal plane on a camera. Because the sample is illuminated with a homogeneous beam, any scattering or fluorescent material in adjacent planes gives rise to an out-of-focus haze that deteriorates image quality.

In confocal detection, the objective is illuminated with a parallel excitation beam that it focuses to a point in the front focal plane. Points in adjacent planes at defocusing δz are also illuminated, but with an intensity that decays quickly as $1/\delta z^2$. The camera in widefield detection is replaced by a "spatial filter", or pinhole, and a non-imaging detector such as a photomultiplier tube or an avalanche photodiode is placed behind the pinhole (see Figure 3.5, right). On reaching the plane of the pinhole, light issuing from out of focus planes of the sample is spread over a disk of radius proportional to δz. If the pinhole is small, typically 30–100 μm diameter, the fraction of light passing the pinhole decays as $1/\delta z^2$. Hence overall, out of focus points of the sample contribute intensities that decay sharply as $1/\delta z^4$. This decay is the origin of the "*optical sectioning*" power of the confocal microscope, leading commonly to crisper images. Note, however, that contrary to common belief, the *resolution* (power to separate close points in the sample) in confocal mode is only slightly better than in widefield mode [PAW 95]. Note furthermore that the confocal mode works well when the sample is inhomogeneous, typically a set of bright more or less compact objects in a "dark" medium. The sectioning power is severely reduced in uniformly scattering or fluorescent materials.

Confocal observation allows the specimen to be imaged one "point" at a time. To generate a complete image, the excitation spot is moved over the specimen and the image built point by point along a given plane, giving an image of the thin slice mentioned above.

Depending on the size of the image and the amount of light available, a laser-scanned frame acquisition may require 0.1–10 s, i.e. much slower than in widefield detection, where the full frame rate may be as high as 100 fps. Scanning options include scanning the incoming beam (laser scanning, usually with galvanometric-mounted mirrors), scanning the objective, or scanning the sample. Laser scanning is the fastest mode. Sample scanning is in general less compatible with the use of temperature- or pressure-controlled cells.

The confocal technique can be combined with other imaging modes, including DIC transmission and reflectance. Figure 3.7 shows an example of simultaneous recording of a transmission image and a confocal reflectance image of a hydrate crust growing over a drop of water on glass, with a surrounding halo, discussed further later in the chapter. From the number of adjacent dark (or bright) fringes ($m \sim 10$) between the edge and the center of the drop, we deduce a height of $h = m \lambda/(2n_w) \sim 2$ µm, with $n_w = 1.33$ the refractive index of water and $\lambda = 532$ nm the wavelength of the laser.

In concluding this section, we note that the blossoming field of light sheet microscopy provides a generally attractive alternative method for optical sectioning. However, its usefulness for studies of hydrates is likely to be restricted by the need to position a second microscope objective to the side of the sample, which may be difficult considering the nature of sample cells for hydrate research.

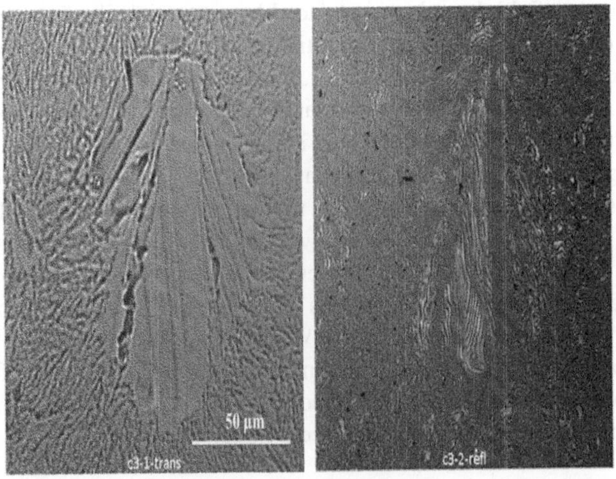

Figure 3.7. *Hydrate halo and crust on glass under cyclopentane. Left: Transmission. Right: Confocal reflectance (from [MAR 16])*

3.2.2. *Brewster angle microscopy*

Brewster angle microscopy is a non-invasive imaging technique based on the reflection properties of interfaces near Brewster's angle of incidence. At this angle, a laser beam polarized parallel to the plane of incidence (p-polarization) has zero reflectance. The presence on the surface of heterogeneities leads to changes in refractive indices and thicknesses, and hence changes in reflectivities, which can be very sensitively detected, providing an image of the topography of the surface. Brewster angle microscopy is thus a vanishing background technique.

3.2.2.1. *Fresnel's laws of reflection from dielectrics and Brewster's angle*

When the angle of incidence i_1 of the laser beam (Figure 3.8, left) does not exceed the limiting angle, there is not only refraction but also a reflection, and the fraction of energy reflected (reflectance) or transmitted (transmittance) depends on the direction of polarization of the light, either in the p direction, composed of the incident plane, or in the s direction, perpendicular to the incident plane.

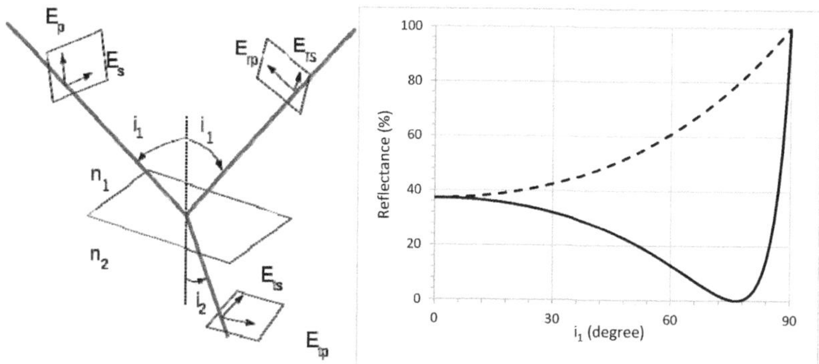

Figure 3.8. *Left: Decomposition of the electrical field vector in the p and s planes. E_p and E_s refer to the incident beam, and subscripts r refer to the reflected beam and t to the refracted beam. Right: Reflectances of the p-wave (full curve) and s-wave (dashed curve) as a function of angle of incidence i_1, calculated for n_1 = 1 (air) and n_2 = 4.1 (silicon). Brewster's angle i_B ~ 78.5°. For a color version of this figure, see www.iste.co.uk/broseta/hydrates1.zip*

The Fresnel equations for the reflection and transmission coefficients are [JAC 98]:

$$r_s = \frac{E_{rs}}{E_s} = \frac{n_1 \cos i_1 - n_2 \cos i_2}{n_1 \cos i_1 + n_2 \cos i_2} \quad [3.1]$$

$$t_s = \frac{E_{ts}}{E_s} = \frac{2 n_1 \cos i_1}{n_1 \cos i_1 + n_2 \cos i_2} \quad [3.2]$$

and

$$r_p = \frac{E_{rp}}{E_p} = \frac{n_2 \cos i_1 - n_1 \cos i_2}{n_2 \cos i_1 + n_1 \cos i_2}$$ [3.3]

$$t_p = \frac{E_{tp}}{E_p} = \frac{2 n_1 \cos i_1}{n_2 \cos i_1 + n_1 \cos i_2}$$ [3.4]

Plots of reflectances, defined as $r_s^2 = R_s^2/E_s^2$ and $r_p^2 = R_p^2/E_p^2$, are shown in Figure 3.8 (right). For an air/silicon interface, r_s varies monotonically from 37% at normal incidence up to 100% at grazing incidence (=π/2). r_p varies between the same limits but passes through zero when i_1 reaches the Brewster's angle i_B given by

$$\tan i_B = \frac{n_2}{n_1}$$ [3.5]

Therefore, an incident beam polarized in the p direction will be completely extinguished because only s polarized rays can be reflected. This phenomenon is exploited in Brewster angle microscopy, where the specimen is observed through the reflection of a p incident beam and the analyzer in the p direction. The non-reflection of p rays by an interface illuminated at Brewster angle makes them disappear while those reflected by a thin layer are generally elliptically polarized and then visible.

3.2.2.2. *Reflection from a thin plane layer*

The thin layers targeted by ellipsometry are generally located at an air/substrate interface. The substrate can be a liquid or a solid.

In the presence of a layer covering the substrate, the reflected beam is the combination of beams due to multiple reflections. In Figure 3.9, the amplitude of each ray is expressed as a function of the transmission coefficient t1 and the reflection coefficients r1 of the air/water interface and r2 of the layer/substrate interface (see equations [3.1]–[3.4]). The complex exponential accounts for the phase shift $\Delta\varphi$ between each reflected ray. Considering a layer of thickness e and refractive index n, this phase shift can be calculated as (see Figure 3.9) $\Delta\varphi = 2\pi[n(AE + ED) - AB]/\lambda$, where λ is the wavelength of the light beam, or:

$$\Delta\varphi = 2\pi \frac{2ne \cos r}{\lambda}$$ [3.6]

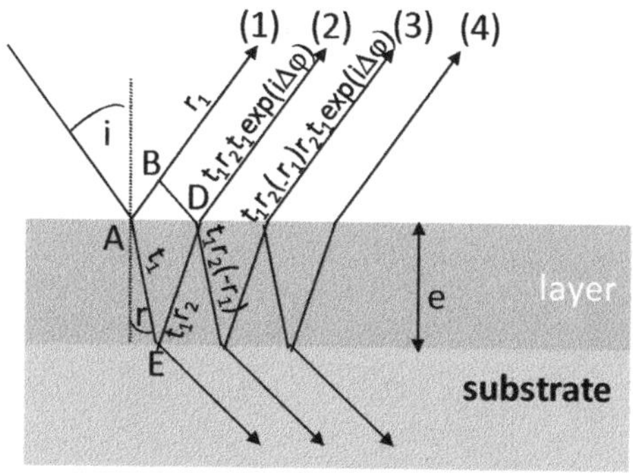

Figure 3.9. *Multiple reflections (and transmissions) from a thin layer of refractive index* n *and thickness* e *deposited on a substrate. The amplitude of the reflected beam is the sum of the amplitudes of rays (1), (2), ... For a color version of this figure, see www.iste.co.uk/broseta/hydrates1.zip*

The amplitude of the reflected ray is obtained by summing the amplitudes of all rays (1), (2), etc., leading to the following reflection coefficient:

$$r_p = \frac{r_{p1}+r_{p2}\exp(i\Delta\varphi)}{1+r_{p1}r_{p2}\exp(i\Delta\varphi)} \quad [3.7]$$

for a polarization along the p direction, and

$$r_s = \frac{r_{s1}+r_{s2}\exp(\Delta\varphi)}{1+r_{s1}r_{s2}\exp(i\,\Delta\varphi)} \quad [3.8]$$

along the s direction ($i^2 = -1$) (see [AZZ 79] for more details).

3.2.2.3. *Principle of a nulling ellipsometry measurement*

Light is reflected by the surface of the sample in the presence of a thin layer. When a polarized light beam strikes a surface, the p and s components suffer an alteration in their phase and amplitude and exhibit different reflectivities. Therefore, a change in the form and size of the ellipse of polarization is observed. This change monitors the properties of the thin layer. Figure 3.10 shows the different components that are necessary to perform an ellipsometric measurement by the nulling technique. A beam of non-polarized light enters a linear polarizer, allowing the light to become polarized in the desired direction P. The p and s components of the electric field vector E have the same phase. This linearly polarized light then arrives at the compensator, a quarter wave plate, where the amplitude and phase of the two components are shifted, converting the linear polarized light into an elliptically polarized one. This beam strikes the sample surface, where the aim is to achieve a linear polarized light after reflection of the light on the thin layer. To achieve this, in general the quarter wave plate is set to an angle C = 45° and the polarizer is turned to the desired P angle, producing a linearly polarized light after reflection from the thin layer.

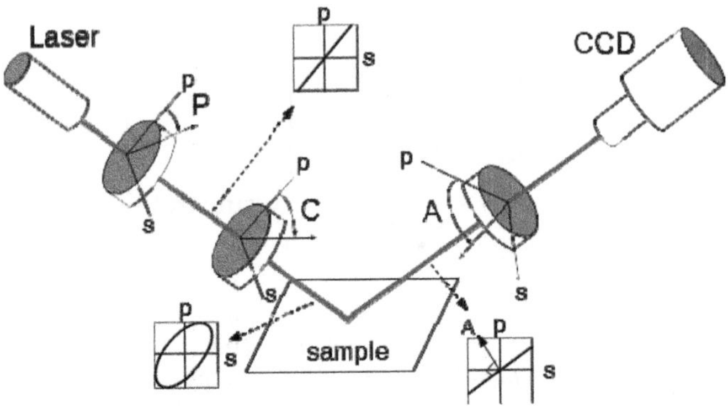

Figure 3.10. *A nulling ellipsometer in action. For a color version of this figure, see www.iste.co.uk/broseta/hydrates1.zip*

An analyzer is placed just before the receptor and turned to an angle A to extinguish the beam in order to determine the direction of polarization of the beam after reflection. This is called "nulling" and corresponds to a minimum of the signal in the photodetector. Knowledge of the three angles P, C and A makes it possible to directly determine the ellipsometric angles ψ and δ, which are related to the ratio of the complex reflection coefficients of p and s polarizations of the thin layer:

$$r_p/r_s = \tan\Psi \exp i\delta = -\tan A \, \frac{\tan C - i\tan(C-P)}{1+i\tan C\tan(C-P)}, \qquad [3.9]$$

where ψ is the angle whose tangent gives the ratio of the change in amplitude for the p and s components and δ is the relative phase shift between the p and s components upon reflection. The knowledge of the ratio r_p/r_s relative to the layer allows the determination of its thickness e using equations [3.7]–[3.9] with just one measurement, knowing *a priori* the refractive index n of the thin layer. In cases where n is unknown, it is possible to determine both n and e starting from multiple nulling performed at different incident angles i around the Brewster angle i_B to ensure a more pronounced sensitivity.

3.3. Selected examples

The examples presented and discussed in sections 3.3.1 and 3.3.2 were recorded using the experimental setup described in Figure 3.11(a–c). At the start of all experiments reported in sections 3.3.1–3.3.3, a drop of water is deposited in a 2 mm path length photometric absorption cell, which is flooded with cyclopentane and placed in a temperature-controlled cell (Figure 3.11). The main difficulty is control of condensation, solved by a slow flow of dry nitrogen through the cell and blowing dry nitrogen over the observation window [MAR 16].

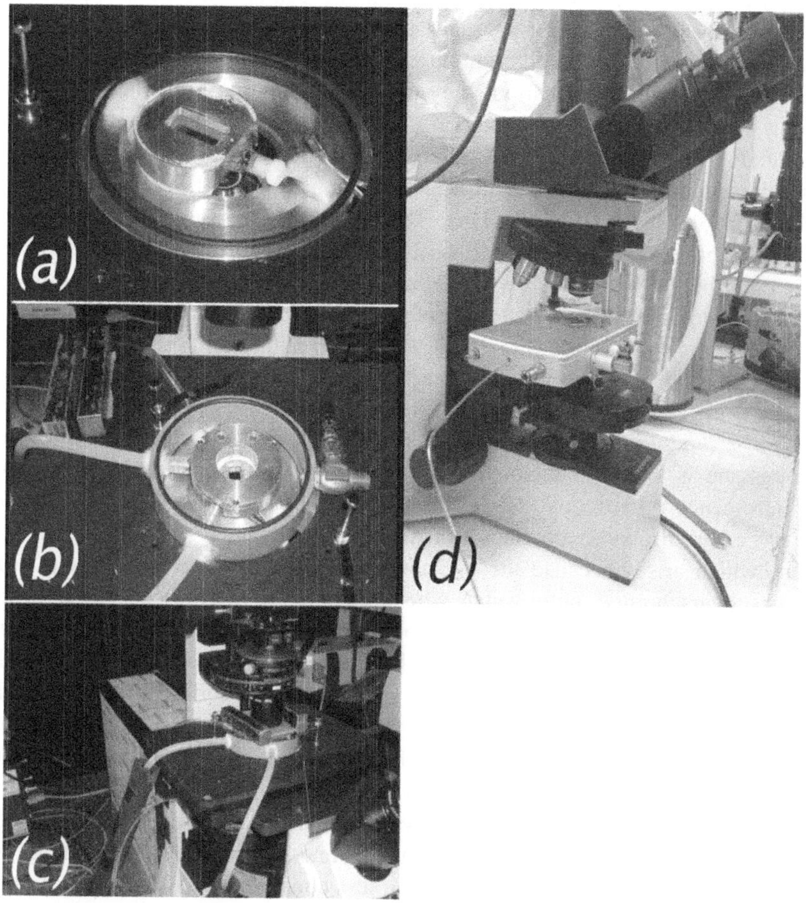

Figure 3.11. *Two examples of compact design for observation of hydrate formation at high resolution. (a–c) The homemade setup used for work at ambient pressure. a) A 2-mm photometric absorption cell (Hellma) is inserted in an aluminum block (resistance thermometer to the right). The block rests on insulating pads on a round cover slip in a rebated stage plate; b) temperature control by a Peltier cooler and radiator. The setup is enclosed in an ad hoc chamber, in which a slow flow of dry nitrogen prevents condensation; c) the chamber, with top window in place, fits between the condenser and objectives on the standard stage of the Ti-Eclipse inverted microscope (Nikon); d) a commercial sample holder for high-pressure work (CAP 500, Linkam) mounted on an B50 upright microscope (Olympus)*

3.3.1. *Hydrate halos growing on glass substrates*

This section illustrates how insights into the structure and growth mechanisms of the halo can be gained from the combined use of the three imaging modes presented above: DIC, reflectance and fluorescence. The halo is the thin hydrate that spreads on a hydrophilic (water-wet) substrate from the contact line of a "reservoir" of water, e.g. a water drop under the guest phase. Hydrate halos are often observed in experiments conducted in cells equipped with see-through windows allowing the observation of hydrate formation: they are seen ascending the (glass or sapphire) windows from the contact line as soon as the water surface has been covered with hydrate [FAN 14, SUN 15]. One proposed halo, growth mechanism is that water flows by capillarity between the substrate and the halo, from the "reservoir" of water to the tip of the halo [BEL 10]. However, the optical microscopy imaging modes presented above reveal somewhat different mechanisms, with other sources of water contributing to halo growth. These results have been recently published in [MAR 16], to which the reader is referred for more details. The experimental configuration is shown in Figure 3.6: a water drop sitting on glass and immersed in the guest phase (cyclopentane).

Prior to the initiation of halo growth, water is present on hydrophilic glass under cyclopentane as a thin precursor film around the contact line of the drop (as shown by fluorescence microscopy, see section 3.2.1.3), but also in the form of a breath figure, i.e. the dropwise condensation of a non-completely wetting liquid (here, water), which arises because the cyclopentane phase (preequilibrated with water) becomes oversaturated with water when temperature decreases. These droplets appear because a stable water film (complete wetting) cannot exist on the substrate. Figure 3.12 shows the images of a breath figure near a large water drop on glass, captured using the various imaging modes: transmission, confocal reflectance, DIC and fluorescence (with DASPI molecules used as fluorophores). The combination of these imaging modes reveals features of the breath figure, such as the size (2–3 μm) and substrate coverage (~2–3%) of the water microdroplets. Note the sensitivity of DIC: the dew droplets are at most a few tens of nanometers high. In fluorescence mode, the water microdroplets appear dimmer than the surrounding glass (Figure 3.12(d)), an effect likely due to the less constrained environment of DASPI molecules. The physical picture that emerges from these observations is depicted in Figure 3.12(e).

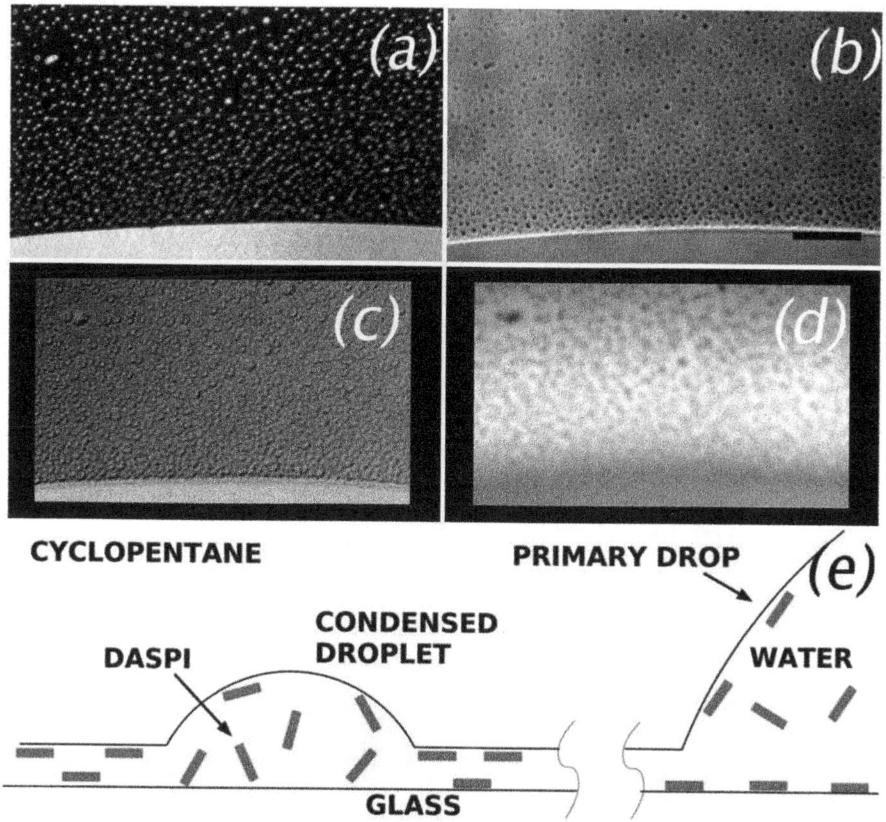

Figure 3.12. *Water condensing on hydrophilic glass as a breath figure near the large water drop, whose contact line is apparent at the lower edges (T = –4 °C). (a and b) Confocal reflectance and transmission pair. (c and d) DIC (inverted contrast for clarity) and fluorescence images in the same region, showing the correspondence between the droplets apparent in (a and c) and the dark (non-fluorescent) spots d). Scale bar: 20 µm. e) The physical picture (not to scale) is shown, with DASPI molecules schematized as red bars (from [MAR 16]). For a color version of this figure, see www.iste.co.uk/broseta/hydrates1.zip*

Halo growth is initiated when the hydrate crust on the drop surface reaches the contact line. The halo then propagates on the substrate as a stable front.

The halo's detailed structure and growth mechanism on the substrate (glass) can be deciphered by combining the fluorescence, DIC and reflectance imaging modes. For example, the drop in intensity of DASPI, coincident with the extreme edge of the halo (see Figure 3.13(b)), shows a travelling wedge of water under the halo fringe. DIC imaging further reveals dew droplets forming under the extreme edge of the halo on apparently bare glass. These droplets, together with the breath figure noticed above, locally enhance the halo thickness, giving rise to the radial texture evident in the DIC images (Figure 3.13(a)). This conclusion is confirmed by the brightness of the extreme edge of the halo in (confocal) reflectance mode [MAR 16].

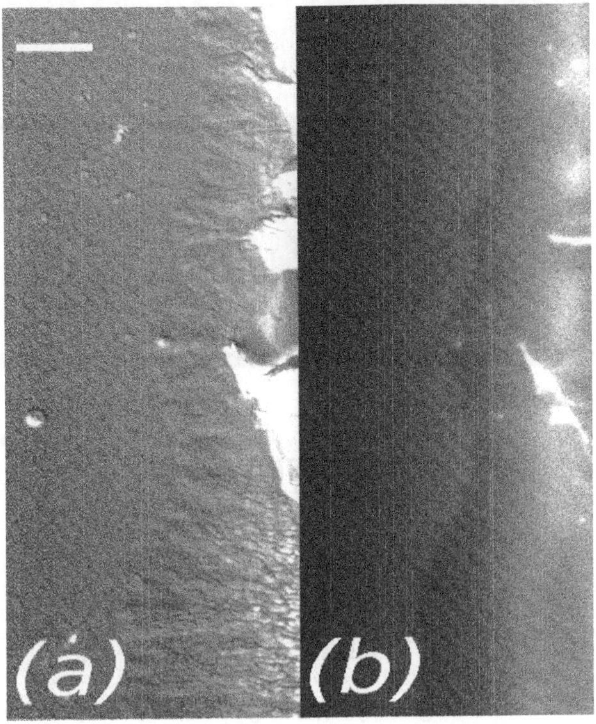

Figure 3.13. *DIC a) and fluorescence b) pair of the growing cyclopentane hydrate halo (T = 0 °C), 90 s after it has emerged from the contact line of the water drop (visible on the right-hand side at the top). The halo displaces DASPI on the glass or in the water precursor film, indicating intimate contact between the halo and the substrate. Scale bar: 20 μm*

Finally, the rate of thickening of the halo can be deduced by counting interference fringes that appear the moment it melts. The fringes arise by interference of light waves reflected off the meltwater–glass interface (a flat reference surface) and the CP–meltwater interface. Thicknesses below the micrometer and thickening rates in the order of 5–10 nm/s are thus obtained for halos formed at T = 0–1 °C on glass. Evidently, this slow thickening is fed by the water present in cyclopentane by ablimation and/or by settling of "fog" droplets – convection of the latter is in fact observed in the measurement cell.

None of the phenomena described above are observed when the substrate is hydrophobic: no halo was observed on silane-treated glass, see [MAR 16] for details.

3.3.2. *Hydrate crystallization in a guest-in-water emulsion*

Most, if not all, published microscopy or macrophotography studies of hydrate crystallization report processes occurring at, or from, the interface between bulk water and guest phases (see, e.g., the review [SUN 10], and the more recent papers [LI 13, LI 14, DAN 15]). Gas hydrate nucleation usually occurs at this interface, which then rapidly covers itself with a polycrystalline hydrate crust. When this interface meets a hydrophilic substrate, a hydrate halo expands from the contact line as a halo under the guest phase (see previous section 3.3.1). The growth rate and texture of the polycrystalline crust on the water/guest interface strongly depend on the subcooling degree. At low subcooling, the crust grows slowly over the water/guest interface and has a rough texture, with some large crystals present. At high subcooling, growth is rapid and the texture, made up of tiny crystallites, is therefore smooth. Gas hydrate crystals may also grow perpendicularly from this crust into the aqueous phase or, less often, into the guest phase [SER 03, SAI 11].

The crystallization processes reported in this section take place in a two-dimensional guest-in-water emulsion: the guest phase (here, cyclopentane) forms a collection of micron-sized droplets dispersed in the (continuous) aqueous phase and trapped beneath glass (see Figure 3.3). The easiest way to prepare this emulsion is to form and melt a hydrate crust over a drop of water hanging from the top of the sample (glass) cell. Although buoyancy pins these droplets to the wall, their Brownian motion suggests minimal interactions with the glass.

Crystallization in this two-dimensional emulsion is triggered by lowering the temperature. Two types of crystal growth are observed: (1) contactless growth of large polygonal plates at the expense of neighboring droplets and (2) irregularly branched crystalline masses growing by interdroplet crystallization with direct contacts.

3.3.2.1. *Growth of large isolated crystals*

In this process, isolated hydrate crystals nucleate in the emulsion and then grow to a macroscopic size in bulk water at the expense of the neighboring guest droplets. The growth mechanism is similar to that which is well known in cloud physics and described by Wegener, Bergeron and Findeisen (WBF) for an ice crystal growing in an assembly of (supercooled) water droplets. In this mechanism, the water vapor pressure is lower over ice than over (metastable) liquid water, so water molecules migrate from over the water droplets to the ice crystal, which thus grows through vapor deposition (or ablimation) at the expense of the surrounding water droplets.

Here, as is apparent in Figures 3.14 and 3.15, the hydrate crystal grows similarly at the expense of the neighboring droplets of the guest phase (here, cyclopentane): the concentration of guest molecules dissolved in water is higher over guest droplets than over the hydrate, hence their migration to the hydrate crystal, which thus grows at the expense of these droplets.

This is illustrated in the following DIC images (Figures 3.14 and 3.15), which show large cyclopentane hydrate crystals growing in the 2d CP-in-water emulsion at the expense of the neighboring cyclopentane droplets. The buildup of a region depleted in cyclopentane droplets near these crystals is emphasized in Figure 3.15.

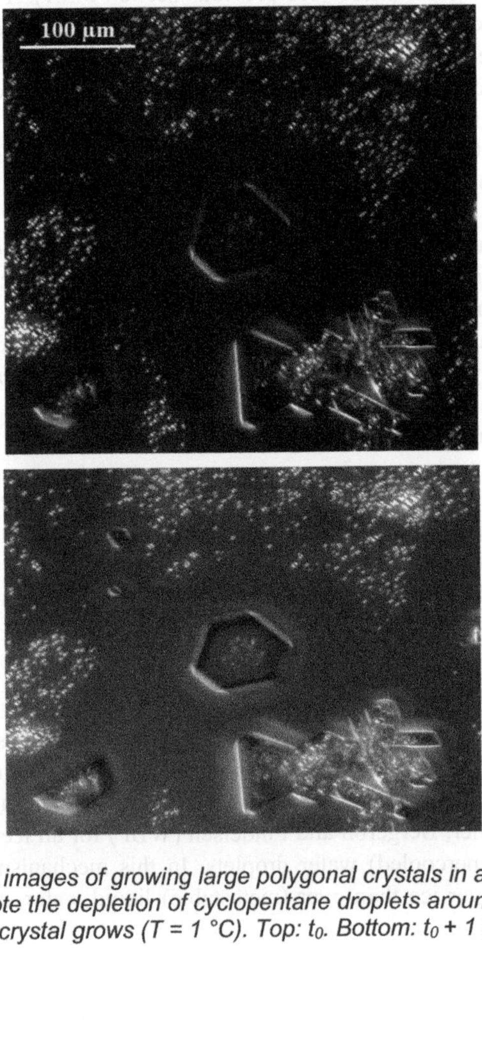

Figure 3.14. *DIC images of growing large polygonal crystals in a cyclopentane-in-water emulsion. Note the depletion of cyclopentane droplets around the crystal as the hydrate crystal grows (T = 1 °C). Top: t_0. Bottom: t_0 + 1 h 30 min*

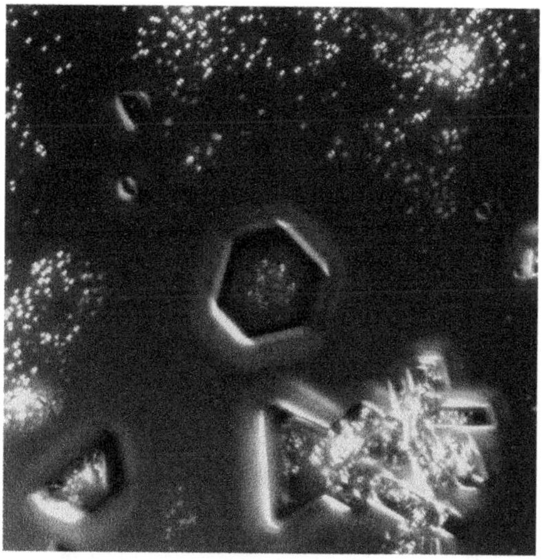

Figure 3.15. *Composite image with times t_0 (Figure 3.14, top) and t_0 + 1 h 30 min (Figure 3.14, bottom) color-coded in red and green, respectively. Red highlights objects that disappear between the views, green highlights those which appear during the waiting time and yellow highlights the common features. The green edges of the hydrate crystals highlight the growth. Note the disappearance of the emulsion closest to the growing crystals. For a color version of this figure, see www.iste.co.uk/broseta/hydrates1.zip*

3.3.2.2. *Growth of percolating polycrystalline masses*

Another type of hydrate crystal growth is observed when the cyclopentane droplets are initially close enough to each other. Crystallization starts in or on one droplet, which elongates. In touching another droplet, it initiates further crystallization, etc. The outcome of this process is a branched crystal structure spanning the emulsion (Figure 3.16). Depending on the density of the emulsion, infilling is also observed, leading to thickening of the branches. The WBF process contributes both to branching and thickening.

This phenomenon is analogous to the in-plane ice frost growth on chilled hydrophobic surfaces recently investigated by many works [GUA 13, PET 14, BOR 16], where one of the supercooled water droplets in an assembly of such droplets (sitting on a hydrophobic substrate) crystallizes and then a “horn” sprouts from it toward a neighboring water drop. The latter droplet starts evaporating and therefore decreases in size. There is a competition between bridge growth and droplet evaporation. If the droplets are close enough, they connect. In the present case, the substrate is hydrophilic, hence oleophobic, establishing the analogy with water droplets on a hydrophobic substrate. The large water drop replaces air as an atmosphere, cyclopentane replaces water and hydrate replaces ice.

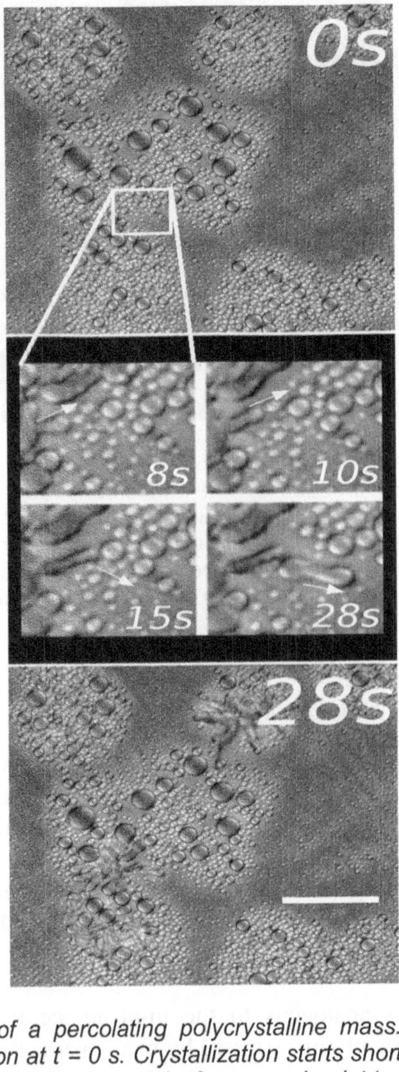

Figure 3.16. *Growth of a percolating polycrystalline mass. No hydrate crystal is apparent in the emulsion at t = 0 s. Crystallization starts shortly after. The zoomed-in views show how crystallization proceeds, from one droplet to another, in the direction indicated by arrows. Two ramified (branched) crystal structures are apparent at t = 28 s. Scale bar: 100 μm*

In some cases, the two types of crystallization presented and discussed above, i.e. the polygonal crystals and percolating crystal masses, are observed simultaneously (see Figure 3.17). A full account is in preparation.

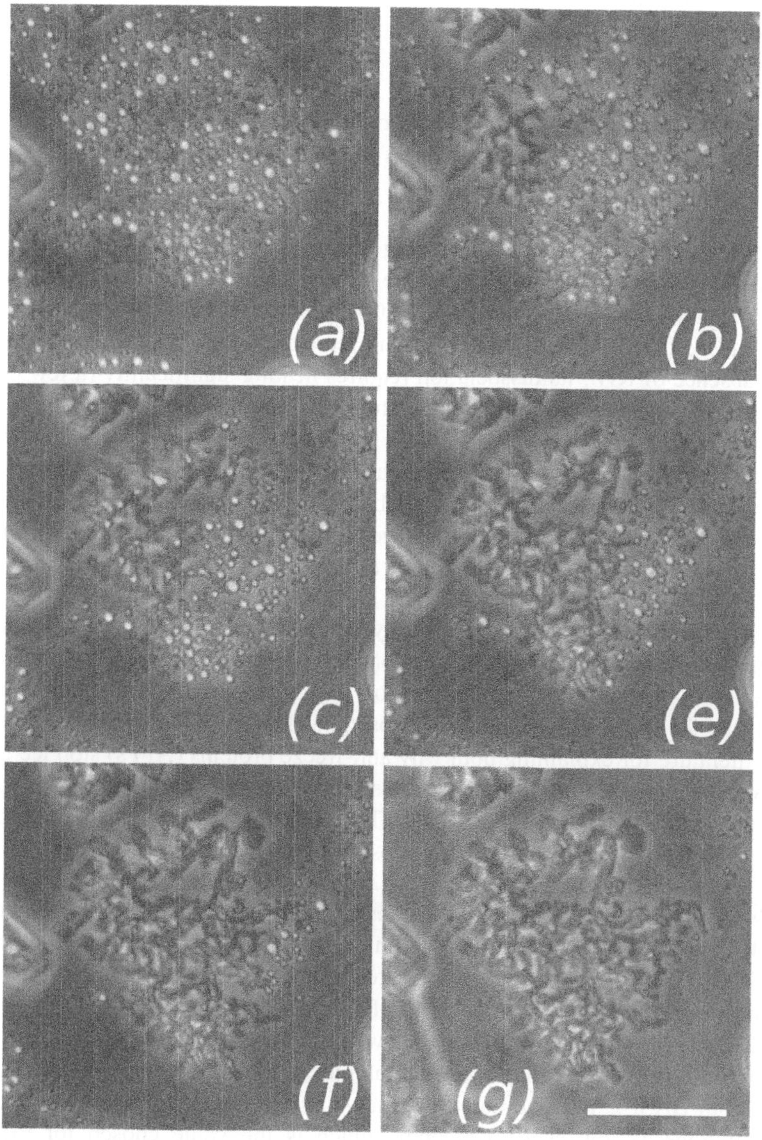

Figure 3.17. *Growth of a percolating crystal mass in an emulsion over a period of 100 s (from a to g). The rightmost edges of a large polygonal crystal are apparent on the left. Dark field mode. Scale bar: 100 µm*

3.3.3. *Adsorption of kinetic hydrate inhibitors*

Kinetic hydrate inhibitors (KHIs) are polymer molecules added to the aqueous phase in small amounts (typically a few hundred ppm) to prevent or at least slow down hydrate formation and/or growth, and thus ensure flow in oil and gas transportation lines (e.g. pipelines). The inhibiting mechanisms are not well understood (see the recent review by Ke and Kelland [KE 16]), although it seems clear that a key role is played by the adsorption of KHIs onto the hydrate crystals. This adsorption of KHIs very likely alters the crystal morphology and/or lowers the crystal growth rate.

Many of these KHIs show a surprising feature: the gas hydrate, whose formation and growth is prevented or delayed by the KHI, remains stable for temperatures that exceed T_{eq} by a few degrees Celsius [DUC 09]. Once the hydrate has been formed, which indeed occurs under a larger driving force and/or a longer waiting time than in the absence of KHI, its dissociation requires heating significantly (a few degrees Celsius) above the equilibrium temperature T_{eq}. Remember that, in the absence of KHI, heating just above T_{eq} suffices to dissociate the hydrate. The hydrates stabilized by KHI adsorption at temperatures higher than T_{eq} are referred to as "superheated" hydrates [DEL 11]. It is debated whether this feature, which seems to be correlated with the inhibiting power of the KHI, is an equilibrium (or thermodynamic) phenomenon or a kinetic effect.

Based on Brewster angle microscopy observations, we argue here that non-uniform (or "patchy") polymer adsorption is responsible for these two apparently contradictory phenomena – inhibition of gas hydrate formation and enhanced stability of the formed hydrate. The proposed mechanism is similar to that proposed in the literature for explaining both why anti-freeze proteins (AFPs) stop ice growth at T well below 0 °C and why ice crystals melt at temperatures significantly above 0 °C up to ~0.5 °C [CEL 10, CEL 13]. This mechanism is presented and discussed below following the Brewster angle microscopy observations.

The hydrate system considered in these experiments is again the cyclopentane hydrate at ambient pressure, which has an equilibrium (or dissociation) temperature T_{eq} ~7 °C. The three polymer KHIs tested (Figure 3.18) are widely used in the industry: poly(N-vinylpyrrolidone) (PVP, M_w = 40,000, Aldrich), polyvinyl-caprolactam (PVCap, low molecular weight, Luvicap EG from BASF) and vinylpyrrolidone: vinyl-caprolactam copolymer (VP/VCap, M_w = 55,000, Luvicap 55 W from BASF). The two latter are known to be good gas hydrate inhibitors for concentrations as low as of 100 ppm, which is the value chosen for the three aqueous solutions.

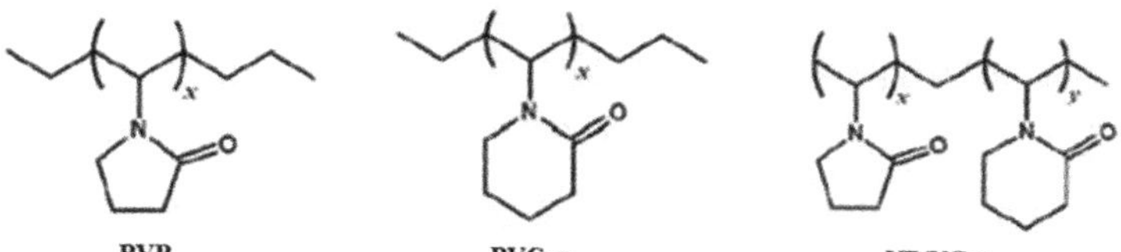

Figure 3.18. *Monomer structures of PVP, PVCap and VP/VCap*

A brief account is given here of the macrophotography observations of millimeter-sized drops of these solutions in cyclopentane using the experimental setup and procedure described in [MAR 15a]. (For more details, see also [MAR 15b].) The formation of cyclopentane hydrate at the surface of these drops requires larger subcooling and hydrate growth over the drop surface is much slower than in the absence of KHI (pure water). The inhibiting efficiency is observed to be in the order: PVP < VP-VCap < PVCap.

The morphologies are different from those found with pure water, i.e. in the absence of KHI: the hydrate that ultimately covers the water drop is less crystalline, thinner and more tenuous. When this hydrate cover has not formed, small white dots are apparent on the drop surface, which persist for tens of minutes at temperatures above T_{eq}. When the hydrate cover has formed the water drop surface at $T < T_{eq}$, it remains stable or a "superheated" hydrate above T_{eq} over a temperature interval, $\Delta T_{superheating}$. The values, reported in Table 3.1 for the three aqueous solutions investigated, show that this interval increases with KHI's inhibiting efficiency.

Inhibitor	$\Delta T_{superheating}$ (°C)
PVP	1 ± 0.5
VP/VCap	3 ± 0.5
PVCap	4 ± 0.5

Table 3.1. *Temperature interval (above T_{eq}) where "superheated" hydrates are observed to be stable*

3.3.3.1. *Adsorption of polymer KHIs at water/hydroxylated surfaces as revealed by ellipsometry*

Our initial project was to carry out Brewster angle microscopy observations of the interface between cyclopentane hydrate and the above aqueous solutions. Unfortunately, we were not able to grow the hydrate crystal with the required planarity and smoothness. We then decided to use an oxidized silicon wafer as an analog of a hydrate crystal face. In fact, the silica (SiO_2) that forms on the wafer surface in air is hydroxylated in water and, similarly to a hydrate crystal surface, bears hydroxyl (-OH) groups pointing outward.

The precleaned silicon wafers were immersed in the solutions containing 100 ppm of the three different KHIs for 15 minutes. The excess polymer solution was drained, the wafers were then rinsed with distilled water and dried before being placed in the ellipsometer.

Figure 3.19(a) shows the raw Brewster angle microscopy images of the oxidized silicon wafers without any inhibitor adsorbed, and after immersion in the different polymer solutions and mild drying (b and c). The color code varies from black (no layer present on the water) to white passing through brown, red and yellow. The brighter the surface, the thicker the adsorbed layer. Figure 3.19(b), which is brighter than Figure 3.19(a) (uncovered surface), shows that PVP has adsorbed on the wafer surface as a thin layer with uniform thickness (the dark and bright horizontal lines are wafer scratches). The bright spots in Figures 3.19(c) and (d) reveal the presence of "grains", probably resulting from a non-homogeneous, "patchy" adsorption; this effect is more pronounced with PVCap.

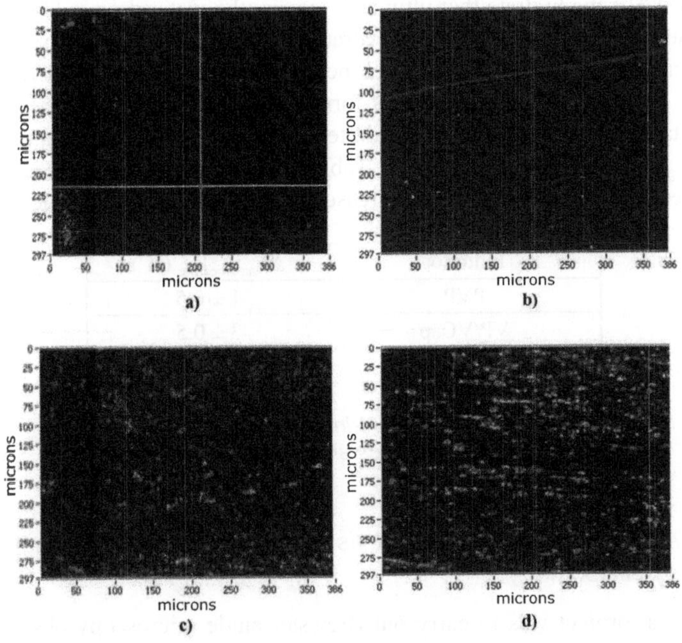

Figure 3.19. *Brewster angle microscopy images of the silicon wafers with a) no inhibitor, b) PVP, c) VP/VCap and d) PVCap. For a color version of this figure, see www.iste.co.uk/broseta/hydrates1.zip*

Ellipsometric measurements also allow thickness maps of the analyzed surfaces to be determined. For this purpose, images are taken at different polarizer and analyzer angles in order to determine the ellipsometric angles δ and ψ of each pixel (see section 3.2.2). The ellipsometer software then produces a 3D map that simulates the thicknesses of the layer deposited on the oxidized silicon surface. The thin layer left after KHI adsorption has taken place is composed of the silica (SiO_2) layer covering the silicon surface and the adsorbed polymer layer, both assumed to have the same refractive index ~1.46.

Figure 3.20 presents the raw data, i.e. the thicknesses of the layers of SiO_2 + polymer. Without any polymer adsorbed, a layer of about 1–2 nm is observed (Figure 3.20(a)), corresponding to the silica (SiO_2) formed on the silicon surface in the presence of water as discussed above. Figure 3.20(b) shows a uniform thickness layer of about 3–4 nm revealing that PVP forms a more or less homogeneous 2 nm thick layer all over the surface (peaks are positioned along the wafer scratches).

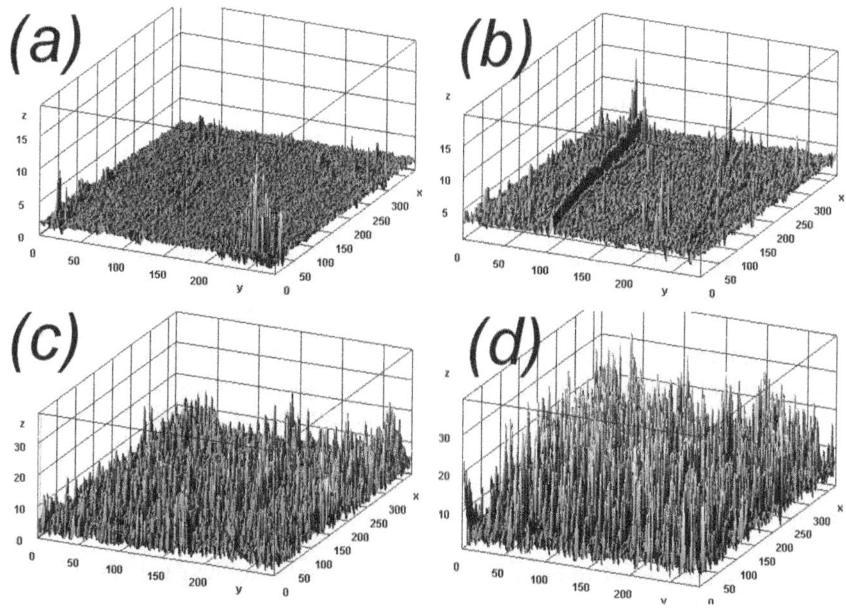

Figure 3.20. *3D maps showing the thicknesses (z, in nanometers) of adsorbed layers on silicon wafers with a) no inhibitor adsorbed, b) PVP adsorbed, c) VP/VCap adsorbed, d) PVCap adsorbed. x and y scales are in micrometer. Note the different z-scales*

In contrast, VP/VCap forms a layer with comparable thickness to that of PVP (about 2 nm) but with local inhomogeneities – or patches – having thicknesses up to

6–8 nm (Figure 3.20(c)). Layers formed from PVCap exhibit thicker patches (about 10–16 nm) (Figure 3.20(d)). With PVCap the layer in-between the patches has more or less the same thickness as that observed for the two other KHIs. Note that the maps shown in Figure 3.20 are smoothed by a numerical procedure, so the patch extents are overestimated in the observation plane. In order to have more realistic information about the lateral extent of these patches, one has to refer to Figure 3.19.

Summarizing these observations, the non-uniformity or "patchiness" of polymer adsorption increases in the order PVP < PV:PVCap < PVCap. This ranking parallels that of the temperature extent above T_{eq} where superheated hydrate crystals are observed (see Table 3.1).

3.3.3.2. *Are superheated hydrate crystals a consequence of patchy polymer adsorption?*

The mechanism proposed and discussed here is adapted from that accounting for the activity of AFPs: these molecules not only stop the growth of small ice crystals well below 0 °C, but they also prevent these ice crystals from melting when the temperature is raised above 0 °C [CEL 10, CEL 13]. Superheated ice crystals have thus been observed at temperatures up to 0.44 °C, with hyperactive AFPs irreversibly adsorbed onto them [CEL 13]. Owing to the extreme AFP dilution in water, only a very small fraction of the ice surface is occupied by the adsorbed AFP: AFP adsorption is thus necessarily "patchy". The chemical structure of these molecules is such that, locally (at the attachment sites), ice growth or melting is suppressed. Between these attachment sites, the interface can be curved, i.e. it can bulge in one direction or the other, which has the effect of displacing the ice freezing and melting temperatures: this is the Gibbs–Thomson effect. The displacement is positive in the case of melting (i.e. the melting temperature is higher than the 0 °C of a flat interface) and negative in the case of freezing, which corresponds to an interface curvature differing in sign. This simple explanation, which can be refined [CEL 10], has the merit of being able to explain both the inhibiting (or anti-freeze) character of these AFPs and the existence of superheated ice crystals well above 0 °C.

The above argument can be straigthforwardly adapted to the gas hydrate system of interest here. Figure 3.21 is a pictorial explanation of why the gas hydrate remains stable at temperatures above T_{eq}. This is because the hydrate/water interface develops some curvature. According to the Gibbs–Thomson equation, the temperature range above T_{eq}, over which superheated hydrates exist, is related to the radius of curvature R (approximately the distance between adsorption patches) as follows:

$$\Delta T_{superheating} = -\frac{4\gamma T_{eq}}{QR\,\rho} \qquad [3.10]$$

where γ is the hydrate/water interfacial tension (approximately the ice/water value ~30 mN/m), T_{eq} ~ 280 K, ρ is the hydrate density (~950 kg/m^3) and Q is the latent heat of hydrate dissociation ~280 kJ/kg [NAK 08]. The factor on the right-hand side of equation [3.10] is for a spherical interface. A value of $\Delta T_{superheating}$ in the range of 1 K (as observed experimentally, see Table 3.1) thus corresponds to a radius R in the submicron range (~0.1 µm). Note that R must be counted negatively for the configuration of Figure 3.21. This range is somewhat smaller than that which is observed experimentally in Figures 3.19 and 3.20. More work is needed to understand the reasons for this difference. Future ellipsometric measurements should be carried out with real hydrate crystals.

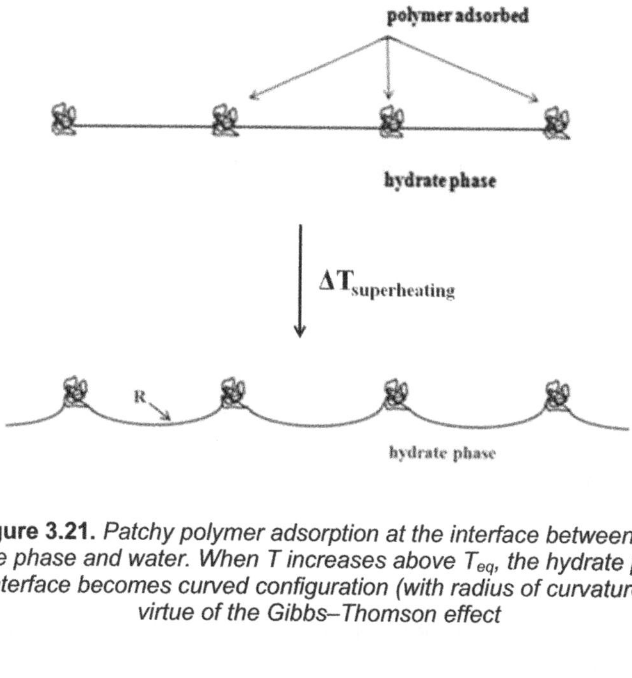

Figure 3.21. *Patchy polymer adsorption at the interface between the hydrate phase and water. When T increases above T_{eq}, the hydrate persists if its interface becomes curved configuration (with radius of curvature R) by virtue of the Gibbs–Thomson effect*

3.4. Concluding remarks

As emphasized in section 3.1, it seemed rather surprising that brightfield optical microscopy overwhelmingly (if not exclusively) dominates in gas hydrate research. Our aim here was to show the potential of other, vanishing background contrast modes to highlight the delicate transparent interfaces present in gas hydrate systems.

In this chapter, high-resolution optical microscopy methods have been reviewed and their value for gas hydrate research is highlighted, with examples addressing a model hydrate system, cyclopentane hydrate. Insights have been obtained into the structure and growth mechanisms of a hydrate halo on a hydrophilic substrate, as well as into the hydrate crystallization processes that occur within a guest-in-water emulsion. The adsorption of common gas hydrate kinetic inhibitors has been investigated on oxidized silicon, an analog of a hydrate crystal, by means of Brewster angle microscopy, revealing the crucial role played by the inhibitor adsorption mode.

The experimental constraints of studying gas hydrates on a microscope are as follows: (1) condensation of ambient humidity and (2) handling high pressures. We have illustrated how to solve the first problem. The second problem could be solved by using microfluidic devices, the simplest device being the cylindrical capillary, which can be inserted in commercial temperature-controlled stages (see Figure 3.11(d)).

3.5. Acknowledgments

The work described in this chapter has been partly supported by the French Agence Nationale de la Recherche (Project HYDRE ANR-15-CE06-008).

3.6. Bibliography

[ALL 69] ALLEN R.D., DAVID G.B., NOMARSKI G., "The Zeiss-Nomarski differential interference equipment for transmitted-light microscopy", *Zeitschrift für Wissenschaftliche Mikroskopie und Mikroskopische Technik*, vol. 69, pp. 193–221, 1969.

[AZZ 79] AZZAM R.M.A., BASHARA N.M., *Ellipsometry and Polarized Light*, North Holland, Amsterdam, 1979.

[BEL 10] BELTRAN J.G., SERVIO P., "Morphological investigations of methane – hydrate films formed on a glass surface", *Crystal Growth & Design*, vol. 10, pp. 4339–4347, 2010.

[BOR 16] BOREYKO J.B., HANSEN R.R., MURPHY K.R. *et al.*, "Controlling condensation and frost growth with chemical micropatterns", *Science Reports*, vol. 6, p. 19131, 2016.

[CEL 10] CELIK Y., GRAHAM L.A., MOK Y.F. *et al.*, "Superheating of ice crystals in antifreeze protein solutions", *Proceedings of the National Academy of Sciences*, vol. 107, pp. 5423–5428, 2010.

[CEL 13] CELIK Y., DRORI, R., PERTAYA-BRAUN N. *et al.*, "Microfluidic experiments reveal that antifreeze proteins bound to ice crystals suffice to prevent their growth", *Proceedings of the National Academy of Sciences*, vol. 110, no. 4, pp. 1309–1314, 2013.

[DAN 15] DANIEL-DAVID D., GUERTON F., DICHARRY C. *et al.*, "Hydrate growth at the interface between water and pure or mixed CO_2/CH4 gases: influence of pressure, temperature, gas composition and water-soluble surfactants", *Chemical Engineering & Science*, vol. 132, pp. 118–127, 2015.

[DEL 11] DEL VILLANO L., KELLAND M., "An investigation into the laboratory method for the evaluation of the performance of kinetic hydrate inhibitors using superheated gas hydrates", *Chemical Engineering and Science*, vol. 66, pp. 1973–1985, 2011.

[DUC 09] DUCHATEAU C., GLENAT P., POU T.E. *et al.*, "Hydrate precursor test method for the laboratory evaluation of kinetic hydrate inhibitors", *Energy and Fuels*, vol. 24, pp. 616–623, 2009.

[DUQ 16] DU QUESNAY J.R., DIAZ POSADA M.C., BELTRAN J.G., "Novel gas hydrate reactor design: 3-in-1 assessment of phase equilibria, morphology and kinetics", *Fluid Phase Equilibria*, vol. 413, no. 15, pp. 148–157, 2016.

[ESM 16] ESMAIL S., BELTRAN J.G., "Methane hydrate propagation on surfaces of varying wettability", *Journal of Natural Gas Science and Engineering*, vol. 35, pp. 1535–1543, 2016.

[FAN 14] FANDINO O., RUFFINE L., "Methane hydrate nucleation and growth from the bulk phase: Further insights into their mechanisms", *Fuel*, vol. 117, pp. 442–449, 2014.

[GUA 13] GUADARRAMA-CETINA J., MONGRUEL A., GONZALEZ-VIÑAS W. *et al.*, "Percolation-induced frost mechanism", *European Physical Letters*, vol. 101, p. 16009, 2013.

[HAU 16] HAUGE L.P., GAUTEPLASS L., HØYLANDA M.D. *et al.*, "Pore-level hydrate formation mechanisms using realistic rock structures in high-pressure silicon micromodels", *International Journal of Greenhouse Gas Control*, vol. 53, pp. 178–186, 2016.

[JAC 98] JACKSON J.D., *Classical Electrodynamics*, 3rd ed., Wiley, New York, 1998.

[KAR 12] KARANJKAR P.U., LEE J.W., MORRIS J.F., "Surfactant effects on hydrate crystallization at the water–oil interface: hollow-conical crystals", *Crystal Growth & Design*, vol. 12, no. 8, pp. 3817–3824, 2012.

[KAT 06] KATSUKI D., OHMURA R., EBINUMA T. *et al.*, "Formation, growth and ageing of clathrate hydrate crystals in a porous medium", *Philosophical Magazine*, vol. 86, pp. 1753–1761, 2006.

[KE 16] KE W., KELLAND M.A., "Kinetic Hydrate Inhibitor Studies for Gas Hydrate Systems: A Review of Experimental Equipment and Test Methods", *Energy and Fuels*, vol. 30, pp. 10015–10028, 2016.

[KIT 13] KITUMURA M., MORI Y.H., "Clathrate hydrate film growth along water/methane phase boundaries – an observational study", *Crystal Research and Technology*, vol. 48, pp. 511–519, 2013.

[LIM 09] LIMOZIN L., SENGUPTA K., "Quantitative reflection interference contrast microscopy (RICM) in soft matter and cell adhesion", *ChemPhysChem*, vol. 10, pp. 2752–2768, 2009.

[MAR 15a] MARTINEZ DE BAÑOS M.L., CARRIER O., BOURIAT P. *et al.*, "Droplet-based millifluidics as a new tool to investigate hydrate crystallization: insights into the memory effect", *Chemical Engineering and Science*, vol. 123, pp. 564–572, 2015.

[MAR 15b] MARTINEZ DE BAÑOS M.L., Mechanisms of formation and dissociation of cyclopentane hydrates, PhD Thesis, University of Pau and Pays de l'Adour, November 2015.

[MAR 16] MARTINEZ DE BAÑOS M.L., HOBEIKA N., BOURIAT P. *et al.*, "How do gas hydrates spread on a substrate", *Crystal Growth and Design*, vol. 16, pp. 4360–4373, 2016.

[MIT 15] MITARAI M., KISHIMOTO M., SUH D. *et al.*, "Surfactant effects on the crystal growth of clathrate hydrate at the interface of water and hydrophobic-guest liquid", *Crystal Growth & Design*, vol. 15, no. 2, pp. 812–821, 2015.

[NAK 08] NAKAJIMA M., OHMURA R., MORI Y.H., "Clathrate hydrate formation from cyclopentane-in-water emulsions", *Industrial & Engineering Chemistry Research*, vol. 47, pp. 8933–8939, 2008.

[PET 14] PETIT J., BONACCURSO E., "General frost growth mechanism on solid substrates with different stiffness", *Langmuir*, vol. 30, pp. 1160–1168, 2014.

[PAW 95] PAWLEY J.B. (ed.), *Handbook of biological confocal microscopy*, 3rd ed., Springer, New York, 1995.

[SAI 11] SAITO K., KISHIMOTO M., TANAKA R. *et al.*, "Crystal growth of clathrate hydrate at the interface between hydrocarbon gas mixture and liquid water", *Crystal Growth & Design*, vol. 11, pp. 295–301, 2011.

[SER 03] SERVIO P., ENGLEZOS P., "Morphology of methane and carbon dioxide hydrates formed from water droplets", *AIChE Journal*, vol. 49, pp. 269–276, 2003.

[SLO 08] SLOAN E.D., KOHC A., *Clathrate hydrates of natural gases*, 3rd ed., CRC Press, Boca Raton, Florida, 2008.

[SME 97] SMELIK E.A., KING H.E. Jr., "Crystal-growth studies of natural gas clathrate hydrates using a pressurized optical cell", *American Mineralogist*, vol. 82, pp. 88–98, 1997.

[SUN 10] SUN C.Y., PENG B.Z., DANDEKAR A. *et al.*, "Studies on hydrate film growth", *Annual Reports on the Progress of Chemistry, Section C (Physical Chemistry)*, vol. 106, pp. 77–100, 2010.

[SUN 15] SUNDRAMOORTHY J.D., SABIL K.M., LAL B. *et al.*, "Catastrophic crystal growth of clathrate hydrate with a simulated natural gas system during a pipeline shut-in condition", *Crystal Growth & Design*, vol. 15, pp. 1233–1241, 2015.

[TOH 01] TOHIDI B., CLENELL M.B., ANDERSON R. *et al.*, "Visual observation of gas-hydrate formation and dissociation in synthetic porous media by means of glass micromodels", *Geology*, vol. 29, p. 867, 2001.

4

Calorimetric Characterization of Clathrate and Semiclathrate Hydrates

4.1. Introduction

A number of thermodynamic and thermophysical properties related to gas hydrates are needed in the various applications involving these materials. On the one hand, the knowledge of pressure–temperature stability limits of hydrates in the presence of chemical inhibitors is required in order to anticipate hydrate formation in oil and gas industry facilities. On the other, the effect of thermodynamic promoters needs to be characterized if hydrates are to be used in new processes such as gas storage or treatment, water purification or cold storage and transfer. The enthalpies of gas hydrate formation or dissociation and the associated heat capacities are needed to assess energy balances in these processes and to develop high-efficiency phase change materials for thermal energy management.

For decades the phase behavior of gas hydrates has been extensively studied, primarily using constant volume reactors equipped with see-through windows and equipped with temperature and pressure sensors. In these studies, the appearance of gas hydrates was evidenced by visual observation and by the detection of a pressure drop and/or a temperature jump. An alternative approach consists of developing and using calorimetry tools. This was essentially used to measure the heat of dissociation and heat capacity of gas hydrates synthesized in a reactor and then transferred into the calorimetric vessel [HAN 84, HAN 86a, HAN 86b, HAN 88, RUE 88, HAN 89, LIE 90, KAN 01].

Chapter written by Didier DALMAZZONE, Luiz Paulo SALES SILVA, Anthony DELAHAYE and Laurence FOURNAISON.

In other investigations, differential scanning calorimetry (DSC) was used on model hydrates in order to assess the efficiency of kinetic inhibitors [KOH 98, SHA 14] or to study the formation of hydrates in water-in-oil emulsions [FOU 99, FOU 01, FOU 02a, FOU 02b, CLA 02, FOU 06]. Model hydrates are formed at ambient pressure from compounds such as tetrahydrofuran [DEV 99] or trichlorofluoromethane [JAK 96]. Model hydrates were thus used to simulate the behavior of natural gas hydrates in ambient pressure DSC experiments.

The use of differential calorimetry with gas hydrates, formed directly inside the calorimetric vessels under a controlled gas pressure, was initiated to study the thermodynamics of hydrates of methane and natural gas in deep offshore drilling fluids [DAL 00, DAL 01a, DAL 01b, DAL 02a, DAL 02b, DAL 02c, DAL 02d, DAL 02e, DAL 03, KHA 03, DAL 04, KHA 04, LE 04]. At the same time, differential thermal analysis (DTA) was implemented for the study of carbon dioxide hydrates as phase change material for refrigeration applications [FOU 04, LIN 08]. These techniques have proven to be not only faster, but also more powerful than, and as precise as, more traditional techniques. Their application was then extended to thermodynamic and thermophysical measurements carried out in the context of the following emerging gas hydrate applications: gas separation, CO_2 capture and biogas valorization [DES 09, BEN 11, BOU 11, SAL 16a, SAL 16b], hydrogen storage [DES 10, KAR 13, KAR 14, FUK 15b] and hydrate-based phase change materials for refrigeration [MAR 08, MAR 09, MAY 10a, MAY 10b, MAY 11, MAY 12, LIN 14, FOU 15].

In this chapter, we first present the principles of the thermal analysis and calorimetry techniques that were used to study gas hydrates. We then focus on the methodologies developed to form hydrates inside the calorimetric vessel and to precisely determine their dissociation temperatures and enthalpies as well as their heat capacities. Finally, we give an overview of the application of DSC to the study of hydrate formation kinetics in a dispersed aqueous phase.

4.2. DTA and differential scanning calorimetry

4.2.1. *Principles of DTA and DSC*

This section briefly introduces the principles of differential thermal and calorimetric methods. The reader is invited to refer to [BRO 98] for further details. DTA consists of measuring the temperature difference between a sample and a reference during a programmed warming or cooling of both the sample and reference holders. This temperature difference depends on [MAC 75]:

– the warming or cooling rate imposed by the furnace;

– the heat capacity of the sample and reference materials, as well as that of their holders (crucibles);

– the enthalpy variation of the sample caused by phase changes and chemical transformations occurring during the temperature program;

– the thermal resistance between the temperature probes used to measure the temperature difference.

In DTA experiments, the sample temperature is directly measured by a thermocouple in contact with the substance under study. It is thus a technique designed to precisely determine the temperature of phase transitions. Even though enthalpy determinations are possible using DTA, they remain approximate because the heat flux exchanged by the sample is not measured [GRA 68].

The first DSC devices have been developed as an improvement of DTA in order to overcome the difficulty of converting temperature differences into enthalpy measurements. These devices, commonly called "plate DSCs", use networks of thermocouples to measure the heat flow exchanged between the bottom of the sample, reference vessels and the thermostat.

Another type of DSC is based on the miniaturization of the Tian–Calvet principle [CAL 56, CAL 58], a highly sensitive heat measurement technique used in microcalorimetry. In such DSC, numerous thermocouples are arranged around the crucible in a cylindrical geometry in order to measure the radial heat flux exchanged with the furnace. Because the contact area of the cell with the heat sensor is much greater than in plate DSC, the heat losses are reduced and this principle gives much more precise measurements. Sophisticated DSCs used for research purposes are usually fitted with these more sensitive and more precise sensors, while plate DSCs are better suited for routine measurements. Figure 4.1 illustrates DTA, plate DSC and Tian–Calvet DSC.

4.2.2. *Examples of pressure-controlled DTA and DSC devices for hydrate studies*

A DTA device was specially developed at Irstea to study the formation of CO_2 hydrates in relation to cold storage and transportation applications [FOU 04]. As can be seen in Figure 4.2, the apparatus is composed of two glass vessels (volume ~40 mL) fitted with magnetic stirrers. Vessel (1) contains the mixture under study; vessel (2) contains an inert solution. The vessels are immersed in a thermostatic bath (3) cooled by means of a cold generation unit (4). A 1,000 mL syringe pump (ISCO, 1,000 D) fed via a CO_2 bottle (6) allows a calibrated amount of gas to be injected into the vessels. The instruments are linked to an acquisition interface (7): one type

T thermocouple inside each vessel (8) and two pressure sensors (9). A DTA (10) sensor made of height thermocouples connected in series measure the differential heat flow between the two vessels. The apparatus may be operated within wide ranges of pressure (0–3 MPa) and temperature (263–303 K). Compared to the µDSC design, the advantage of this equipment is that it enables simultaneous heat exchange measurements and visual observation of the hydrates formation and dissociation.

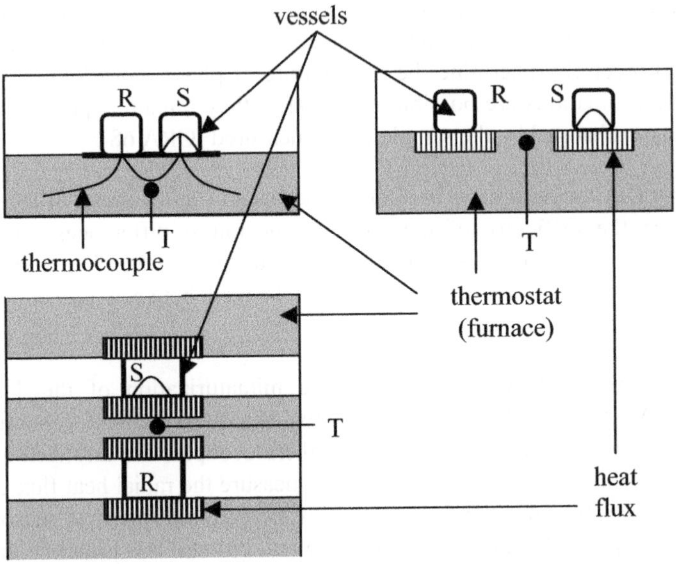

Figure 4.1. *General design of equipment for differential thermal and calorimetric measurements. Upper left: ATD; upper right: plate DSC; bottom left: Tian–Calvet DSC. S: sample holder; R: reference crucible (often left empty); T: temperature probe*

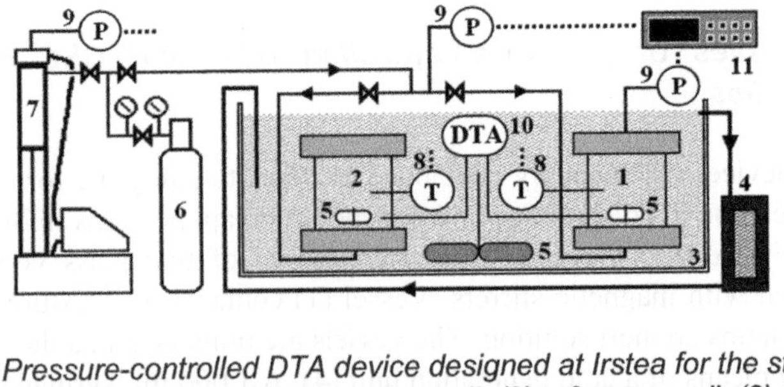

Figure 4.2. *Pressure-controlled DTA device designed at Irstea for the study of CO_2 hydrates. Experimental device: (1) Measuring cell; (2) reference cell; (3) temperature-controlled bath; (4) cooling/heating unit; (5) stirrers; (6) CO_2 bottle; (7) syringe pump; (8) thermocouples; (9) pressure gauges; (10) DTA sensor; (11) acquisition interface*

The vessels are first submitted to a cooling ramp of -1 K·min^{-1} in order to form the hydrates. When the temperature is stabilized at a value of 273 K to avoid ice crystallization, hydrate crystallization is initiated by stirring. The exothermal crystallization is evidenced by a differential heat flow peak detected by the DTA sensor (see Figure 4.3) and a temperature peak (not presented). Hydrate formation also causes a pressure drop, due to gas consumption. In the closed system, the formation stops when all excess CO_2 has been consumed to reach L_w-H-V equilibrium.

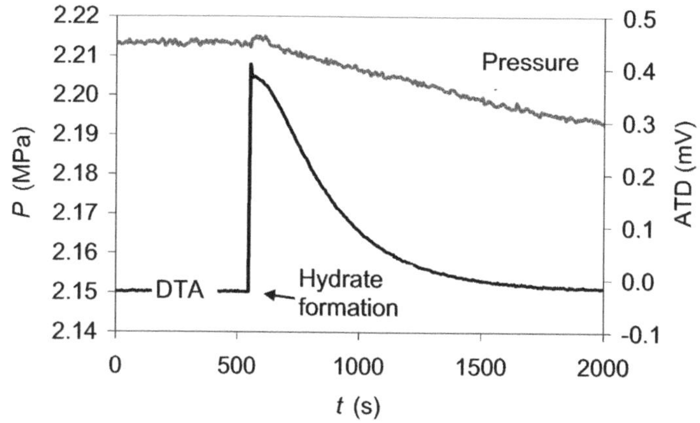

Figure 4.3. *DTA signal obtained during the formation of CO_2 hydrates at a temperature of 274 K and an initial CO_2 pressure of 2.213 MPa. For a color version of this figure, see www.iste.co.uk/broseta/hydrates1.zip*

After the hydrates have been formed and the system stabilized, the vessels are submitted to a temperature ramp at a slow warming rate of 0.05–0.2 K·min^{-1} (Figure 4.4). When the dissociation of hydrates takes place, a DTA peak is observed. The maximum amplitude of the DTA peak is used to measure the temperature of L_w-H-V equilibrium. By integrating this peak, it is also possible to determine the dissociation enthalpy, after calibrating the heat flow sensor (for example by comparison with the heat of melting of an ice sample of known mass).

Several commercial DSC devices may be adapted to pressure-controlled experiments by using specially designed vessels. Figure 4.5 shows the various equipment that have been used for over 15 years at ENSTA ParisTech to study a large variety of hydrates: for details, see [DAL 00] and the references which follow. The sample volume is much smaller than in the DTA device described above, typically between 20 and 50 µL, allowing DSC experiments to be run at much faster rates. Even if moderate rates are recommended to get accurate thermodynamic measurements, DSC remains a fast and powerful technique. On the other hand, unlike Irstea's DTA equipment, it provides neither stirring nor visualizing of the vessels' content.

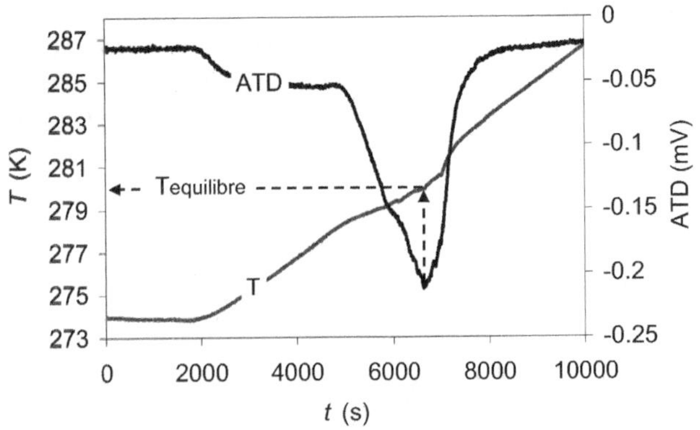

Figure 4.4. *DTA signal obtained during the dissociation of CO_2 hydrates formed in the conditions in Figure 4.3. For a color version of this figure, see www.iste.co.uk/broseta/hydrates1.zip*

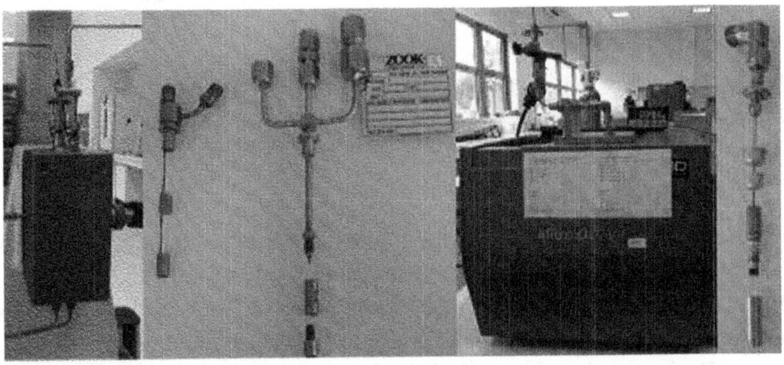

Figure 4.5. *Equipment used at ENSTA ParisTech for calorimetric study of gas hydrates. From left to right: furnace of the DSC 111 Tian–Calvet calorimeter (SETARAM); 100 and 400 bar pressure-controlled cells for the DSC 111; µDSC VII calorimeter (SETARAM); pressure-controlled cell for the µDSC VII. For a color version of this figure, see www.iste.co.uk/broseta/hydrates1.zip*

Measurements are generally performed at constant controlled pressure. Crystallization requires several degrees of undercooling and for that reason crystallization heat flow signals may not be used for equilibrium determinations. Melting (dissociation) experiments are used for that purpose, as discussed in the methodological section 1.3. Figure 4.6 presents some examples of thermograms obtained with various gas hydrate systems.

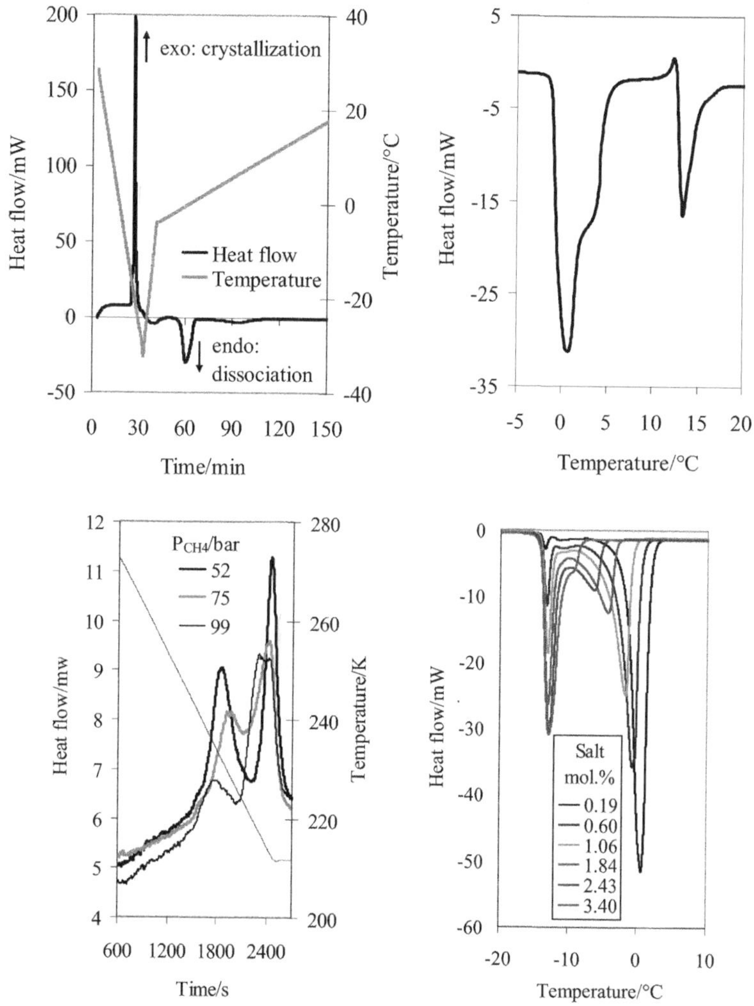

Figure 4.6. *Examples of thermograms obtained with hydrates. Upper left: 15 wt.% tetrabutylphosphonium bromide in water at ambient pressure, complete formation/dissociation cycle, temperature and heat flow versus time; upper right: 100 bar CH_4 in pure water, dissociation thermogram (heat flow versus temperature) showing successive melting of ice and methane hydrates; bottom left: formation of methane hydrates in a water-in-oil emulsion; bottom right: dissociation of tributylmethylammonium chloride semiclathrate hydrates at various compositions. For a color version of this figure, see www.iste.co.uk/broseta/hydrates1.zip*

4.2.3. *Temperature calibration of DSC*

Unlike in DTA, there is no temperature probe in contact with the sample in a DSC crucible. The temperature sensor is located in the furnace, outside the crucibles, and the difference between the experimental temperature and the exact temperature of the sample holder has to be determined indirectly. This correction is crucial because DSC is usually used in a dynamic way, at a non-zero heating rate. The heat exchanges that continuously take place in the system may then lead to unacceptable errors if a correct protocol is not applied. First, a proper temperature calibration must be performed in order to correct for the temperature lag between the probe and the calorimetric vessel. This is done by recording the melting thermograms of known substances. Preferably, metals or selected organic compounds can be used as calorimetric standards [SAB 99]. Samples of high purity must be used for that purpose. The onset point of the melting peak (see Figure 4.7) is taken as the melting temperature of the standard. The difference between the measured temperature and the known melting point of the standard serves to determine the correction function. At least two standards must be used to repeat the calibration at different temperatures in order to account for a possible deviation of the probe. It is advisable to perform the calibration in a temperature range adapted to the experimental conditions of the study. For example, for studies on gas hydrates a calibration with mercury (T_m = –38.8 °C) and gallium (T_m = 29.8 °C) covers the range of interest. Finally, the calibration has to be performed for each temperature at various heating rates, this parameter having a great influence on the temperature correction. The correction function applied to obtain the temperature inside the sample cell from the measured temperature is given by the following expression:

$$T_{Sample} - T_{meas} = AT_{meas} + B_1\beta + B_2\beta^2 + C \qquad [4.1]$$

where $\beta = dT_p/dt$ represents the programmed scanning rate. Note that for correct calibration, the melting must take place when the actual rate of temperature increase is rigorously equal to the programmed scanning rate, i.e. when $dT_{meas}/dt = \beta$, as shown in Figure 4.7. The constants A, B_1, B_2 and C are then obtained by linear regression of the calibration results.

It must be noted that the parameters B_1 and B_2 in equation [4.1] depend on the heat capacity and thermal conductivity of the sample holder. This means that temperature calibration should be performed for each new DSC cell being used. For practical reasons, cells of the same model, provided by the same manufacturer, having the same dimensions, same weight, made of the same material, etc., can be used with the same correction function. Yet it is a good precaution to check the calibration by performing a simple experiment, consisting of melting a sample, ice for example, at two different heating rates. The result should always stand within 0.1 or 0.2 °C of the expected value.

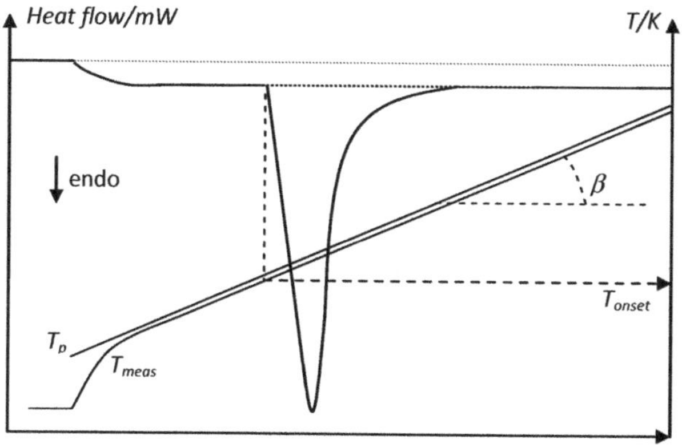

Figure 4.7. *Heat flow and temperature versus time during the melting of a metal sample for calibration purpose. T_p: programmed temperature; T_{meas}: measured temperature; β: temperature scanning rate*

The same simple experiment can also be used to check that the measured melting heat stands within 1–2% of the expected value. The calibration of heat flow sensitivity, as performed by the DSC manufacturer, is generally worthy of confidence and does not require adjustment when changing the model of sample holder, unlike temperature calibration.

4.3. Phase equilibrium determination in hydrate systems using pressure-controlled TDA and DSC

4.3.1. *Proper exploitation of DSC thermograms*

As already mentioned, the DSC technique has been designed to measure heat flows, not temperatures. Nevertheless, accurate measurement of phase change temperatures may be derived from DSC results, provided that extreme care is exercised in exploiting the thermograms.

After careful calibration of the DSC, as described above, the measured temperature is automatically corrected so that the recorded temperature is equal, or very close, to that of the sample holder. Nevertheless, uncertainties remain about the actual temperature of the sample itself. As can be seen in Figure 4.8, the temperature of interest, namely the melting temperature T_m, may differ from the recorded temperature T_c, due to the heat flow that has to be exchanged, through the liquid phase, between the solid–liquid interface and the wall of the crucible.

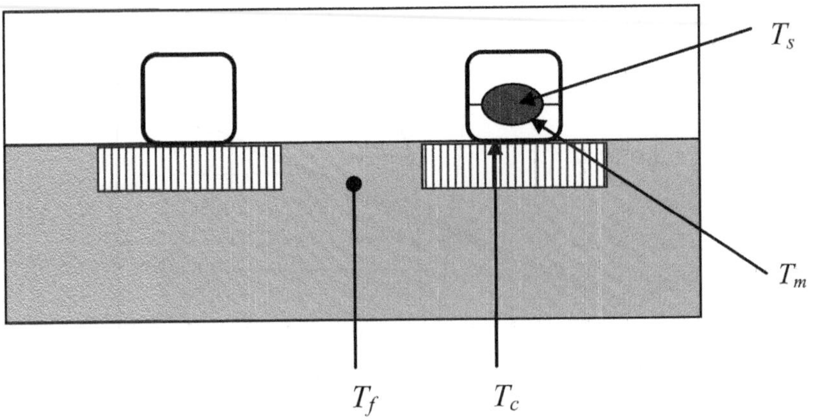

Figure 4.8. *Temperature profile inside a DSC crucible during sample melting. T_f : furnace temperature, measured by the probe (T_f = T_{meas}); T_c: crucible temperature, obtained after correction using equation [4.1]; T_m: melting temperature at the solid–liquid interface; T_s: temperature inside the solid sample. For a color version of this figure, see www.iste.co.uk/broseta/hydrates1.zip*

The resulting thermal gap strongly depends on the amount of heat to be transferred and the thermal resistance of the liquid phase. Two kinds of phase changes have to be considered:

– constant temperature phase change, corresponding to any invariant equilibrium situation: melting of a substance in equilibrium with pure liquid phase, melting of a eutectic or a peritectic mixture;

– progressive phase change, occurring when the variance of the equilibrium system is greater than zero: for example, melting of a substance in the presence of a solute in the liquid phase.

In the first case, the melting starts as the temperature of the crucible reaches the melting point. The onset temperature (see Figure 4.9) thus corresponds to the melting point with an uncertainty of typically 0.1–0.2 K. The evolution of the pressure signal shows that the onset point precisely corresponds to the starting of gas release due to hydrate dissociation. Similarly, the end of gas release and the inflexion point (maximum of the slope) of the heat flux signal coincide. Note, however, that the "end temperature" may not be considered as an equilibrium point, since a significant heat flow is still being transferred to the sample vessel at that point.

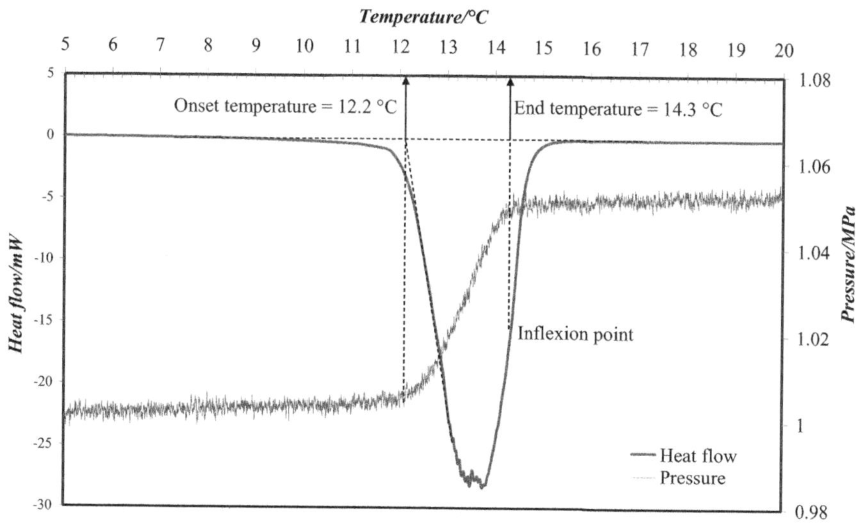

Figure 4.9. *Measurement of hydrates dissociation temperature using continuous heating, H_2O + CO_2 + CH_4 + tetrahydropyran (unpublished results). For a color version of this figure, see www.iste.co.uk/broseta/hydrates1.zip*

In the case of a progressive phase change, as illustrated in Figure 4.10, the thermodynamic property of interest is the temperature at the end of melting. It has to be measured at the end of the melting peak, when the heat flux is close to its maximum, i.e. far from equilibrium. It is commonly accepted that using DSC at a sufficiently low heating rate assures a quasi-homogeneous temperature profile inside the sample holder, thus reducing the measurement uncertainties to acceptable values. However, recent works have evidenced that the usual way of determining the melting temperatures can lead to errors exceeding 1 K, even at very low heating rate, in the case of progressive melting [LIN 13].

The most precise way of measuring phase change temperatures in this case is the stepwise method [LAF 12, LIN 13]: the sample is heated by small steps of temperature until complete melting. The temperature resolution depends on the step that is applied, for example 0.1 K as in Figure 4.11. The end of hydrate dissociation is evidenced by both the heat flow and the pressure signals. With such tiny temperature increases, the driving force for melting is very low and the time required to reach equilibrium may be very long. In the case presented here, steps of

4 h were necessary to achieve equilibrium and the complete experiment lasted more than 36 h.

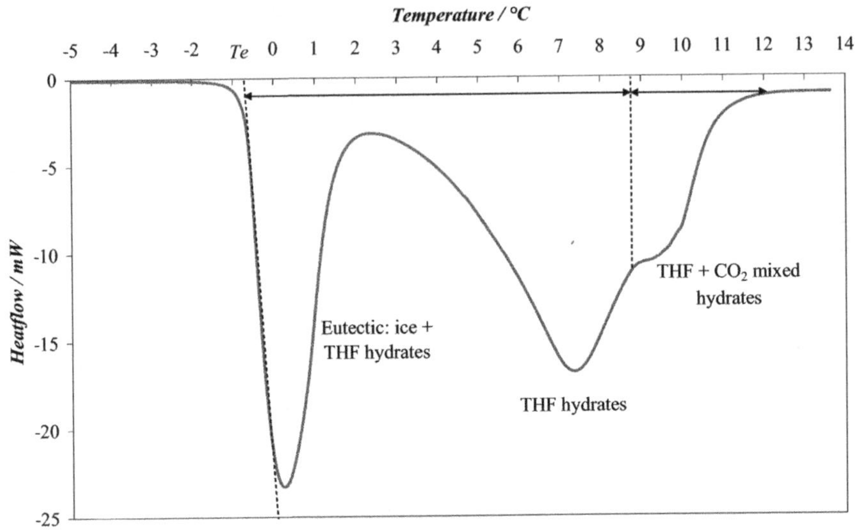

Figure 4.10. *Sample thermogram with constant-temperature and progressive melting. Aqueous solution of THF at 11% in weight + 3.7 bar CO_2 (unpublished results). Eutectic mixture melts at constant temperature Te. Arrows show the ranges of dissociation of THF hydrates and (CO_2 + THF) mixed hydrates. For a color version of this figure, see www.iste.co.uk/broseta/hydrates1.zip*

Another, more rapid, method consists of performing two melting runs at different heating rates in order to extrapolate the results to isothermal condition. As shown in Figure 4.12, thermograms recorded at 0.2 and 0.5 K/min with the same sample are used to extrapolate the final melting temperature, that is the temperature at the inflexion point of the peak [KOU 06, KOU 07, LIN 13]. Uncertainties of the order of 0.2 K may be obtained by this method, taking far less time than the stepwise method. Logically, the two methods (extrapolated final temperature and onset point) give the same result for a constant-temperature melting (eutectic in Figure 4.12). The extrapolated method is sometimes impracticable, especially if two peaks are overlapping as in Figure 4.10. In such a case, the stepwise method should be used instead.

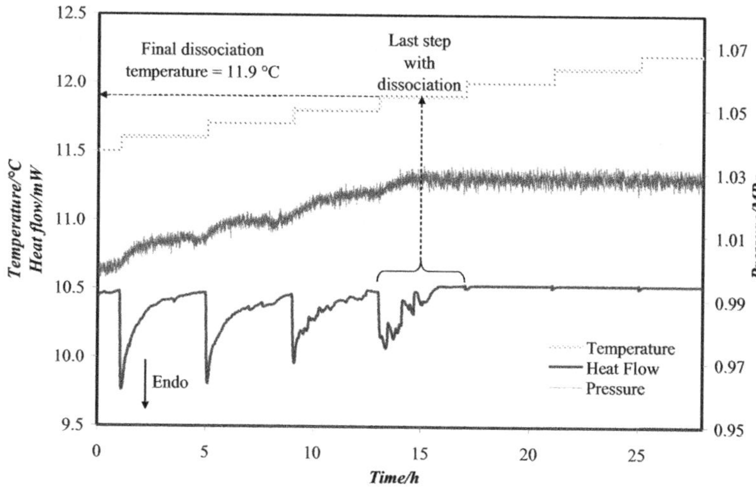

Figure 4.11. *Measurement of the phase change temperature using the stepwise method, H_2O + CO_2 + CH_4 + tetrahydropyran (unpublished results; heat flow signal was rescaled). For a color version of this figure, see www.iste.co.uk/broseta/hydrates1.zip*

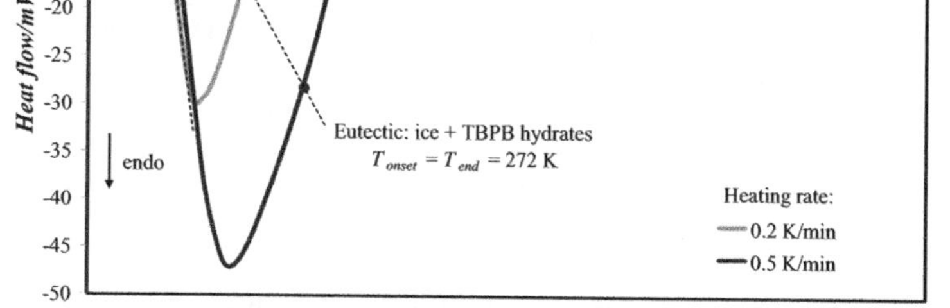

Figure 4.12. *The extrapolated method is used to measure the melting temperatures H_2O + 15 wt.% tetrabutylphosphonium bromide (unpublished results). The eutectic mixture melts at a constant temperature of 272 K; excess hydrates undergo progressive melting in the range 272–280.5 K. For a color version of this figure, see www.iste.co.uk/broseta/hydrates1.zip*

These two methods are used, often in combination, for comparison and validation purposes for the determination of a large variety of hydrates' phase diagrams, as summarized in Table 4.1.

Hydrate forming species/additives	Application	Reference
CCl_3F, CH_4, natural gas (C1–C4 + CO_2 + N_2 mixture) with/without inhibitors ($NaCl/CaCl_2$)	Drilling fluids	[FOU 99, FOU 01, FOU 02b]
CO_2 CO_2 + tetrahydrofuran CO_2 + TBPB	Refrigeration	[DEL 06] [DEL 06, MAR 08, MAR 09] [MAY 10a, MAY 10b, MAY 11, LIN 13]
$CO_2 + N_2$ + TBAB $CO_2 + CH_4$ + TBPB $CO_2 + CH_4$ + TBPO	Gas separation	[LIN 08, DES 09] [SAL 16a] [SAL 16b]
$H_2 + C_3H_8$ H_2 + TBAOH H_2 + $TBABH_4$ H_2 + TBAC/TBPB	H_2 storage	[KAR 13] [KAR 14] [DES 10, FUK 15b]

TBAB, tetrabutylammonium bromide; TBAC, tetrabutylammonium chloride; TBAOH, tetrabutylammonium hydroxide; $TBABH_4$, tetrabutylammonium borohydride; TBMAC, tetrabutylmethylammonium chloride; TBPO, tributylphosphine oxide.

Table 4.1. *Systems studied using DTA/DSC for phase diagram determinations*

4.4. Measuring the heat of dissociation and heat capacity of gas hydrates

DSC is particularly well suited to measure heats of phase transition and heat capacities. The principle of these measurements is illustrated in Figure 4.13 for the general case of a pure substance.

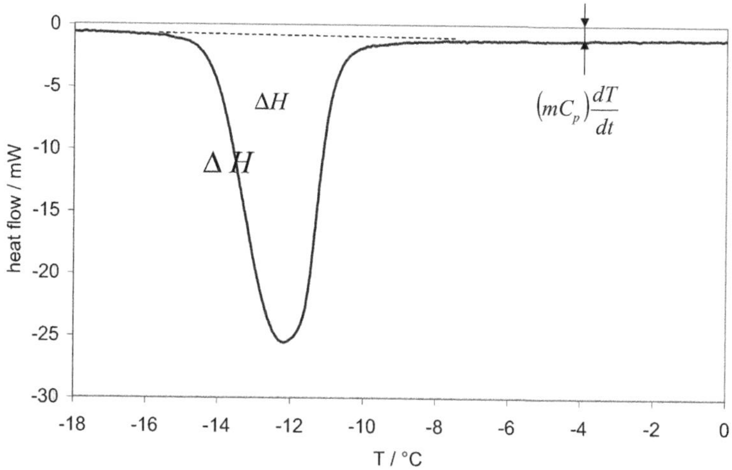

Figure 4.13. *Principle of the measurement of phase change enthalpy ΔH and heat capacities C_p by DSC. For a color version of this figure, see www.iste.co.uk/broseta/hydrates1.zip*

The phase change enthalpy results from the integration of the melting peak. The heat flow measured in the absence of phase change is related to the heat capacities of the phases present by:

$$\dot{Q} = \left(c_{p,M} + m_S C_{p,S} - c_{p,R}\right)\frac{dT}{dt} \quad [4.2]$$

where $c_{p,M}$ and $c_{p,R}$ refer to the heat capacities of the measurement and reference vessels, respectively, and $C_{p,S}$ to the massic heat capacity of the sample of mass m_S. To obtain accurate C_p measurements, a blank experiment must be performed with both vessels empty. The corresponding heat flow:

$$\dot{Q}_{\text{blank}} = \left(c_{p,M} - c_{p,R}\right)\frac{dT}{dt} \quad [4.3]$$

is then subtracted from the measurement to finally obtain:

$$C_{p,S} = \frac{\dot{Q} - \dot{Q}_{\text{blank}}}{m_S}\left(\frac{dT}{dt}\right)^{-1} \quad [4.4]$$

4.4.1. *Quantitative* in situ *hydrate formation*

In the case of gas hydrates, the main difficulty associated with these quantitative measurements arises from the uncertainty on the amount of hydrates present in the crucible. The first calorimetric experiments were carried out with samples of hydrates formed *ex situ* and then transferred to the calorimetric crucible [HAN 84, HAN 86a, HAN 86b, HAN 88]. The transfer of such unstable phases from a pressurized crystallization reactor to the calorimetric crucible is a delicate operation; it is in fact very difficult to precisely weigh the sample and ensure that no hydrates have dissociated during the filling and pressurizing of the crucible.

Forming the hydrates by directly reacting water and gas at proper pressure and temperature in the DSC crucible may appear more convenient but this procedure also bears large uncertainties. First, as there is no stirring inside the cell, the contact area between the liquid phase and the poorly soluble gas is limited. Second, hydrates crystallization usually requires large super cooling, it is therefore in competition with the much easier crystallization of metastable gasless solids, such as ice. Finally, because there is no means of analyzing or observing the sample inside the crucible, the proportion of the sample converted into hydrates after a certain time is not measurable.

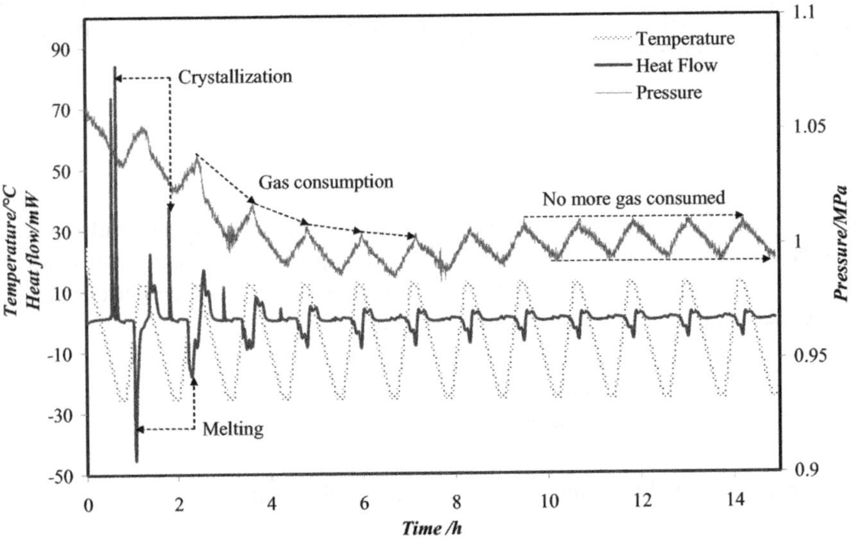

Figure 4.14. *Formation of hydrates during multicycle program (unpublished results). For a color version of this figure, see www.iste.co.uk/broseta/hydrates1.zip*

The best method to ensure complete conversion of water by *in situ* liquid–gas reaction consists of multicycle programming, as described in [DEL 06]. As can be seen in Figure 4.14, successive cooling and warming cycles allow the water consumption to be monitored by the decrease in the signals associated with the crystallization and melting of metastable phases. At the same time, the pressure evolution shows a consumption of gas at each cycle, thus evidencing an accumulation of hydrates. When the pressure and heat flow evolutions remain constant over successive cycles, this is evidence that any metastable phase has disappeared.

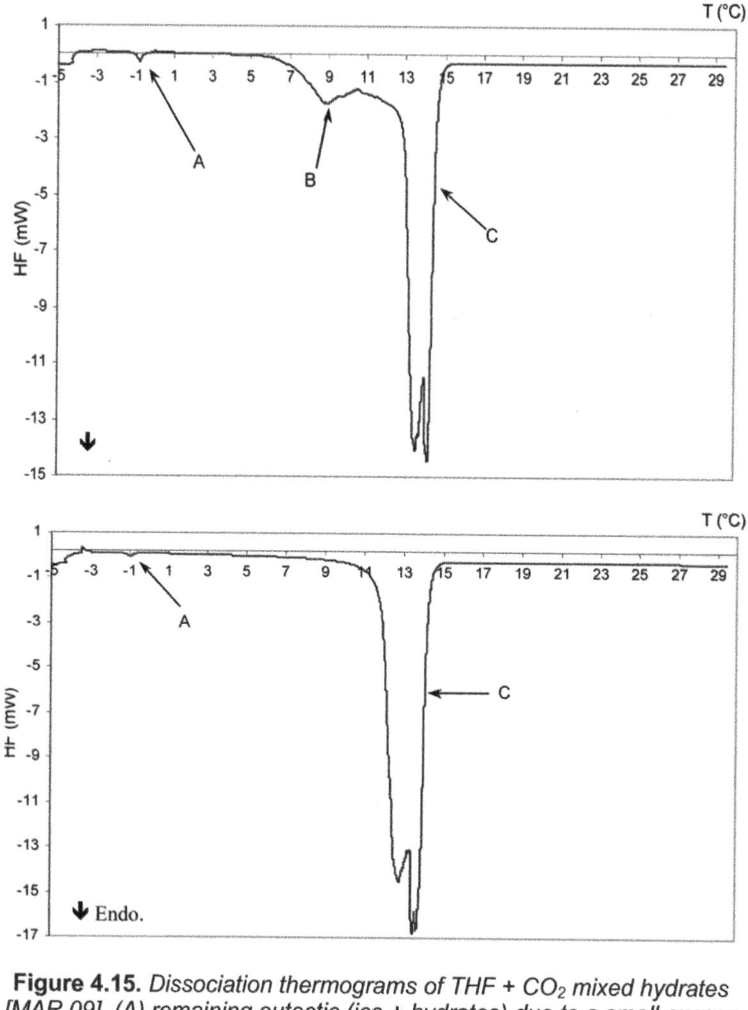

Figure 4.15. *Dissociation thermograms of THF + CO_2 mixed hydrates [MAR 09]. (A) remaining eutectic (ice + hydrates) due to a small excess of water; (B) metastable hydrates; (C) stable hydrates. For a color version of this figure, see www.iste.co.uk/ broseta/hydrates1.zip*

Note that metastable phases may yet subsist, at temperatures higher than the upper limit of the cycles. This is a common problem in the study of mixed hydrates formed with a gas and a thermodynamic promoter, such as THF [MAR 08] or peralkylonium salts [BOU 11]. For example, with THF + CO_2 + water at least three solid phases may appear: ice, gas-free THF hydrates (THF·17H_2O), and CO_2 + THF mixed hydrates (the phase of interest). But it must be considered that the latter is a non-stoichiometric phase with respect to CO_2. Its chemical formula can be expressed as: THF·17H_2O·$x$$CO_2$, where the variable x is maximal when the cage occupancy reaches its nominal value at the given gas pressure, thus ensuring maximal stabilization of the clathrate structure. Because pure THF hydrate is stable even in the absence of any gas, there is a continuous range of metastable solids that can be present in a sample. These metastable phases will dissociate in a large range of temperatures and, as the dissociation enthalpy of hydrates depends on cage occupancy, the result will be subject to large uncertainties. Figure 4.15 illustrates the difference between the melting of incompletely filled CO_2 + THF mixed hydrates and a well-formed dissociation peak suitable for accurate enthalpy determination. It often requires a very long cycling procedure to get this result, making the measurement of dissociation enthalpy of gas hydrates a long and difficult process.

4.4.2. *Indirect enthalpy measurement and gas content evaluation*

It is often easier to calculate the dissociation enthalpies of hydrates based on *p-T* equilibrium data by applying the Clapeyron equation:

$$\Delta_{\text{diss}} H = T \Delta_{\text{diss}} V \frac{dp}{dT} \qquad [4.5]$$

where $\Delta_{\text{diss}} V$ stands for the volume variation induced by the dissociation process:

$$\Delta_{\text{diss}} V = V_V + V_L - V_H \qquad [4.6]$$

In most cases, the volume of condensed phases is neglected and the volume variation is approximated as the volume of gas species released upon hydrate dissociation, thus leading to the Clausius–Clapeyron equation:

$$\Delta_{\text{diss}} H \approx -ZR \frac{d \ln p}{d(1/T)} \qquad [4.7]$$

The validity and accuracy of this approximation has been discussed [AND 03] as it takes into account neither the gas solubility in water nor the volume of liquid and hydrate phases. Another aspect of this expression is that it gives access to the dissociation enthalpy on the basis of one mole of gas. It is thus impossible to express the enthalpy on the basis of a given amount of hydrates, unless the precise composition of the hydrates is known. On the other hand, when measuring dissociation enthalpies using DSC, the result is expressed with respect to the amount of sample, that is the mass of water or solution placed in the DSC vessel. If now we consider that both methods give the same energy value, the calculated one being expressed per mole of gas and the measured one per mole of water, it is possible to approximate the hydrates composition by combining these two enthalpies [MAR 08]. If the hydrates formed by a given gas G are represented by the formula $G{\cdot}n_H H_2O$, the hydration number n_H may be approximated by:

$$n_H = \frac{n_{H_2O}}{n_G} \approx \frac{\Delta_{\text{diss}} H_{(C-C)}\left(\text{kJ}\times\text{mol}_G^{-1}\right)}{\Delta_{\text{diss}} H_{(\text{DSC})}\left(\text{kJ}\times\text{mol}_{H_2O}^{-1}\right)} \qquad [4.8]$$

where subscripts (*C-C*) and (DSC) refer to dissociation enthalpies determined using the Clausius–Clapeyron equation and DSC experiments, respectively. This approximation was applied to various (gas + promoter) mixed hydrates: CO_2 + THF [MAR 08]; CO_2 + tetrabutylammonium chloride or tetrabutylammonium nitrate or tetrabutylphosphonium bromide [MAY 10a]; H_2 + tetrabutylammonium chloride or tetrabutylammonium bromide or tetrabutylphosphonium bromide [DES 10]. Figure 4.16 gathers the Clapeyron diagrams of these systems. The compositions of some CO_2 + additive mixed hydrates calculated from equation [4.8] are reported in Figure 4.17. The extreme variability of gas uptake potential, ranging from 2 wt.% CO_2 trapped in some saline semiclathrates to more than 40 wt.% in THF clathrate hydrates, is illustrated.

The validity and uncertainty of these estimates have been discussed in [DES 10]. The major source of error is the assumption of negligible volume of condensed phases, especially for hydrates containing low amounts of gas and for relatively soluble gases. In these cases, the amount of gas released by the hydrates' dissociation that remains dissolved in the liquid phase should be considered for better accuracy, as well as the volume expansion of the liquid due to gas dissolution.

4.4.3. *Heat capacity measurement*

Few experimental data of heat capacities have been published for gas hydrates. This is probably due to the fact that heat capacities are more difficult to measure and more prone to uncertainties than dissociation heats. For these measurements, the same tedious and time consuming process is required to assure quantitative formation of hydrates. In addition, the amplitude of the heat flow signal is very low and may thus be perturbed by tiny variations in the experimental conditions. The DSC device must be placed in a perfectly regulated thermal ambient and the experimental gas pressure must be strictly regulated. Moreover, being non-stoichiometric compounds, hydrates are subject to composition variations, not only with gas pressure as seen in Figure 4.17, but also, very likely, with temperature.

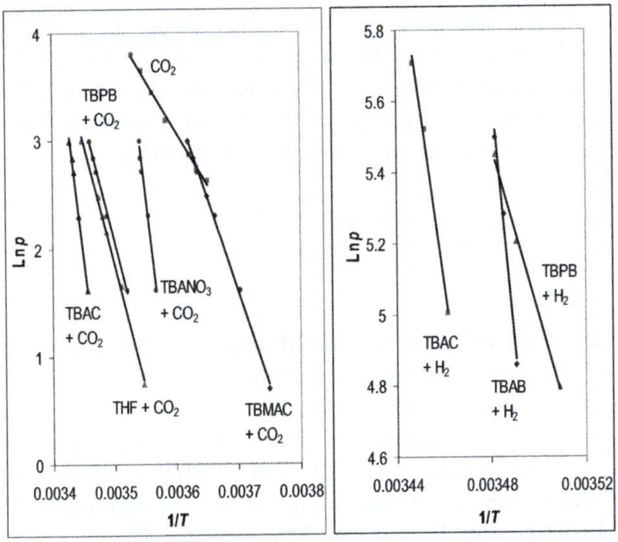

Figure 4.16. *Clapeyron diagrams of pure and mixed hydrates: left: CO_2, CO_2 + tetrahydrofuran (THF), CO_2 + tributylmethylammonium chloride (TBMAC), CO_2 + tetrabutylammonium chloride (TBAC), CO_2 + tetrabutylammonium bromide (TBAB), CO_2 + tetrabutylphosphonium bromide (TBPB); right: H_2 + TBAC, H_2 + TBAB, H_2 + TBPB. For a color version of this figure, see www.iste.co.uk/broseta/hydrates1.zip*

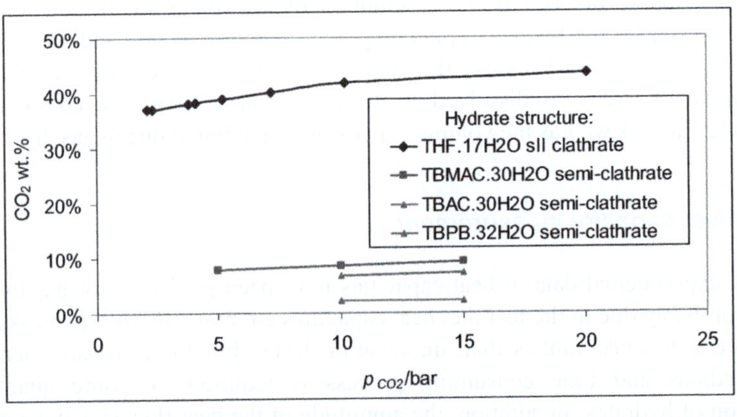

Figure 4.17. *Composition of various mixed hydrates estimated using enthalpy values. For a color version of this figure, see www.iste.co.uk/broseta/hydrates1.zip*

Figure 4.18 presents the heat capacities of THF hydrates and mixed THF + CO_2 hydrates measured using DSC [MAR 09]. These results are compared to heat capacities of ice and krypton hydrates. There is a good accordance in the results obtained by various authors for pure THF hydrates, with a heat capacity slightly lower than that of ice. In the presence of CO_2 at low pressure, the C_p per gram of hydrates is reduced, due to the increased molar mass of the solid, which contains between 35 and 45% in mass of CO_2 according to the evaluation presented in Figure 4.17. The very low C_p of heavy krypton hydrates may be noticed as a comparison. The C_p of CO_2 + THF hydrates then regularly increases with gas pressure, showing increasing interactions between the gas and the solid lattice. As can be seen for mixed hydrates at 0.95 MPa CO_2 pressure, the measurement may be subject to unexpected variations, probably due to the melting of remaining metastable phases.

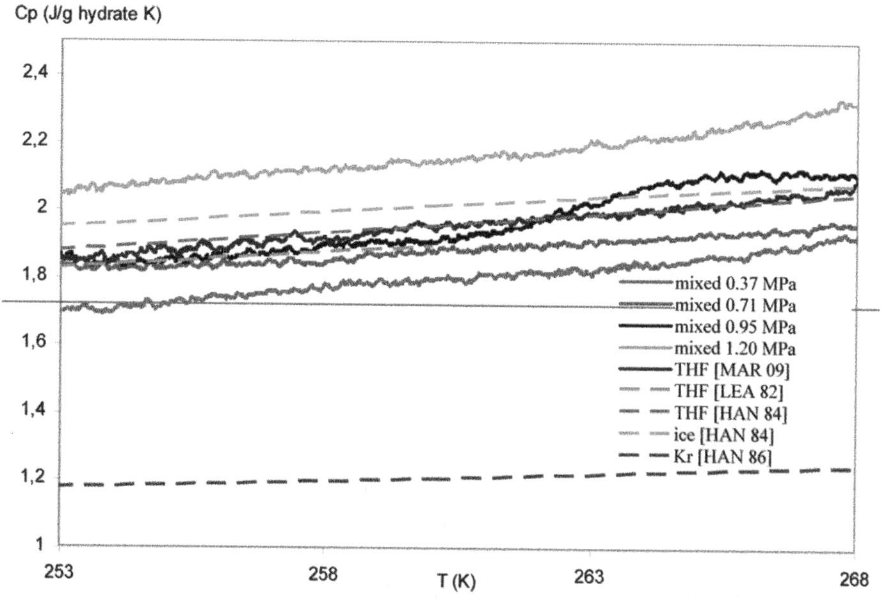

Figure 4.18. *Heat capacities of various hydrates compared to ice: krypton; tetrahydrofuran (THF); mixed CO_2 + THF at various CO_2 pressures [MAR 09]. For a color version of this figure, see www.iste.co.uk/broseta/hydrates1.zip*

To conclude this section, it is worth noting that the numerous experimental data collected were recently exploited in the development of novel modeling methods: the HLV equilibria and, for the first time, dissociation enthalpies of mixed THF + CO_2 hydrates were represented in [MAR 08, MAR 09]. The modeling of semiclathrate hydrates of (TBAB + CO_2) was proposed in [PAR 11] and then successfully extended to a variety of (salt + gas) couples in [FUK 14, FUK15a, FUK15b].

From a practical point of view, the difficulty of quantitatively converting the water into hydrates in the sample holder was pointed out earlier as a major hindrance. For this reason, interesting progress is expected for hydrates calorimetric studies with the development of stirred high-pressure DSC and calorimetric vessels [MAR 13, PLA 13, TOR 15].

4.5. Measuring the kinetics of hydrate formation

DSC has been used as a tool to compare and assess kinetic inhibitors designed to prevent the formation of gas hydrates in oil and gas processing industries [HES 08, LAC 09, SHA 14]. In [DAV 10], measurements of the mass transfer resistance through a film of methane hydrate as a function of time obtained using HP-DSC were reported. However, the crystallization of small samples usually requires much longer induction times, or higher super cooling degrees, than in macroscopic reactors. With samples of a few tens of microliters in volume and no agitation to enhance gas–liquid mixing, hydrate crystallization in a DSC vessel requires super cooling often exceeding 20 K to take place at a reasonable rate. At such low temperatures, the crystallization of metastable ice and gas-free hydrates may not be avoided, requiring a multicycle program to reach quantitative formation as described above. For these reasons, DSC is not suitable for a realistic investigation of the kinetics of crystallization of hydrates in the most general case.

If the water phase is dispersed in the form of a water-in-oil emulsion, DSC is the ideal tool. In such cases, each water droplet behaves as a tiny individual reactor. The huge area of interface between the water phase and the gas-rich oil phase compensates the size effect, which is thus less limiting. In addition, the lack of stirring is balanced by the high solubility of gas in the oil phase. Finally, the heat release measured by the DSC is a more sensitive and immediate way of detecting hydrates formation than the pressure drop due to gas consumption. As an example, the kinetics of model hydrate $CCl_3F\cdot 17H_2O$ formation in water-in-oil emulsions, which was first investigated using dielectric spectroscopy [JAK 96], has been thoroughly characterized by means of ambient pressure DSC [FOU 99, FOU 01, CLA 02, FOU 02a, FOU 02b, FOU 06], including for its emulsion and stability properties [DAL 01c, CLA 05].

Gas hydrate formation in water-in-oil emulsions is of industrial interest, for example in the drilling industry where more and more complex fluids are used as “drilling muds”. Drilling fluids have to comply with increasingly severe requirements as drilling conditions are getting more difficult: deep off shore drilling, directional drilling, etc. Gas kicks in deep offshore wells during drilling operations are a severe safety issue for surface equipment and operators [BAR 89, LAI 89, OUA 92, EBE 97].

Following the model studies cited above, pressure-controlled DSC was first applied to investigate the thermodynamic conditions for the potential formation of gas hydrates in the water phase of oil-based drilling fluids [DAL 00, DAL 01a, DAL 01b, DAL 02a, DAL 02b, c, DAL 02d, DAL 02e, DAL 03, KHA 03, KHA 04]. These studies demonstrated the possibility of methane or natural gas hydrate formation, in spite of the high concentration of thermodynamic inhibitors used to prevent their occurence [DAL 04, DAL 05]. In subsequent studies, the kinetics of formation of methane hydrates was investigated in a pressure range that enabled the conditions of very deep offshore operations to be simulated (up to 40 MPa, or a height of the water column of 4,000 m) [HAM 06, DAL 06, DAL 09]. Figure 4.19 presents thermograms obtained during the isothermal formation of methane hydrates in a water-in-oil emulsion at two different CH_4 pressures. The (dispersed) water phase in the emulsion was a concentrated $CaCl_2$ brine similar to those used in deep offshore drilling fluids. The oil phase and emulsifying system were also representative of industrial oil-based fluids. For each pressure, the experiments were repeated at different temperatures. The subcooling degree ΔT was calculated as the difference between the operating temperature and the HLV equilibrium temperature.

It can be observed that the hydrate formation peaks exhibit a regular, symmetric shape characteristic of a normal distribution function and that the peak width strongly depends on ΔT. This was interpreted as a confirmation that the crystallization of hydrates in each droplet takes place independently, the global formation process being thus randomly distributed. This complies with the hypothesis of a primary nucleation in each droplet considered as an independent reactor. A model, based on a normal distribution of induction times, coupled to a crystal growth equation applied locally to each droplet of the population, was proposed to represent these results [DAL 09].

A recent study [SEM 15] on methane hydrate crystallization in a water-in-crude oil emulsion evidenced a more complex behavior than in oil-based drilling fluids. Contrary to the symmetric patterns shown in Figure 4.19, these authors observed that hydrate formation in the studied emulsions occurred at different times within the sample and that each crystallization episode seemed to obey a collective process. They concluded that local secondary nucleation, probably induced by the contact of a drop of water with the neighboring hydrate particle, was responsible for this phenomenon. Given the great differences of compositions of the studied systems, these opposite results are not surprising.

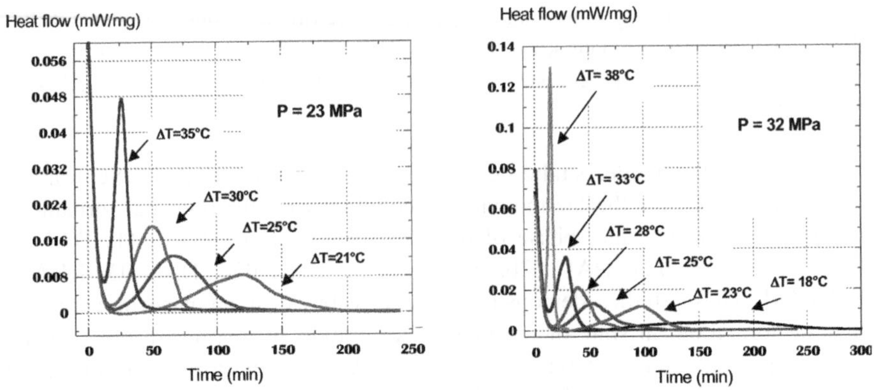

Figure 4.19. *Effect of subcooling degree ΔT on the kinetics of formation of methane hydrate in a water-in-oil emulsion at CH_4 pressures of 23 MPa (left) and 32 MPa (right). For a color version of this figure, see www.iste.co.uk/broseta/hydrates1.zip*

Therefore, it should be emphasized that DSC is a unique tool for the study of crystallization phenomena in complex fluids as it allows questions that could not be resolved by any other simple technique to be easily elucidated. The principles of emulsion microcalorimetry have been presented in detail in [DAL 01c]. DSC was successfully used by various researchers to study the complex behavior of dispersed water + (oil and gas) systems in hydrate-forming conditions, which is a hot topic in connection with flow assurance in the oil industry. Hydrate agglomeration in oil-based fluids was investigated in [PAL 05] and [COL 09]. The effect of hydrate formation and dissociation on the stability of water-in-oil emulsions was presented in [DAL 05, LAC 05]. In these works, HP-DSC was used to evidence that repeated cycles of formation/dissociation of methane hydrate destabilize water-in-oil emulsions. A model for hydrate-induced destabilization was proposed in [LAC 05].

4.6. Conclusion

DTA and DSC have long been considered as powerful techniques, yet fast and simple, for determining thermodynamic properties (melting points, phase diagrams, enthalpy change, heat capacities, etc.), thermochemical properties (heats of reaction, thermal decomposition, process safety) and kinetic behaviors. This diversity of possibilities has been widely employed for characterizing pharmaceuticals, alloys, energetic materials, polymers, petroleum and many other kinds of materials. In the case of gas hydrates, however, a limited number of researchers have applied calorimetric methods. In this chapter, we tried to present some aspects of the possible applications of thermal and calorimetric analysis, especially high-pressure DSC, to the study of gas hydrates.

As a matter of fact, when gas hydrates are concerned, measurements often reveal to be neither as simple, nor as fast as anticipated. All the limitations of DSC that have been pointed out, such as the absence of stirring, the very small sample volume or the impossibility of observing the vessel content, impose the use of new or unusual methods. The complex behavior of multiphase systems requires not only rigorous interpretation skills and a good insight into "well known" and commonly admitted principles, but also a good sense of the observation combined with some imagination to adapt these principles to such complicated systems. But this would be of no use without respecting the basics: a properly calibrated device and a suitable programming strategy. For all these reasons, the methodological description has been a major part of the above presentation. Intended to help any scientist willing to start calorimetric studies on hydrates, this chapter might thus also provide useful advice to experienced calorimetry specialists.

4.7. Bibliography

[AND 03] ANDERSON G.K., "Enthalpy of dissociation and hydration number of carbon dioxide hydrate from the Clapeyron equation", *Journal of Chemical Thermodynamics*, vol. 35, pp. 1171–1183, 2003.

[BAR 89] BARKER J.W., GOMEZ R.K., "Formation of hydrates during deepwater drilling operations", *Journal of Petroleum Technology*, vol. 41, pp. 297, 1989.

[BEN 11] BEN ATTOUCHE SFAXI I., Etude thermodynamique d'un procédé de captage du CO_2 par formation d'hydrates appliquée aux fumées de postcombustion, PhD Thesis, Mines ParisTech, France, July 13, 2011.

[BOU 11] BOUCHAFAA W., Mesure et modélisation des conditions de dissociation d'hydrates de gaz stabilisés en vue de l'application au captage du CO_2, PhD Thesis, Ecole Polytechnique, France, November 22, 2011.

[BRO 98] BROWN M.E., GALLAGHER P.K. (eds), *Handbook of Thermal Analysis and Calorimetry. Vol. 1: Principles and Practice*, Elsevier Science B.V., Amsterdam, 1998.

[CAL 56] CALVET E., PRAT H., *Microcalorimétrie; applications physico-chimiques et biologiques*, Masson, Paris, 1956.

[CAL 58] CALVET E., PRAT H., *Récents progrès en microcalorimétrie*, Dunod, Paris, 1958.

[CLA 02] CLAUSSE D., DALMAZZONE D., FOUCONNIER B. *et al.*, "Etude de la cinétique de formation de divers hydrates dans les émulsions eau dans huile", *Third World Congress on Emulsions*, Lyon, France, September 24–27, 2002.

[CLA 05] CLAUSSE D., GOMEZ F., PEZRON I. *et al.*, "Morphology characterization of emulsions by differential scanning calorimetry", *Advances in Colloid and Interface Science*, vol. 117, pp. 59–74, 2005.

[COL 09] COLOMBEL E., GATEAU P., BARRE F. *et al.*, "Discussion of agglomeration mechanisms between hydrate particles in water in oil emulsions", *Oil & Gas Science and Technology – Rev. IFP*, vol. 64, no. 5, pp. 629–636, 2009.

[DAL 00] DALMAZZONE C., DALMAZZONE D., HERZHAFT B., "Differential scanning calorimetry: a new technique to characterize hydrate formation in drilling muds", *Society of Petroleum Engineers Annual Technical Conference and Exhibition*, Dallas, Texas, October 1–4, 2000.

[DAL 01a] DALMAZZONE D., LACHET V., CLAUSSE D., "Caractérisation des équilibres de dissociation des hydrates dans les milieux complexes par les techniques PVT et DSC", *32èmes Journées de Calorimétrie et d'Analyse Thermique*, Hammamet, Tunisie, May 12–14, 2001.

[DAL 01b] DALMAZZONE D., KHARRAT M., LACHET V., "Etude des hydrates de méthane et de gaz naturel dans les émulsions eau dans huile par PVT et DSC", *Colloque sur l'apport de l'analyse thermique et de la calorimétrie dans les industries chimiques, pharmaceutiques et cosmétiques*, Lyon, France, November 14–16, 2001.

[DAL 01c] DALMAZZONE C., CLAUSSE D., "Microcalorimetry" in SJÖBLOM J.E. (ed.), *Encyclopedic Handbook of Emulsion Technology*, Marcel Decker, New York, 2001.

[DAL 02a] DALMAZZONE D., KHARRAT M., LACHET V. *et al.*, "DSC and PVT measurements of methane and trichlorofluoromethane hydrate dissociation equilibria in highly concentrated calcium chloride solutions and water-in-oil emulsions", *Journal of Thermal Analysis and Calorimetry*, vol. 70, pp. 493–505, 2002.

[DAL 02b] DALMAZZONE D., KHARRAT M., CLAUSSE D. *et al.*, "Thermodynamic study of several hydrates in dispersed systems using differential scanning calorimetry", *4th International Conference on Gas Hydrates*, Yokohama, Japan, May 19–23, 2002.

[DAL 02c] DALMAZZONE C., HERZHAFT B., DALMAZZONE D., "Characterization of hydrate formation in drilling muds using differential scanning calorimetry (DSC)", *4th International Conference on Gas Hydrates*, Yokohama, Japan, May 19–23, 2002.

[DAL 02d] DALMAZZONE C., HERZHAFT B., DALMAZZONE D. *et al.*, "Etude par DSC de la formation d'hydrates de gaz dans les fluides de forage à base d'huile", *Third World Congress on Emulsions*, Lyon, France, September 24–27, 2002.

[DAL 02e] DALMAZZONE D., DALMAZZONE C., HERZHAFT B., "Differential scanning calorimetry: a new technique to characterize hydrate formation in drilling muds", *SPE Journal*, vol. 7, no. 2, pp. 196–202, 2002.

[DAL 03] DALMAZZONE C., HERZHAFT B., ROUSSEAU L. *et al.*, "Prediction of gas hydrates formation with DSC technique", *SPE Annual Technical Conference and Exhibition*, 5–8 October, 2003.

[DAL 04] DALMAZZONE D., CLAUSSE D., DALMAZZONE C. *et al.*, "The stability of methane hydrates in highly concentrated electrolyte solutions by differential scanning calorimetry and theoretical computation", *The American Mineralogist*, vol. 89, nos. 8–9, pp. 1183–1191, 2004.

[DAL 05] DALMAZZONE D., HAMED N., CLAUSSE D. *et al.*, "The use of DSC in the study of the thermodynamics and kinetics of formation of model and gas hydrates", *5th International Conference on Gas Hydrates*, Trondheim, Norway, June 13–16, 2005.

[DAL 06] DALMAZZONE D., HAMED N., DALMAZZONE C. *et al.*, "Application of high pressure DSC to the kinetics of formation of methane hydrate in water-in-oil emulsion", *Journal of Thermal Analysis and Calorimetry*, vol. 85, no. 2, pp. 361–368, 2006.

[DAL 09] DALMAZZONE D., HAMED N., DALMAZZONE C. "DSC measurements and modelling of the kinetics of methane hydrate formation in water-in-oil emulsion", *Chemical Engineering Science*, vol. 64, pp. 2020–2026, 2009.

[DAV 10] DAVIES S.R., LACHANCE J.W., SLOAN E.D. *et al.*, "High-pressure differential scanning calorimetry measurements of the mass transfer resistance across a methane hydrate film as a function of time and subcooling", *Industrial & Engineering Chemistry Research*, vol. 49, no. 23, pp. 12319–12326, 2010.

[DEL 06] DELAHAYE A., FOURNAISON L., MARINHAS S. *et al.*, "Effect of THF on equilibrium pressure and dissociation enthalpy of CO_2 hydrates applied to secondary refrigeration", *Industrial & Engineering Chemistry Research*, vol. 45, pp. 391–397, 2006.

[DES 09] DESCHAMPS J., DALMAZZONE D., "Dissociation enthalpies and phase equilibrium for TBAB semi–clathrate hydrates of N_2, CO_2, N_2 + CO_2 and CH_4 + CO_2", *Journal of Thermal Analysis and Calorimetry*, vol. 98, pp. 113–118, 2009.

[DES 10] DESCHAMPS J., DALMAZZONE D., "Hydrogen storage in semiclathrate hydrates of tetrabutyl ammonium chloride and tetrabutyl phosphonium bromide", *Journal of Chemical & Engineering Data*, vol. 55, no. 9, pp. 3395–3399, 2010.

[DEV 99] DEVARACONDA S., GROYSMAN A., MYERSON S. "THF-water hydrate crystallization: an experimental investigation", *Journal of Crystal Growth*, vol. 204, pp. 525–538, 1999.

[EBE 97] EBELTOFT H., YOUSIF M., SOERGAARD E., "Hydrate control during deep water drilling: overview and new drilling fluids", *SPE 38567, SPE Annual Technical Conference and Exhibition*, San Antonio, Texas, October 5–8, 1997.

[FOU 99] FOUCONNIER B., LEGRAND V., KOMUNJER L. *et al.*, "Formation of trichlorofluoromethane hydrate in w/o emulsions studied by differential scanning calorimetry", *Progress in Colloid and Polymer Sciences*, vol. 112, pp. 105–108, 1999.

[FOU 01] FOUCONNIER B., MANISSOL Y., DALMAZZONE D. *et al.*, "Study of trichlorofluoromethane hydrate formation in W/O emulsions: dissociation energy and equilibria with salt + water solutions", *7èmes Journées Européennes Thermodynamique Contemporaine*, Mons, Belgium, August 27–31, 2001.

[FOU 02a] FOUCONNIER B., MANISSOL Y., DALMAZZONE D. *et al.*, "Determination of trichlorofluoromethane hydrate dissociation energy using the shell model", *Entropie*, vol. 38, no. 239/240, pp. 72–77, 2002.

[FOU 02b] FOUCONNIER B., Etude par calorimétrie de la formation de l'hydrate de trichlorofluorométhane en émulsion: système modèle pour l'étude des hydrates de gaz, PhD Thesis, University of Technology of Compiègne, France, July 18, 2002.

[FOU 04] FOURNAISON L., DELAHAYE A., CHATTI I. *et al.*, "CO_2 hydrates in refrigeration processes", *Industrial & Engineering Chemistry Research*, vol. 43, no. 20, pp. 6521–6526, 2004.

[FOU 06] FOUCONNIER B., KOMUNJER L., OLLIVON M. *et al.*, "Study of CCl_3F hydrate formation and dissociation in W/O emulsion by differential scanning calorimetry and X-ray diffraction", *Fluid Phase Equilibria*, vol. 250, pp. 76–82, 2006.

[FOU 15] FOURNAISON L., CLAIN P., DELAHAYE A. *et al.*, Matériau de stockage d'énergie par hydrate mixte, French Patent no. 3 008 709, 2015.

[FUK 14] FUKUMOTO A., PARICAUD P., DALMAZZONE D. *et al.*, "Modeling the dissociation conditions of carbon dioxide + TBAB, TBAC, TBAF, and TBPB semiclathrate hydrates", *Journal of Chemical and Engineering Data*, vol. 59, no. 10, pp. 3193–3204, 2014.

[FUK 15a] FUKUMOTO A., SALES SILVA L.P., PARICAUD P. *et al.*, "Modeling of the dissociation conditions of H_2 + CO_2 semiclathrate hydrate formed with TBAB, TBAC, TBAF, TBPB, and TBNO3 salts. Application to CO_2 capture from syngas", *International Journal of Hydrogen Energy*, vol. 40, no. 30, pp. 9254–9266, 2015.

[FUK 15b] FUKUMOTO A., DALMAZZONE D., PARICAUD P. *et al.*, "Experimental measurements and modeling of the dissociation conditions of tetrabutylammonium chloride semiclathrate hydrates in the presence of hydrogen", *Journal of Chemical & Engineering Data*, vol. 60, no. 2, pp. 343–350, 2015.

[GRA 68] GRAY A.P., PORTER R.J., JOHNSON J.F., "A simple generalized theory for the analysis of dynamic thermal measurement", in *Analytical Calorimetry*, Plenum Press, New York, vol. 1, 1968.

[HAM 06] HAMED N., Etude de la cinétique de formation des hydrates de méthane dans les fluides de forage off-shore par analyse calorimétrique différentielle haute pression, PhD Thesis, Ecole des Mines de Paris, France, November 10, 2006.

[HAN 84] HANDA Y.P., HAWKINS R.E., MURRAY J.J., "Calibration and testing of a Tian–Calvet heat-flow calorimeter. Enthalpies of fusion and heat capacities for ice and tetrahydrofuran hydrate in the range 85 to 270 K", *Journal of Chemical Thermodynamics*, vol. 16, pp. 623–632, 1984.

[HAN 86a] HANDA Y.P., "Calorimetric determinations of the compositions, enthalpies of dissociation, and heat capacities in the range 85 to 270 K for clathrate hydrates of xenon and krypton", *Journal of Chemical Thermodynamics*, vol. 18, pp. 891–902, 1986.

[HAN 86b] HANDA Y.P., "Compositions, enthalpies of dissociation, and heat capacities in the range 85 to 270 K for clathrate hydrates of methane, ethane, and propane, and enthalpy of dissociation of isobutane hydrate as determined by heat-flow calorimeter", *Journal of Chemical Thermodynamics*, vol. 18, p. 915, 1986.

[HAN 88] HANDA Y.P., "A calorimetric study of naturally occurring gas", *Industrial & Engineering Chemistry Research*, vol. 27, pp. 872–874, 1988.

[HAN 89] HANDA Y.P., YAMAMURO O., OGUNI M. *et al.*, "Low temperature heat capacities of xenon and krypton clathrate hydrates", *Journal of Chemical Thermodynamics*, vol. 21, pp. 1249–1262, 1989.

[HES 08] HESTER K.C., DAVIES S.R., LACHANCE J.W. *et al.*, "Hydrate nucleation measurements using high pressure differential scanning calorimetry", *Proceedings of the 6th International Conference on Gas Hydrates (ICGH 2008)*, Vancouver, British Columbia, Canada, July 6–10, 2008.

[JAK 96] JAKOBSEN T., SJÖBLOM J., RUOFF P., "Kinetics of gas hydrate formation in W/O emulsions. The model system trichlorofluoromethane/water/non-ionic surfactant studied by means of dielectric spectroscopy", *Colloids and Surfaces*, vol. 112, p. 73, 1996.

[KAN 01] KANG S.P., LEE H., RYU B.J., "Enthalpies of dissociation of clathrate hydrates", *Journal of Chemical Thermodynamics*, vol. 33, no. 5, pp. 513–521, 2001.

[KAR 13] KARIMI A., Synthèse et caractérisation thermodynamique d'hydrates de gaz contenant de l'hydrogène, PhD Thesis, Ecole Polytechnique, France, November 22, 2013.

[KAR 14] KARIMI A., DOLOTKO O., DALMAZZONE D., "Hydrate phase equilibria data and hydrogen storage capacity measurement of the system H_2 + tetrabutylammonium hydroxide + H_2O", *Fluid Phase Equilibria*, vol. 361, pp. 175–180, 2014.

[KHA 03] KHARRAT M., DALMAZZONE D., "Experimental determination of stability conditions of methane hydrate in aqueous calcium chloride solutions using high pressure differential scanning calorimetry", *Journal of Chemical Thermodynamics*, vol. 35, pp. 1489–1505, 2003.

[KHA 04] KHARRAT M., Etude des conditions de formation et de stabilité des hydrates de gaz dans les fluides de forage, PhD Thesis, Ecole des Mines de Paris, France, October 26, 2004.

[KOH 98] KOH C.A., WESTACOTT R.E., HURACHAND K. *et al.*, "Low dosage natural gas hydrate inhibitor evaluation, *Proceedings of the International Gas Research Conference*, San Diego, California, vol. 1, pp. 194–200, 1998.

[KOU 06] KOUSKSOU T., JAMIL A., ZERAOULI Y. *et al.*, "DSC study and computer modelling of the melting process in ice slurry", *Thermochimica Acta*, vol. 448, pp. 123–129, 2006.

[KOU 07] KOUSKSOU T., JAMIL A., ZERAOULI Y. *et al.*, "Equilibrium liquidus temperatures of binary mixtures from differential scanning calorimetry", *Chemical Engineering Science* vol. 62, pp. 6516–6523, 2007.

[LAC 08] LACHANCE J.W., SLOAN E.D., KOH C.A., "Effect of hydrate formation/dissociation on emulsion stability using DSC and visual techniques", *Chemical Engineering Science*, vol. 63, pp. 3942–3947, 2008.

[LAC 09] LACHANCE J.W., SLOAN E.D., KOH C.A., "Determining gas hydrate kinetic inhibitor effectiveness susing emulsions", *Chemical Engineering Science*, vol. 64, pp. 180–184, 2009.

[LAF 12] LAFOND P.G., OLCOTT K.A., SLOAN E.D. *et al.*, "Measurements of methane hydrate equilibrium in systems inhibited with NaCl and methanol", *Journal of Chemical Thermodynamics*, vol. 48, pp. 1–6, 2012.

[LAI 89] LAI D.T., DZIALOWSKI A.K., "Investigation of natural gas hydrates in various drilling fluids", *SPE/IADC 18637, SPE/IADC Drilling Conference*, New-Orleans, Louisiana, February 28–March 3, 1989.

[LEA 82] LEAIST D.G., MURRAY J.J., POST M.L. *et al.*, "Enthalpies of decomposition and heat capacities of ethylene oxide and tetrahydrofuran hydrates", *Journal of Physical Chemistry*, vol. 86, pp. 4175–4178, 1982.

[LEP 04] LE PARLOUER P., DALMAZZONE C., HERZHAFT B. *et al.*, "Characterisation of gas hydrates formation using a new high pressure Micro-DSC", *Journal of Thermal Analysis and Calorimetry*, vol. 78, no. 1, pp. 165–172, 2004.

[LIE 90] LIEVOIS J.S., PERKINS R., MARTIN R.J. *et al.*, "Development of an automated high pressure heat flux calorimeter and its application to measure the heat of dissociation and hydrate numbers of methane hydrate", *Fluid Phase Equilibria*, vol. 59, pp. 73–97, 1990.

[LIN 08] LIN W., DELAHAYE A., FOURNAISON L., "Phase equilibrium and dissociation enthalpy for semi-clathrate hydrate of CO_2 + TBAB", *Fluid Phase Equilibria*, vol. 264, nos. 1–2, pp. 220–227, 2008.

[LIN 13] LIN W., DALMAZZONE D., FÜRST W. *et al.*, "Accurate DSC measurement of phase transition temperature in the TBPB-water system", *Journal of Chemical Thermodynamics*, vol. 61, pp. 132–137, 2013.

[LIN 14] LIN W., DALMAZZONE D., FÜRST W. *et al.*, "Thermodynamic properties of semiclathrate hydrates formed from TBAB + TBPB + water and CO_2 + TBAB + TBPB + water systems", *Fluid Phase Equilibria*, vol. 372, pp. 63–68, 2014.

[MAC 75] MAC NAUGHTON J.L., MORTIMER C.T., *Differential Scanning Calorimetry*, Perkin Elmer, Butterworth, 1975.

[MAR 08] MARTINEZ M.C., DALMAZZONE D., FÜRST W. *et al.*, "Thermodynamic properties of THF + CO_2 hydrates in relation with refrigeration applications", *AIChE Journal*, vol. 54, no. 4, pp. 1088–1095, 2008.

[MAR 09] MARTINEZ M.C., Etude des coulis d'hydrates de CO_2 en présence d'additifs pour la réfrigération secondaire, PhD Thesis, Ecole des Mines de Paris, France, April 21, 2009.

[MAR 13] MARLIN L., PLANTIER F., MISSIMA D. *et al.*, "Prototype de cellule calorimétrique sous pression, agitée mécaniquement et avec contrôle dynamique de pression: application à la détermination de propriétés thermodynamiques d'hydrates de gaz pour la capture du CO_2", *44ème Journées de Calorimétrie et d'Analyse Thermique JCAT 2013*, Lyon, France, May 2013.

[MAY 10a] MAYOUFI N., DALMAZZONE D., FÜRST W. *et al.*, "CO_2 enclathration in hydrates of peralkyl-(ammonium/phosphonium) salts: stability conditions and dissociation enthalpies", *Journal of Chemical & Engineering Data*, vol. 55, no. 3, pp. 1271–1275, 2010.

[MAY 10b] MAYOUFI N., Caractérisation de coulis d'hydrates contenant du CO_2 appliqué à des systèmes frigorifiques, PhD Thesis, ENSTA ParisTech, France, December 20, 2010.

[MAY 11] MAYOUFI N., DALMAZZONE D., DELAHAYE A. *et al.*, "Experimental data on phase behavior of simple TBPB and mixed CO_2 + TBPB semiclathrate hydrates", *Journal of Chemical & Engineering Data*, vol. 56, no. 6, pp. 2987–2993, 2011.

[MAY 12] MAYOUFI N., DALMAZZONE D., FÜRST W. *et al.*, "Phase behaviour of tri-n-butylmethylammonium chloride hydrates in the presence of carbon dioxide", *Journal of Thermal Analysis and Calorimetry*, vol. 109, pp. 481–486, 2012.

[OUA 92] OUAR H., CHA S.B., WILDEMAN T.R. *et al.*, "The formation of natural gas hydrates in water-based drilling fluids", *Transactions of the Institution of Chemical Engineers*, vol. 70A, pp. 48, 1992.

[PAL 05] PALERMO T., BORREGALES M., ARLA D. *et al.*, "Study of the agglomeration between hydrate particles in oil using differential scanning calorimetry (DSC)", *Proceedings of the Fifth International Conference on Gas Hydrates*, Trondheim, Norway, June 12–16, 2005.

[PAR 11] PARICAUD P. "Modeling the dissociation conditions of salt hydrates and gas semiclathrate hydrates: application to lithium bromide, hydrogen iodide, and tetra-n-butylammonium bromide plus carbon dioxide systems", *Journal of Physical Chemistry B*, vol. 115, no. 2, pp. 288–299, 2011.

[PLA 13] PLANTIER F., MARLIN L., MISSIMA D. *et al.*, "Development of a new type of high pressure calorimetric cell, mechanically agitated and equipped with a dynamic pressure control system: Application to the characterization of gas hydrates", *Review of Scientific Instruments*, vol. 84, no. 12, p.125107, 2013.

[RUE 88] RUEFF R.M., SLOAN E.D., YESAYAGE V.F., "Heat capacity and heat of dissociation of methane hydrates", *AIChE Journal*, vol. 34, no. 9, pp. 1468–1476, 1988.

[SAB 99] SABBAH R. (ed.), "Reference materials for calorimetry and differential thermal analysis", *Thermochimica Acta (special issue)*, vol. 331, no. 2, pp. 121–166, 1999.

[SAL 16a] SALES SILVA L.P., DALMAZZONE D., STAMBOULI M. *et al.*, "Phase equilibria of semi-clathrate hydrates of tetra-n-butyl phosphonium bromide at atmospheric pressure and in presence of CH_4 and CO_2 + CH_4", *Fluid Phase Equilibria*, vol. 413, pp. 28–35, 2016.

[SAL 16b] SALES SILVA L.P., DALMAZZONE D., STAMBOULI M. *et al.*, "Phase Behavior of Simple Tetrabutylphosphine Oxide (TBPO) and Mixed Gas (CO_2, CH_4 and CO_2+CH_4) + TBPO Semiclathrate Hydrates" *Journal of Chemical Thermodynamics*, vol. 102, pp. 293-302, 2016.

[SEM 15] SEMENOV M.E., MANAKOV A.Y., SHITZ E.Y. *et al.*, "DSC and thermal imaging studies of methane hydrate formation and dissociation in water emulsions in crude oils", *Journal of Thermal Analysis and Calorimetry*, vol. 119, no. 1, pp. 757–767, 2015.

[TOR 15] TORRE J.-P., PLANTIER F., MARLIN L., High-pressure colorimetric measurement cell, US Patent no. US 20150204735 A1, 2015.

5

Thermodynamic Modeling of Solid–Fluid Equilibria: From Pure Solid Phases to Gas Semiclathrate Hydrates

5.1. Introduction

Solid phases are often encountered in the chemical industry. For some applications, solids are intentionally formed or dissociated during the industrial process, as they can be either final products of the process or intermediate products used for the separation or storage of compounds. For some other applications, the formation of solids is problematic, and engineers try to avoid it in the process, as solids can block pipes and unit operations. For all these cases, it is necessary to know the conditions of formation of solids. These conditions are directly related to the thermodynamic properties of solid and fluid phases. The development of reliable thermodynamic models for solid-liquid phase equilibrium calculation is then crucial for the design of many processes and separation units.

Clathrates represent one specific class of solid phases. Their structure is composed of a network of host molecules that form cages and can encapsulate gas molecules [SLO 08, CAR 09]. These gas molecules stabilize the solid structure as they interact with the host molecules composing the cages, and the gas content in the clathrate phase depends on temperature, pressure and global composition. Gas hydrates are a particular type of clathrates for which the host molecules are water molecules. Other types of clathrates involve organic molecules such as hydroquinone. Such clathrates are called organic clathrates and have been considered as promising materials to store or separate gas molecules [LEE 15, COU 16, CON 16].

Chapter written by Patrice PARICAUD.

In the field of gas hydrates, many thermodynamic models have been proposed to predict the state conditions under which gas hydrates can form or dissociate. Such models have often been used in the gas and petroleum industries: one major issue is the possible formation of gas hydrates in pipelines, and engineers need models that can predict the formation conditions of these hydrates. These models are able to quantitatively predict the effect of hydrate inhibitors on the melting points of gas hydrates. Most thermodynamic approaches for classical gas hydrates are based on the well-known van der Waals and Platteeuw (vdW-P) model proposed in 1959 [VAN 59]. This model was developed by considering several reasonable hypotheses such as the single occupancy of cages, the non-distortion of the hydrate lattice under pressure and the assumption that interactions between guest molecules and quantum effects are negligible. The solid–liquid equilibrium condition is written as the equality of the chemical potential of water in both phases. The properties of fluid phases are described by a thermodynamic model (equation of state (EoS) or activity coefficient model for the liquid phase), while the chemical potential of water in the hydrate phase is written as the sum of two terms: the first term is the Gibbs free energy of an empty lattice of host water molecules, and the second term is the vdW-P term that quantifies the decrease in free energy due to the presence of gas molecules encapsulated in the cages of the hydrate. Many versions of the vdW-P model have been proposed in the literature and the differences between these versions correspond to different assumptions made or different thermodynamic models used to describe the properties of the fluid phases. The reader is directed to reviews by Sloan and Koh [SLO 08] and Caroll [CAR 09] on applications of the vdW-P model in the literature. We can also refer to other models for hydrates based on equations of state [YOK 05] that can deal with the three types of phases (gas, liquid and solid). One particularity of gas hydrates compared to classical solids is the fact that the composition of the gas hydrate varies dramatically with the pressure and temperature of the system. This feature is taken into account within the vdW-P term that is very similar to the Langmuir adsorption model.

Gas semiclathrate hydrates represent another class of hydrates that have recently received much attention from both industrials and researchers, as they can be used as materials for gas separation [DUC 07, LI 09] and storage, as well as in cooling systems. The structure of gas semiclathrate hydrates has some common features with classical gas hydrates, as they also exhibit cages formed by water molecules, which can encapsulate gas molecules. Moreover, the gas composition in the semiclathrate hydrate phase also depends on pressure and temperature. However, the main difference is the fact that the structure of semiclathrate hydrates contains anions while large cages contain tetraalkyl ammonium or phosphonium cations. Moreover, semiclathrate hydrates can be stable at atmospheric pressure even without the presence of gas molecules. These hydrates behave as salt hydrates and exhibit different phases corresponding to different hydration numbers. Fowler *et al.* [FOW 40] were among the first to form semiclathrate hydrates of tetraalkymammonium salts. The solid–liquid phase diagrams of tetraalkylammonium halide salts + water binary systems can be rather complex and have been reviewed by Dyadin and Udashin [DYA 84] and Aladko *et al.* [ALA 03]. Gas semiclathrate hydrates of tetra-*n*-butylammonium bromide (TBAB) have been extensively studied because of the availability of TBAB.

While the thermodynamic modeling for classical gas hydrate is now well established, the theoretical developments for gas semiclathrate hydrates are rather recent. We can distinguish three main classes of approaches for these systems: the predicting tool proposed by Mohammadi *et al.* [MOH 10] and based on neural network algorithm, the model of Eslamimanesh *et al.* [ESL 12] and the approach proposed by Paricaud [PAR 11], which is based on the Gibbs free energy minimization. The aim of this chapter is to present the thermodynamic modeling of solid–liquid and solid–fluid equilibria with a focus on salt hydrates and semiclathrate hydrates. It is organized as follows: we first recall the expressions to solve a classical solid–liquid equilibrium problem by assuming that the solid phase is pure. We then present the modeling of solid–liquid phase equilibria in the presence of solid solutions and complex solids. We briefly remind the formalism used to write the chemical potential of species in electrolyte solutions, and discuss the phase equilibria of salt hydrates and semiclathrate hydrates. We finally recall the working expressions of the vdW-P model and show how it can be combined with the model of salt hydrate to describe the dissociation conditions of gas semiclathrate hydrates. Different versions of this model are discussed.

5.2. Solid–fluid equilibrium between a fluid mixture and a pure solid phase

5.2.1. *Solid–liquid equilibrium condition*

Before discussing the modeling of solid–liquid equilibria (SLE) in the presence of complex solids and semiclathrate hydrates, we first recall the usual procedure to calculate SLE for engineering applications [PRA 99, PRA 00]. This approach assumes that solid phases are pure. Let us consider a multicomponent liquid mixture of n_c compounds in equilibrium with a pure solid phase S_i^* only composed of molecules of type i (the symbol * denotes a pure phase through all this chapter). The determination of the melting temperature T of the pure solid phase S_i^* in equilibrium with a liquid mixture L at pressure P is equivalent to the determination of the solubility of molecule i in the liquid mixture at temperature T and pressure P. The condition of equilibrium can be obtained from the fact that the total Gibbs free energy of the system G is at its minimum value, i.e. the differential dG of the full system is equal to zero. The total Gibbs free energy G of the system can be expressed as the sum of the Gibbs free energy of the liquid mixture (G^L) and the Gibbs free energy of the pure solid *phase* S_i^* ($G^{S_i^*}$),

$$G = G^L + G^{S_i^*} , \qquad [5.1]$$

$$G^L = \sum_{j=1}^{n_c} n_j^L \mu_j^L , \qquad [5.2]$$

and

$$G^{S_i^*} = n_i^{S_i^*} \mu_i^{S_i^*} . \qquad [5.3]$$

where n_j^L, μ_j^L are the number of moles and chemical potential of component j in the liquid phase, and $n_i^{S_i^*}$, $\mu_i^{S_i^*}$ are the number of moles and chemical potential component i in the pure solid phase. Note that component j can either be a molecule or an ion. The condition of solid–liquid equilibrium correspond to $dG = 0$ at fixed T and P, which leads to

$$dG = \sum_{j=1}^{n_c} \mu_j^L dn_j^L + \mu_i^{S_i^*} dn_i^{S_i^*} = 0 . \qquad [5.4]$$

As the solid phase S_i^* is pure, all changes in moles dn_j^L are equal to zero for component $j \neq i$. As a result, equation [5.4] can be reduced to

$$\mu_i^L dn_i^L + \mu_i^{S_i^*} dn_i^{S_i^*} = 0 \cdot \qquad [5.5]$$

If there is no chemical reaction, the total number of moles of molecule i is conserved ($dn_i^L = -dn_i^{S_i^*}$) and equation [5.5] simplifies to

$$\left(\mu_i^L - \mu_i^{S_i^*}\right)dn_i^L = 0 \cdot \qquad [5.6]$$

Since dn_i^L can take any value, the necessary condition of solid–liquid equilibrium is

$$\mu_i^L = \mu_i^{S_i^*}, \qquad [5.7]$$

which is the well-known condition of the equality of the chemical potential of component i in the coexistent phases. This equation can be solved by expressing the chemical potentials in both phases. In the liquid phase, the chemical potential of component i is usually expressed as

$$\mu_i^L = \mu_i^{L*} + RT\ln(x_i\gamma_i), \qquad [5.8]$$

where R is the ideal gas constant; x_i and γ_i are the mole fraction and activity coefficient of component i in the liquid phase, and μ_i^{L*} is the chemical potential of pure liquid i at T and P. Since T is expected to be lower than the melting point of pure component i, μ_i^{L*} is actually the chemical potential of the supercooled pure liquid i [CHE 82]. Equation [5.7] can be written as

$$\Delta g_{dis} = \mu_i^L - \mu_i^{S_i^*} = \Delta g_i + RT\ln(x_i\gamma_i) = 0, \qquad [5.9]$$

where Δg_{dis} is the dissociation (fusion) Gibbs free energy that must be zero at equilibrium; $\Delta g_i = \mu_i^{L*} - \mu_i^{S_i^*}$ is only a function of T and P. We can express Δg_i at T, P with respect to a reference state at T_{ref}, P_{ref}, which is well known experimentally, by considering a thermodynamic cycle [PRA 99]. For the reference state, it is convenient to consider the experimental melting point $T_{m,i}$ of the pure component i at a reference pressure P_{ref}, which can be either the standard pressure (1 bar), the atmospheric (1 atm) or the triple point pressure. Since the pure liquid i is equilibrated with the pure solid at the melting point $T_{m,i}$ of component i

$$\Delta g_i(T_{m,i}, P_{ref}) = \mu_i^{L*}(T_{m,i}, P_{ref}) - \mu_i^{S_i^*}(T_{m,i}, P_{ref}) = 0, \qquad [5.10]$$

The temperature dependence of $\Delta g_i(T, P_{ref})$ is then expressed by using the exact thermodynamic relation (Gibbs–Helmholtz equation):

$$\left(\frac{\partial(\mu_i / T)}{\partial T}\right)_{P=P_0} = -\frac{h_i}{T^2}, \qquad [5.11]$$

where h_i is the partial molar enthalpy of component i at temperature T and pressure P_{ref}, which can be expressed as $h_i(T, P_{ref}) = h_i(T_{ref}, P_{ref}) + \int_{T_{ref}}^{T} c_{p,i}\, dT$, where $c_{p,i}$ is the molar heat capacity of component i. By assuming that $c_{p,i}$ is constant, h_i is a linear function of T, and the integration of equation [5.11] leads to

$$\begin{aligned} \frac{\mu_i(T)}{T} - \frac{\mu_i(T_{ref})}{T_{ref}} &= -\int_{T_{ref}}^{T} \frac{h_i}{T^2} dT = -\int_{T_{ref}}^{T} \frac{h_i(T_{ref}, P_{ref}) + c_{p,i}(T - T_{ref})}{T^2} dT \\ &= \frac{h_i(T_{ref}, P_{ref})}{T}\left(1 - \frac{T}{T_{ref}}\right) - c_{p,i}\left(\frac{T_{ref}}{T} - 1\right) + c_{p,i} \ln\left(\frac{T_{ref}}{T}\right) \end{aligned} \qquad [5.12]$$

By applying equation [5.12] to the chemical potentials of component i in the pure solid and liquid phases, and using $T_{ref} = T_{m,i}$, we can express $\Delta g_i(T, P_{ref})$ as

$$\begin{aligned} \frac{\Delta g_i(T, P_{ref})}{T} &= \frac{\Delta g_i(T_{ref}, P_{ref})}{T_{ref}} + \frac{\Delta h_i(T_{ref}, P_{ref})}{T}\left(1 - \frac{T}{T_{ref}}\right) + \Delta c_{p,i}\left(\frac{T_{ref}}{T} - 1\right) + \Delta c_{p,i} \ln\left(\frac{T_{ref}}{T}\right), \\ &= \frac{\Delta h_{m,i}}{T}\left(1 - \frac{T}{T_{m,i}}\right) - \Delta c_{p,i}\left(\frac{T_{m,i}}{T} - 1\right) + \Delta c_{p,i} \ln\left(\frac{T_{m,i}}{T}\right) \end{aligned} \qquad [5.13]$$

where $\Delta h_{m,i} = h_i^{L*} - h_i^{S_i^*}$ is the melting enthalpy of pure component i at $T_{m,i}$, P_{ref}. $\Delta c_{p,i} = c_{p,i}^{L*} - c_{p,i}^{S_i^*}$, where $c_{p,i}^{L*}$ and $c_{p,i}^{S_i^*}$ are the heat capacities of the pure liquid and solid phases. We can use the experimental values of $c_{p,i}^{L*}$ and $c_{p,i}^{S_i^*}$ at $T_{m,i}$, P_{ref}. The dependence of Δg_i with respect to pressure can be obtained by using the following thermodynamic relation,

$$\left(\frac{\partial \mu_i}{\partial P}\right)_T = v_i, \qquad [5.14]$$

where v_i is the partial molar volume of component i at T, P. By neglecting the dependence of v_i with respect to pressure (incompressible liquid and solid phases) and integrating equation [5.14], we can express Δg_i as $\Delta g_i(T,P) = \Delta g_i(T,P_{ref}) + \Delta v_i(P - P_{ref})$, and the solid–liquid equilibrium condition becomes

$$\frac{\Delta h_{m,i}}{T}\left(1-\frac{T}{T_{m,i}}\right) - \Delta c_{p,i}\left(\frac{T_{m,i}}{T}-1\right) + \Delta c_{p,i}\ln\left(\frac{T_{m,i}}{T}\right) + \frac{\Delta v_i}{T}\left(P-P_{ref}\right) + R\ln(x_i\gamma_i) = 0 \qquad [5.15]$$

where $\Delta v_i = v_i^{L*} - v_i^{S_i^*}$ is the difference between the molar volumes of pure component i in the pure liquid and solid phases at temperature T and pressure P_{ref}. Equation [5.15] is often used to estimate the solubility of a molecule in a given solvent or to predict solid–liquid phase diagrams in binary systems. By further assuming that $\Delta c_{p,i} = 0$ and $\gamma_i = 1$ (ideal solution), the solubility of molecule i at T and P_{ref} can be analytically estimated as

$$x_i = \exp\left(-\frac{\Delta h_{m,i}}{RT}\left(1-\frac{T}{T_{m,i}}\right)\right). \qquad [5.16]$$

It can be easily seen from equation [5.16] that the melting temperature of a pure solid phase in contact with a solution decreases as the mole fraction of component i in the liquid phase is decreased. It is recommended to use an accurate thermodynamic model for the liquid phase to correctly predict the activity coefficients, as large deviations from the solid–liquid experimental data can be observed if the ideal solution is assumed. It is also recommended to consider a non-zero $\Delta c_{p,i}$ term if the temperature range of the SLE curve is large. As an example, the solid–liquid phase diagram of the ethanol + acetone mixture (Figure 5.1) has been predicted by using equation [5.15] with two different thermodynamic models for the activity coefficients: the ideal solution and the UNIFAC model [FRE 75]. In this diagram, equation [5.15] has been used for both the pure ethanol and the pure acetone solid phases: the two corresponding SLE curves cross at the eutectic point. It can be seen that the UNIFAC model leads to a much better prediction of the experimental data than the ideal solution model, showing that the choice of the thermodynamic model for the liquid phase is crucial.

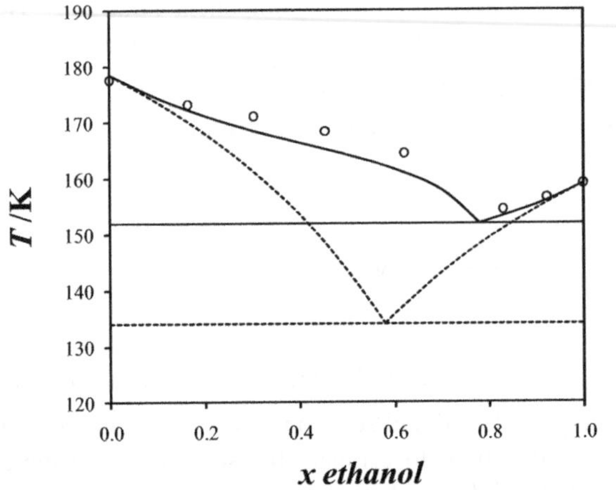

Figure 5.1. *Solid–liquid phase diagram of the ethanol + acetone mixture. The symbols denote the experimental data [SAP 29]. The dashed lines are the SLE curves predicted with the ideal solution model, and the solid lines are the SLE curves predicted with the UNIFAC thermodynamic model [FRE 75]. The horizontal lines are the calculated solid–solid–liquid three phase lines*

Equation [5.15] is suitable for excess Gibbs free energy (activity coefficient) models. However, most excess Gibbs free energy models do not depend on pressure and are only suitable for low and moderate pressures. At high pressures, it is recommended to use an EoS to predict the properties of the liquid phase. In this case, we can express equation [5.8] by considering the fugacity of component i instead of the activity coefficient. The chemical potential of component i in the liquid mixture can be expressed in terms of the fugacity f_i^L of component i in the liquid mixture as

$$\mu_i^L = \mu_i^0 + RT\ln(f_i^L / P_0), \qquad [5.17]$$

where μ_i^0 is only a function of T and P_0 =1 bar. The fugacity f_i^L is written as $f_i^L = Px_i\phi_i^L$, where ϕ_i^L is the fugacity coefficient of component i in the liquid mixture, calculated with the EoS. Equation [5.17] is also used for the pure supercooled liquid i, leading to

$$\mu_i^{L*} = \mu_i^0 + RT\ln(f_i^{L*} / P_0), \qquad [5.18]$$

where $f_i^{L*} = P\phi_i^{L*}$. By combining equations [5.8], [5.17] and [5.18], we can express the activity coefficient with respect to the fugacity coefficients as $\gamma_i = \phi_i^L / \phi_i^{L*}$, and equation [5.15] becomes

$$\frac{\Delta h_{m,i}}{T}\left(1-\frac{T}{T_{m,i}}\right)-\Delta c_{p,i}\left(\frac{T_{m,i}}{T}-1\right)+\Delta c_{p,i}\ln\left(\frac{T_{m,i}}{T}\right)+\frac{\Delta v_i}{T}\left(P-P_{ref}\right)+R\ln\left(x_i\varphi_i^L/\varphi_i^{L*}\right)=0 \cdot \quad [5.19]$$

Equation [5.19] is equivalent to the solid–fluid equilibrium condition proposed by Seiler *et al.* [SEI 01]. It has been used by many authors [SEI 01, NIK 16, PRI 12, RUT 11, PAR 10, OKU 16, TSI 09, DE 05] to describe SLE as well as solid–fluid equilibria involving supercritical fluids. Another approach for solid–fluid calculation considers the sublimation temperature of pure component i as the reference temperature, and is discussed in section 5.2.3.

5.2.2. *SLE in the presence of electrolyte solutions*

One application of equation [5.15] is the prediction of the melting point of ice in the presence of an electrolyte solution. Let us consider a strong electrolyte $C_{\nu_C}A_{\nu_A}$ that fully dissociates into water as $C_{\nu_C}A_{\nu_A} \rightarrow \nu_C C + \nu_A A$, where C and A are the cation and anion of the salt, and ν_C , ν_A are the corresponding stoichiometric coefficients. According to equation [5.15], the activity coefficient of water in the electrolyte solution must be known for an accurate prediction of the melting point of ice. Two formalisms can be used to express the chemical potentials of the ions: we can use either the mole fraction scale or the molality scale. For both formalisms the activity coefficients of the ions are defined with respect to the infinite dilution in the solvent (water). For an electrolyte solution, the activity of water is related to the osmotic coefficient ϕ as [ROB 02, LEE 08] $\phi = -1000\ln(x_w\gamma_w)/(m_s M_w(\nu_C+\nu_A))$ where x_w, γ_w and M_w are the mole fraction, the activity coefficient and the molecular weight of water, respectively; m_s is the molality of the salt defined as the number of moles of salt per kilogram of solvent (water). The chemical potential of ion i in the liquid phase can be expressed in the mole fraction scale as

$$\mu_i^L = \mu_i^{ref(x)} + RT\ln\left(x_i\gamma_i\right), \quad [5.20]$$

where γ_i the activity coefficient of chemical species *i* defined in the mole fraction scale, and x_i the mole fraction of chemical species *i* in the liquid phase; $\mu_i^{ref(x)}$ is the reference chemical potential of ion *i* in the mole fraction scale, and is only a function of *T* and *P*. The reference state is the infinite dilution in water; μ_i^L can also be expressed in the molality scale as

$$\mu_i^L = \mu_i^{ref(m)} + RT\ln\left(m_i\gamma_i^{(m)}\right), \qquad [5.21]$$

where m_i is the molality of ion *i*; $\gamma_i^{(m)}$ and $\mu_i^{ref(m)}$ are the activity coefficient and the reference chemical potential of ion *i* in the molality scale. The molality of the ion (m_i) is related to x_i as $m_i = x_i/(x_w M_w)$ where M_w is the molecular weight of water expressed in kg mol^{-1}. Since μ_i^L has the same value in the mole fraction and molality scales, it can be shown that

$$\mu_i^{ref(m)} = \mu_i^{ref(x)} + RT\ln M_w, \qquad [5.22]$$

and $\gamma_i^{(m)} = \gamma_i x_w$. The mean activity coefficient of the salt γ^{+-} is defined as [ROB 02, LEE 08] $\gamma^{+-} = \left(\gamma_C^{(m)\nu_C}\gamma_A^{(m)\nu_A}\right)^{1/(\nu_C+\nu_A)}$, where $\gamma_C^{(m)}$ and $\gamma_A^{(m)}$ are the activity coefficients of the cation and anion in the molality scale. Further details about the thermodynamics of electrolytes can be found elsewhere [ROB 02, LEE 08, KON 10]. To determine the activity coefficients of all chemical species, a thermodynamic model specific to electrolyte systems must be used. We can use either an activity coefficient model such as eNRTL [CHE 82, CHE 04] or Pitzer's model [PIT 73], or an electrolyte EoS such as Statistical Associating Fluid Theory with Variable Range for Electrolytes (SAFT-VRE) [GAL 98, GAL 99]. The reader is directed to an excellent review of thermodynamic models for electrolyte solutions by Kontogeorgis and Folas [KON 10].

Let us consider an aqueous solution of hydriodic acid (HI) as an example. Paricaud *et al.* [PAR 10] used the SAFT-VRE EoS [GAL 98, GAL 99] to describe the properties of HI electrolyte solutions. The SAFT-VRE model is a molecular-based EoS that can be used to describe the properties of both liquid and vapor phases, and its parameters have physical meanings. An excellent description of the mean activity coefficient and osmotic coefficient of HI can be obtained by fitting the diameter of the ions and the water-ion solvation energy, as shown in Figure 5.2.

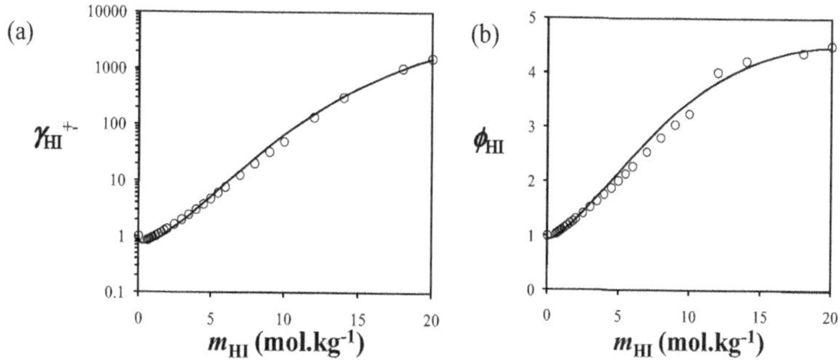

Figure 5.2. *Mean activity coefficient a) and osmotic coefficient b) of hydriodic acid solutions (HI + water mixture) at atmospheric pressure and T = 298.15 K. The symbols denote the experimental data [HAA 63, HAM 72] and the solid lines are calculated with the SAFT-VRE model [PAR 10]*

The ice–liquid equilibrium was predicted by using equation [5.15] and the following parameters for water and ice: $\Delta h_i = 6.01$ kJ mol^{-1}, $T_{m,I} = 273.15$ K, $\Delta v_i = -1.6$ cm^3·mol^{-1}, and $\Delta c_{p,i} = 40.9$ J mol^{-1} K^{-1}. The predictions of the SLE data are excellent, as shown in Figure 5.3. We can observe that the melting point of ice significantly decreases as the amount of HI is increased.

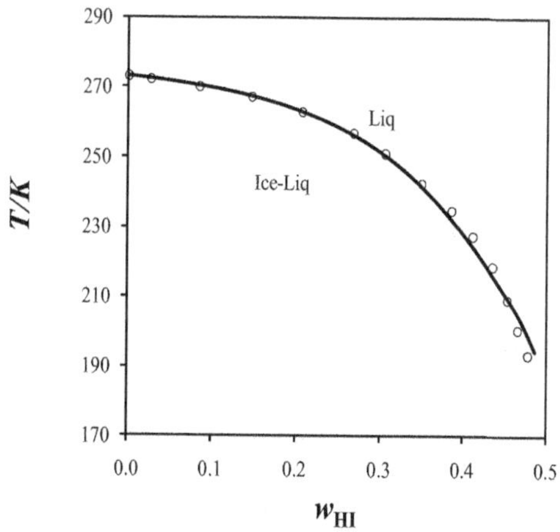

Figure 5.3. *Melting point of ice in contact with an aqueous solution of hydriodic acid (HI + water mixture) at atmospheric pressure. The symbols denote the experimental data [PIC 93 , PAS 60] and the solid lines are the SLE curves calculated by solving equation [5.15] and using the SAFT-VRE thermodynamic model [GAL 99] for the calculation of the activity coefficients*

5.2.3. *Solid–fluid equilibrium condition*

In some cases, one may be interested in knowing the solubility of a given solute in a gas or a supercritical fluid. A well-known application of solid–fluid equilibria is the extraction of solid solutes by using supercritical carbon dioxide. In this case, the solid–fluid equilibrium condition involving component i and a pure solid phase becomes

$$\mu_i^F = \mu_i^{S_i^*}, \quad [5.23]$$

where μ_i^F is the chemical potential of component i in the fluid mixture (gas, liquid or supercritical fluid); μ_i^F can be expressed with respect to the fugacity f_i^F of component i in the fluid mixture as [PRA 00]

$$\mu_i^F = \mu_i^0 + RT\ln(f_i^F / P_0), \quad [5.24]$$

where P_0 = 1 bar and μ_i^0 only depends on T. The chemical potential $\mu_i^{S_i^*}$ of component i in the pure solid phase can be expressed by considering a different thermodynamic cycle from the one used in the previous sections. To express $\mu_i^{S_i^*}$, we first consider a reference point at temperature T along the sublimation curve of pure component i. Let us assume that the sublimation curve $P_{sub,i}(T)$ (sublimation pressure vs. temperature) is known as a correlation of experimental data (Antoine's law or DIPPR like expression [PRO 13]). The correlation $P_{sub,i}(T)$ must be well behaved over a wide temperature range, even above the triple point of pure component i. At temperature T along the sublimation curve, the solid–gas equilibrium condition for pure component i is given by

$$\mu_i^{F*} = \mu_i^0 + RT\ln\left(f_i^{F*} / P_0\right) = \mu_i^{S_i^*}(T, P_{sub,i}(T)), \quad [5.25]$$

where f_i^{F*} is the fugacity of pure component i at temperature T and pressure $P_{sub,i}(T)$; f_i^{F*} is calculated as $f_i^{F*} = P_{sub,i}(T)\phi_i^{F*}$, where ϕ_i^{F*} is the fugacity coefficient of pure component i at $T, P_{sub,i}(T)$; ϕ_i^{F*} can be calculated with an EoS for fluids. By extrapolating equation [5.25] to temperatures above the triple point and integrating equation [5.14], we can express the chemical potential of the pure solid phase as

$$\mu_i^{S_i^*}(T,P) = \mu_i^0 + RT\ln\left(\phi_i^{F*}P_{sub,i}(T)/P_0\right) + \int_{P_{sub,i}}^{P} v_i^{S_i^*}\,dP\,, \qquad [5.26]$$

where $v_i^{S_i^*}$ is the molar volume of the pure solid phase at T. By assuming that the molar volume of the solid phase does not depend on pressure, the equilibrium condition (equation [5.23]) can be expressed as the equality of the fugacity of component i in the liquid and solid phases ($f_i^F = f_i^{S_i^*}$), where the liquid phase fugacity is $f_i^F = \phi_i^F x_i P$, and the solid phase fugacity is given by

$$f_i^{S_i^*} = \phi_i^{F*}P_{sub,i}(T)\exp\left[\frac{v_i^{S_i^*}\left(P - P_{sub,i}(T)\right)}{RT}\right]. \qquad [5.27]$$

The expression of the solid fugacity based on the sublimation pressure can also be used for gas hydrate phases (see section 5.5.2).

5.3. Solid–liquid equilibrium between a liquid mixture and a solid solution

A solid solution is composed of a fixed number of compounds: it contains some of the compounds of the liquid but it does not have to contain all of them. It has a well-defined crystalline structure but the nature of the molecules (or ions) at each site of the solid structure is assumed to be random. Moreover, the composition of a solid solution varies continuously with respect to temperature, pressure and global composition of the system. Solid solutions are usually observed when the compounds are very similar in terms of size and interactions. For instance, they are encountered in mixtures of the components of air and rare gases. To derive the equilibrium conditions, let us consider a liquid phase of n_c compounds in equilibrium with a solid solution S composed of n_c^S compounds that are all present in the liquid phase ($n_c^S \leq n_c$). The components j for $1 \leq j \leq n_c^S$ are present in both the liquid phase and the solid solution, while the components j for $n_c^S < j \leq n_c$ are only present in the liquid phase. The total Gibbs free energy of the system is given by

$$G = \sum_{j=1}^{n_c} n_j^L \mu_j^L + \sum_{j=1}^{n_c^S} n_j^S \mu_j^S , \quad [5.28]$$

where n_j^S and μ_j^S are the number of moles and chemical potential of component *j* in the solid solution *S*. At equilibrium, *G* is at its minimum value, thus

$$dG = \sum_{j=1}^{n_c} \mu_j^L dn_j^L + \sum_{j=1}^{n_c^S} \mu_j^S dn_j^S = 0 . \quad [5.29]$$

Since the components *j* for $n_c^S < j \le n_c$ are not observed in the solid phase, their number of moles in the liquid phase does not change ($dn^L_{j, n_c^S < j \le n_c} = 0$). Each molecule (or ion) *j* leaving the liquid phase goes into the solid phase ($dn_j^L = -dn_j^S ,\ 1 \le j \le n_c^S$). Equation [5.29] can then be reduced to

$$\sum_{j=1}^{n_c^S} \left(\mu_j^L - \mu_j^S\right) dn_j^L = 0 \cdot \quad [5.30]$$

As the infinitesimal changes dn_j^L can take any value, the n_c^S conditions of the solid–liquid phase equilibria are

$$\mu_j^L = \mu_j^S , \quad 1 \le j \le n_c^S . \quad [5.31]$$

Equation [5.7] is a specific case of equation [5.31]. In the general case of a solid solution, the phase equilibrium calculation requires a specific thermodynamic model for the solid solution in order to express the chemical potentials of the species in the solid solution. At moderate pressures, it is common to use an activity coefficient model for both the solid and liquid phases, and the chemical potentials μ_j^S are expressed with respect to a reference state, which can be a pure solid phase S_j^* containing only component *j*. μ_j^S are then written as

$$\mu_j^S = \mu_j^{S_j^*} + RT \ln\left(x_j^S \gamma_j^S\right) , \quad [5.32]$$

where x_j^S and γ_j^S are the mole fraction and chemical potential of component *j* in the solid solution *S*. If another activity coefficient model for the liquid phase is used, the equilibrium conditions can be written as

$$\Delta g_j + RT\ln\left(x_j^L\gamma_j^L\right) - RT\ln\left(x_j^S\gamma_j^S\right) = 0, \qquad [5.33]$$

where $\Delta g_j = \mu_j^{L*} - \mu_j^{S_j^*}$ is only a function of temperature *T* and pressure *P*. By analogy with procedure for pure solid phases (section 5.2), we can express Δg_j as

$$\begin{aligned}\frac{\Delta g_j(T,P)}{T} &= \frac{\Delta g_j(T_{ref,j},P_{ref,j})}{T_{ref,j}} + \frac{\Delta h_j}{T}\left(1-\frac{T}{T_{ref,j}}\right) - \Delta c_{p,j}\left(\frac{T_{ref,j}}{T}-1\right), \\ &+ \Delta c_{p,j}\ln\left(\frac{T_{ref,j}}{T}\right) + \frac{\Delta v_j}{T}\left(P-P_{ref,j}\right)\end{aligned} \qquad [5.34]$$

where $T_{ref,j}$, $P_{ref,j}$, Δh_j, $\Delta c_{p,j}$ and $\Delta g_j(T_{ref,j},P_{ref,j})$ are parameters that are specific to component *j* and the solid solution. $\Delta g_j(T_{ref,j},P_{ref,j})$ can be calculated with the activity coefficient model by applying equation [5.33] at $T = T_{ref,j}$, $P = P_{ref,j}$. The combination of all the parameters with two activity coefficient models (one for the liquid and one for the solid solution) represents a very flexible model that can describe complex solid–liquid phase diagrams observed experimentally. This approach can be used to calculate both solid–liquid and solid–solid phase equilibria.

Some approximations can be made when solid–liquid experimental data are not available. By assuming that the pure solid phase S_j^* has the same structure as the solid solution *S*, the parameters $T_{ref,j}$, $P_{ref,j}$, Δh_j, $\Delta c_{p,j}$ and $\Delta g_j(T_{ref,j},P_{ref,j})$ can be estimated at the melting point of component *j*: $P_{ref,j}$ = 1 atm, $T_{ref,j} = T_{m,j}$, $\Delta h_j = \Delta h_{m,j}$, $\Delta c_{p,j} = \Delta c_{p,j}$, $\Delta v_j = \Delta v_j$. We can further assume that the solid solution is ideal, meaning that the activity coefficients in the solid solution are equal to 1. This approximation cannot be applied to describe solid–solid equilibrium or azeotropic behavior.

Other approaches such as equations of states have been proposed to model the thermodynamic properties of solid solutions. We can refer to the EoS of Yokozeki [YOK 03], which can be applied to both fluid and solid phases. This EoS is given by

$$P = \frac{RT}{v-b}\frac{v-d}{v-c} - \frac{a}{v^2}, \qquad [5.35]$$

where a, b, c and d are the parameters of the EoS for a pure compound. These parameters are determined by forcing the EoS to go through the experimental critical and triple points of pure compounds. The EoS is a modification of the van der Waals EoS [VAN 04]: it has an additional term $(v-d)/(v-c)$ that diverges at $v \to c$ and leads to an additional density root at fixed T and P, which corresponds to the solid phase density. This EoS can be extended to mixtures by using simple one-fluid mixing rules and introducing some binary interaction parameters. The Yokozeki EoS has been used to describe complex phase diagrams involving solid–fluid as well as solid–solid phase equilibria over a wide pressure range [YOK 04]. It has also been used to represent the phase behavior of gas hydrates [YOK 05]. Stringari and Campestrini [STR 13] have developed some mixing rules for the Yokozeki EoS to predict the global phase diagram of Lennard-Jones mixtures. They applied the model to real systems such as biogas [RIV 14], natural gas [STR 14] and cryogenic mixtures [CAM 14]. Yokozeki EoS assumes that all solid phases are solid solutions, which is not always the case in practice. However, it can be used to describe rather complex diagrams by adjusting the binary parameters.

5.4. SLE between a liquid mixture and a solid compound

5.4.1. *Solid–liquid equilibrium with salt hydrates*

A solid compound (also called solid complex) is a solid phase formed from a mixture of chemical species of fixed composition. Solid compounds are observed as intermediate phases between the pure solid phases in the solid–liquid temperature composition diagram of a binary mixture. The formation of a solid compound from the liquid phase can be viewed as a chemical reaction, because its composition at thermodynamic equilibrium is stoichiometric. In this section, we focus on solid complexes formed from aqueous electrolyte solutions. However, the expressions for such systems can be easily extended to solid compounds formed from non-electrolyte systems. Solid compounds formed from aqueous solutions and containing water are called “hydrates”, but they should not be confused with gas hydrates. In this chapter, solid complexes formed water and ions are called “salt hydrates” in order to distinguish them from gas hydrates.

Let us consider a salt $C_{\nu_C}A_{\nu_A}$ composed of V_C cations C and V_A anions A. This salt is assumed to be fully dissociated in water. Several salt hydrate phases may form in such a system. Such hydrates should be neutral so they should contain ν_A anions for V_C cations. We can consider a salt hydrate phase (H) having the following relative composition $C_{\nu_C}A_{\nu_A}(H_2O)_{\nu_w}$ in equilibrium with the electrolyte solution; ν_W is the number of molecules of water per molecule of salt and is called the hydration number. The total Gibbs energy G of the system (salt hydrate + liquid phases) is given by

$$G = n_w^L\mu_w^L + n_C^L\mu_C^L + n_A^L\mu_A^L + n_w^H\mu_w^H + n_C^H\mu_C^H + n_A^H\mu_A^H, \qquad [5.36]$$

where (n_w^L, n_C^L, n_A^L) and (n_w^H, n_C^H, n_A^H) are the numbers of moles of water molecules, cations and anions in the liquid and hydrate phases, respectively; (μ_w^L, μ_C^L, μ_A^L) and (μ_w^H, μ_C^H, μ_A^H) are the corresponding chemical potentials. The SLE condition can be derived by expressing the differential of G. The stoichiometry requirement for the composition of the hydrate phase leads to $dn_w^L/\nu_w = dn_C^L/\nu_C = dn_A^L/\nu_A$ and $dn_w^H/\nu_w = dn_C^H/\nu_C = dn_A^H/\nu_A$. The mass balance equations between the liquid and hydrate phases lead to $dn_i^L = -dn_i^H$ for all species i. As a result, the differential of G is given by

$$dG = \frac{dn_w^H}{\nu_w}\left(\nu_w\mu_w^H + \nu_C\mu_C^H + \nu_A\mu_A^H - \left(\nu_w\mu_w^L + \nu_C\mu_C^L + \nu_A\mu_A^L\right)\right). \qquad [5.37]$$

Since $dG = 0$ and dn_w^H can take any value at equilibrium, the equilibrium condition at fixed T and P is given by

$$\Delta g_{dis} = \nu_w\mu_w^L + \nu_C\mu_C^L + \nu_A\mu_A^L - \nu_w\mu_w^H - \nu_C\mu_C^H - \nu_A\mu_A^H = 0, \qquad [5.38]$$

where Δg_{dis} is the dissociation Gibbs free energy. In equation [5.28], Δg_{dis} is zero at equilibrium, but the corresponding dissociation enthalpy is different from zero and is calculated with the Gibbs–Helmholtz equation as $\Delta h_{dis} = -T^2\partial(\Delta g_{dis}/T)/\partial T$. Δh_{dis} is the energy required for melting the hydrate phase, and is expressed in Joules per mole of salt. It can be measured experimentally by differential scanning calorimetry. Equation [5.38] still holds for systems containing gas semiclathrate hydrates (see section 5.4.2). Let us define the molar Gibbs free energy of the hydrate phase (g^H) as $g^H = \nu_w\mu_w^H + \nu_C\mu_C^H + \nu_A\mu_A^H$; g^H is expressed in Joules per mole of salt. Since the salt hydrate composition is fixed, g^H

only depends on T and P. By expressing the chemical potentials in the liquid phases with respect to the activity coefficients, the equilibrium condition (equation [5.38]) can be written as:

$$\Delta g_{dis} = \Delta g^H + RT\left(\nu_C \ln(x_C\gamma_C) + \nu_A \ln(x_A\gamma_A) + \nu_w \ln(x_w\gamma_w)\right) = 0 \quad [5.39]$$

where $\Delta g^H = \nu_w \mu_w^{ref(x)} + \nu_C \mu_C^{ref(x)} + \nu_A \mu_A^{ref(x)} - g^H$ only depends on T and P. Note that the equilibrium condition [5.39] is not the equality of the chemical potential of water in both phases. This equation can be seen as the condition for the following pseudo-chemical equilibrium,

$$\nu_C \mathrm{C} + \nu_A \mathrm{A}_{\nu_A} + \nu_w \mathrm{H_2O} \leftrightarrows \mathrm{C}_{\nu_C}\mathrm{A}_{\nu_A}(\mathrm{H_2O})_{\nu_w} \,. \quad [5.40]$$

Most authors, like Tumakaka *et al.* [TUM 07] or Kleiner *et al.* [KLE 09], have solved SLE with complex solids by considering a chemical equilibrium similar to equation [5.40] and an equilibrium constant. The equilibrium constant $K^{(x)}$ in the mole fraction scale corresponding to equation [5.40] can be written from equation [5.39] as [TUM 07, KLE 09]

$$K^{(x)} = \exp\left(-\frac{\Delta g^H}{RT}\right) = (x_C\gamma_C)^{\nu_C}(x_A\gamma_A)^{\nu_A}(x_w\gamma_w)^{\nu_w}. \quad [5.41]$$

Equation [5.41] is obtained by assuming that the activity of the salt hydrate is equal to 1. This assumption is equivalent to the fact that Δg^H only depends on T and P. To solve equation [5.39] or [5.41], we need to express Δg^H with respect to T and P. By analogy with the SLE condition for a pure solid phase (equation [5.19]), we can express Δg^H with respect to T and P and a reference state (T_{ref}, P_{ref}) as

$$\frac{\Delta g^H}{RT} = \frac{\Delta h^H}{RT}\left(1 - \frac{T}{T_{ref}}\right) - \frac{\Delta c_p^H}{R}\left(\frac{T_{ref} - T}{T}\right) + \frac{\Delta c_p^H}{R}\ln\left(\frac{T_{ref}}{T}\right) + \frac{\Delta v^H}{RT}(P - P_{ref}) + \frac{\Delta g^H(T_{ref}, P_{ref})}{RT}, \quad [5.42]$$

where Δh^H, Δv^H and Δc_p^H are parameters characterizing the hydrate phase of relative composition $C_{\nu_C}A_{\nu_A}(H_2O)_{\nu_w}$; $\Delta g^H(T_{ref}, P_{ref})$ can be calculated with the activity coefficient model by applying the equilibrium condition (equation [5.39]) at the reference state (T_{ref}, P_{ref}). The equilibrium condition becomes

$$-\frac{\Delta h^H}{RT}\left(1-\frac{T}{T_{ref}}\right)+\frac{\Delta c_p^H}{R}\left(\frac{T_{ref}-T}{T}\right)-\frac{\Delta c_p^H}{R}\ln\left(\frac{T_{ref}}{T}\right)$$
$$-\frac{\Delta v^H}{RT}(P-P_{ref})-\frac{\Delta g^H(T_{ref},P_{ref})}{RT}=\nu_C\ln(x_C\gamma_C)+\nu_C\ln(x_A\gamma_A)+\nu_w\ln(x_w\gamma_w). \quad [5.43]$$

The enthalpic parameter Δh^H can be estimated by using the experimental dissociation enthalpy of the hydrate measured at one point of the SLE curve. We can also define an entropic parameter, Δs^H, such that $\Delta h^H = T_{ref}\Delta s^H$. Several choices can be made for the reference state. When the SLE curve of the hydrate exhibits congruent melting, a relevant choice for the reference state is to consider the congruent melting point at atmospheric pressure for the reference state [TUM 07]. Choosing this reference state ensures that the calculated SLE curve goes through the experimental congruent melting point. Let T_{ref} be the congruent melting point of the hydrate at atmospheric pressure (P_{ref} = 1 atm). By applying equation [5.39] at the stoichiometric composition of the hydrate phase ($x_w^{(sto)}$, $x_C^{(sto)}$, $x_A^{(sto)}$), we can determine $\Delta g^H(T_{ref}, P_{ref})$ as

$$\frac{\Delta g^H(T_{ref},P_{ref})}{RT_{ref}}=-\nu_C\ln\left(x_C^{(sto)}\gamma_C^{(sto)}\right)-\nu_A\ln\left(x_A^{(sto)}\gamma_A^{(sto)}\right)-\nu_w\ln\left(x_w^{(sto)}\gamma_w^{(sto)}\right), \quad [5.44]$$

where $x_w^{(sto)}/\nu_w = x_C^{(sto)}/\nu_C = x_A^{(sto)}/\nu_A = 1/(\nu_w+\nu_C+\nu_A)$. For a given salt + water system, several hydrates of various hydration numbers may be observed. The dissociation enthalpy Δh_{dis} of the hydrate per mole of salt is obtained as $\Delta h_{dis} = -T^2\partial(\Delta g_{dis}/T)/\partial T$. The approach requires that the composition of the hydrate (ν_w, ν_C, ν_A) is known.

The equilibrium condition [5.41] can be expressed in terms of molalities rather than mole fractions, as

$$K^{(m)} = \exp\left(-\frac{\Delta g^{H(m)}}{RT}\right) = \left(m_C \gamma_C^{(m)}\right)^{\nu_C} \left(m_A \gamma_A^{(m)}\right)^{\nu_A} a_w^{\nu_w}. \quad [5.45]$$

where m_C and m_A are the molalities of the cations and anions, respectively, and $a_w = x_w \gamma_w$ is the activity of water. We can show from equations [5.21], [5.22], [5.41] and [5.42] that $K^{(x)} = K^{(m)} M_w^{\nu_C+\nu_A}$. The constant $K^{(m)}$ can be viewed as the solubility product of the salt hydrate. Many authors have performed calculations of the melting points of salt hydrates. In most studies, a chemical equilibrium is considered for the SLE condition and the equilibrium constant is expressed in the molality scale (equation [5.45]). Since most studies of salt hydrates were done at atmospheric pressure, the parameter Δv^H is assumed to be zero, and the parameters $\Delta g^{H(m)}(T_{ref}, P_{ref})$, $\Delta h^{H(m)}$, and $\Delta c_p^{H(m)}$ in the molality scale are calculated as [LI 11]

$$\begin{aligned} \Delta g^{H(m)}(T_{ref}, P_{ref}) &= \nu_C g_{f,C}^0 + \nu_A g_{f,A}^0 + \nu_w g_{f,w}^0 - g_{f,hyd}^0, \\ \Delta h^{H(m)} &= \nu_C h_{f,C}^0 + \nu_A h_{f,A}^0 + \nu_w h_{f,w}^0 - h_{f,H}^0, \\ \Delta c_p^{H(m)} &= \nu_C c_{p,C}^0 + \nu_A c_{p,A}^0 + \nu_w c_{p,w}^0 - c_{p,H}^0, \end{aligned} \quad [5.46]$$

where g_f^0, h_f^0, s^0 and c_p^0 are standard Gibbs free energies of formation, absolute entropies, standard enthalpies of formation and standard heat capacities of the chemical species and hydrate *H*. These constants can be found in the NBS database [WAG 82]. Monnin *et al.* [MON 02] modeled the solid–liquid phase behavior of the LiCl + water mixture by considering a chemical equilibrium similar to equation [5.45] and using the mean spherical approximation model [BLU 75, BLU 80] to describe the activity coefficients of the species. An excellent description of the SLE data could be obtained over the entire composition range. Besides, a linear relation was found between the formation enthalpy of the hydrate ($h_{f,H}^0$) and the hydrate number (ν_w). Li *et al.* [LI 11] also used an equilibrium constant (solubility product) combined with the LIQUAC activity coefficient model to compute the SLE in electrolyte systems. They considered the standard formation enthalpies and absolute entropies of the species and the hydrates, which are reported in the NBS database [WAG 82]. A similar approach based on Pitzer's model [PIT 73] for electrolyte systems was used by Lach *et al.* [LAC 15], and an excellent prediction of the SLE of NaOH + water system was obtained.

Paricaud [PAR 11] used the SAFT-VRE EoS [GAL 98, GAL 99] to describe the properties of different strong electrolyte solutions and predict the SLE in the presence of salt hydrates. In the SAFT-VRE approach, water molecules are modeled as spheres interacting via a square-well (SW) potential and four association sites are considered on each sphere to represent hydrogen bonding. Ions are modeled as charged hard spheres. The solvation (ion-water) interactions are described with a short-range SW potential, and the ionic parameters are fitted to the experimental osmotic and mean activity coefficients. The molecular models within the SAFT approach are depicted in Figure 5.4.

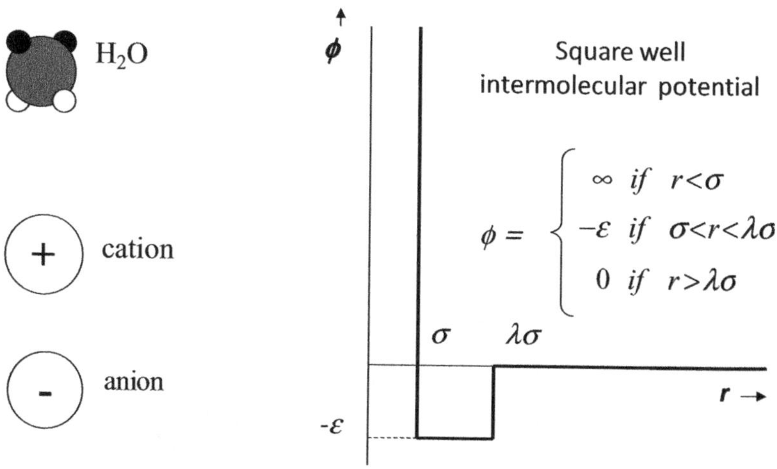

Figure 5.4. *Molecular models used in the SAFT-VRE approach [GAL 98, GAL 99] for aqueous electrolyte solutions: water is modeled as a sphere with four association sites, while ions are represented by charged hard spheres. The solvation (ion-water) interactions are described with short-range square-well potentials*

The SLE was solved from equation [5.39] and using the congruent melting point as the reference temperature. Two parameters (Δh^H, T_{ref}) were adjusted to SLE experimental data, while the heat capacity parameter (Δc_p^H) was neglected. A good description of SLE for the LiBr + water and HI + water systems was obtained over a wide composition range (Figure 5.5).

The computation of the solubility of a salt in a given solvent (water) corresponds to a particular case of equation [5.45], where the hydration number is zero ($\nu_w = 0$). The solubility product of salt can then be expressed in the molality scale as $K^{(m)} = \left(m_C \gamma_C^{(m)}\right)^{\nu_C} \left(m_A \gamma_A^{(m)}\right)^{\nu_A}$. The computation of solubility of salts has been performed by many authors [FUR 82, MGA 91, FAR 93, FAR 04, MAS 04, FAR 05, PIN 05, LI 07, HU 15], and the main difference between all the approaches is the thermodynamic model used to predict the ionic activity coefficients.

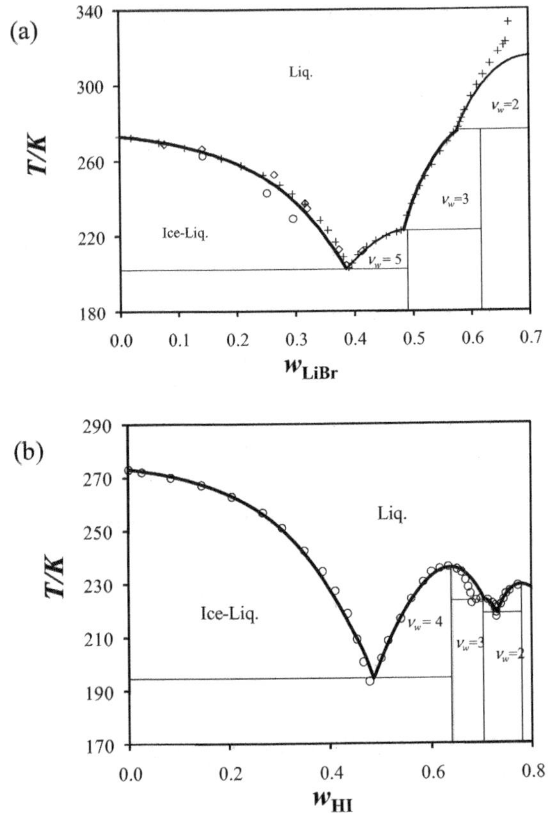

Figure 5.5. *Solid–liquid phase diagram of the LiBr + water a) and HI + water b) binary mixtures. The symbols denote the experimental data [PIC 93, PAS 60, STE 27, JON 04, PÁT 06] and the solid lines are the SLE curves calculated by solving equation [5.44] and using the SAFT-VRE thermodynamic model [GAL 99] for the calculation of the activity coefficients*

5.4.2. *Solid–liquid equilibrium with semiclathrate hydrates*

Semiclathrate hydrates can be considered as one particular type of salt hydrate. These solids can form at atmospheric pressure from aqueous solutions of tetraalkylammonium or tetraalkylphosphonium salts. Their structure is made up of water molecules and ions: the anions belong to the crystalline framework, while the alkyammonium cations are encapsulated in large cages formed by water molecules. The solid–liquid phase behavior of semiclathrate hydrates is complex and the crystalline structures and hydration numbers of the different hydrates are still not perfectly known. Dyadin and co-workers [DYA 84, LIP 02] have reviewed the phase behavior of these systems and published phase diagrams as well as a couple of properties (congruent melting points, structure data, densities, hydration numbers, etc.) for a broad variety of systems.

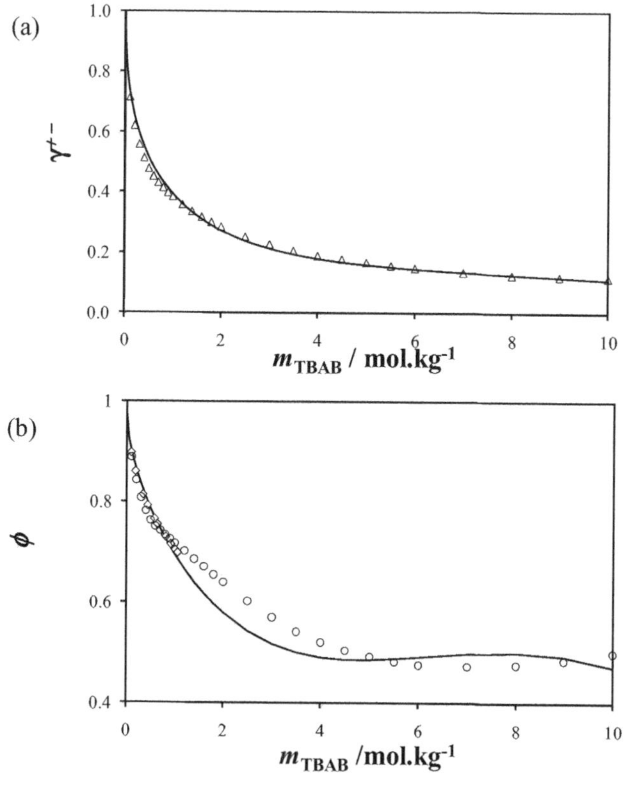

Figure 5.6. *Thermodynamic properties of TBAB + H_2O solutions at T = 298.15 K and atmospheric pressure [PAR 11]. The solid lines correspond to the predictions of the SAFT-VRE model: a) mean activity coefficients; (b) osmotic coefficients. The symbols denote the experimental data (circles [LIN 64], diamonds [AMA 05])*

The TBAB + water binary system has probably been the most studied mixture among these systems, because TBAB is cheap as it is used in industry for various applications (phase transfer catalyst, intermediate compound for the production of other alkylammonium salts, etc.). Let H_{ν_w} be a salt hydrate of hydration number ν_w. For the TBAB + water system, four different hydrates (H_{36}, H_{32}, H_{26} and H_{24}) were found for TBAB weight fractions inferior to 0.6 by Dyadin and co-workers [DYA 84, LIP 02], while Shimada and co-workers found only two hydrates H_{26} and H_{38} denoted as type A and type B, respectively. Numerous SLE data from different sources are available for the TBAB + H_2O system [DYA 84, LIP 02, DAR 05, SHI 05, OYA 05, ARJ 07, SUN 08, DES 09] for TBAB weight fractions below 0.5.

Concerning the modeling of the TBAB aqueous phases, it is well known that TBAB salt exhibits ion pairing even at moderate concentrations [SLU 97]. Since the composition range of interest for the study of semiclathrate hydrates is the dilute region, we can assume that TBAB is fully dissociated in water and does not form any ion pairing. This approach has been used to aqueous solutions of alkylammonium salts by several authors [PAR 11, AMA 05, BEL 04, KWA 11, HER 14], and a good description of the experimental mean activity coefficients and osmotic coefficients can be obtained by adjusting the parameters of the electrolyte thermodynamic model (Figure 5.6). Ion pairing can be taken into account, either by considering a chemical equilibrium between the ions forming ion pairs [NAJ 15], or by using an associating thermodynamic model for electrolyte solutions [PAP 15].

Once the parameters of the thermodynamic model for the electrolyte solution have been regressed, the SLE of the TBAB + water mixture can be described by using equation [5.39]. Following the work of Shimada *et al.* [SHI 05, SHI 03], Paricaud [PAR 11] proposed to model the solid–liquid phase behavior of the TBAB + water system by considering only two hydrates: H_{26} type A and H_{38} type B. The heat capacity parameter, Δc_p^H, was neglected as the temperature range of the SLE is rather narrow. The congruent melting point of H_{26} was set to the experimental value obtained by Oyama *et al.* [OYA 05] (T_{ref} = 285.15 K), while the congruent melting point (metastable) of H_{38} was fixed to 283.5 K. As shown in Figure 5.7, the model based on Equation [5.39] can accurately describe the SLE data at atmospheric pressure for TBAB weight fractions up to 0.5. The predicted dissociation enthalpies by Paricaud's model are equal to 193.02 $kJ \cdot g^{-1}$ for type A hydrate at the congruent composition (w_{TBAB} = 0.4077) and 199.17 $kJ \cdot g^{-1}$ for type B hydrate at the triple point composition (w_{TBAB} = 0.174). These values are close to the experimental data [OYA 05] (193.18 and 199.59 $kJ \cdot g^{-1}$). The parameter $\Delta v^H = -30 cm^3$ per mole of TBAB for type B hydrate (H_{38}) was optimized on the experimental dissociation temperatures at different pressures reported by Arjmandi *et al.* [ARJ 07]. A very good agreement between the predictions of the model and the experimental data was obtained, as shown in Figure 5.8. Since there is no such data for type A hydrate (H_{26}), we can use the same parameter $\Delta v^H = -30 \text{cm}^3.\text{mol}^{-1}$ for both type A and type B hydrates. Kwaterski and Herri [KWA 11] obtained very similar results by considering the NRTL electrolyte model instead of the SAFT-VRE model, showing that the approach to compute SLE for salt hydrates is relevant for the TBAB + water mixture as long as the activity coefficients of the species are well described. The introduction of ion pairing for the modeling of the liquid phase leads to similar descriptions of mean activity and osmotic coefficients, and no significant improvement was obtained for the description of the SLE [NAJ 15].

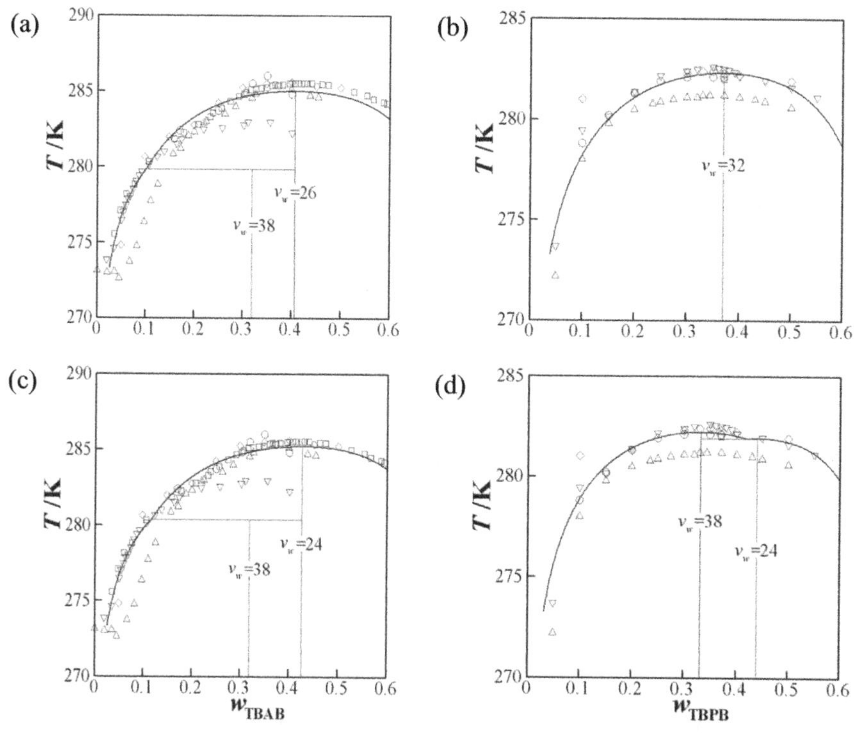

Figure 5.7. *Temperature composition diagrams of TBAB + H_2O mixture calculated by considering the hydrate phases H_{38} and H_{26}: a) and the hydrate phases H_{38} and H_{24}; b) Temperature composition diagrams of the TBPB + H_2O mixture calculated by considering the hydrate phase H_{32}; c) and the hydrate phases H_{38} and H_{24}; d) The composition is expressed in terms of the salt weight fraction. The symbols denote the experimental data. (a and b) Squares [LIP 02]; down triangles v_w = 38 from [OYA 05]; up triangles v_w = 26 from [OYA 05]; diamonds [DAR 05]; circles [DES 09]. (c and d) Squares [DYA 87]; up triangles [MAY 10]; diamonds [SUG 12]; circles [LIN 13]; down triangles [ZHA 13]. The solid lines are the calculated SLE curves*

The thermodynamic approach has been extended to other alkylammonium salts [FUK 14, BAB 16]: tetrabutylammonium chloride (TBAC), tetrabutylammonium fluoride (TBAF), tetrabutylammonium nitrate ($TBANO_3$) and tetrabutylphosphonium bromide (TBPB). An excellent description of the solid–liquid phase diagrams of the salt + water systems could be obtained over a wide composition range [FUK 14], as shown in Figures 5.7 and 5.8. Different assumptions have been made for the hydration numbers of the TBAB and TBPB salts. It is observed that the use of type A hydrate with hydration number equal to 24 leads to a better description of the SLE data (Figure 5.7) for TBAB and TBPB. A group contribution method has been proposed by Fukumoto *et al.* [FUK 14] to predict the parameter Δh^H and the congruent melting points (T_{ref}) of semiclathrate hydrates.

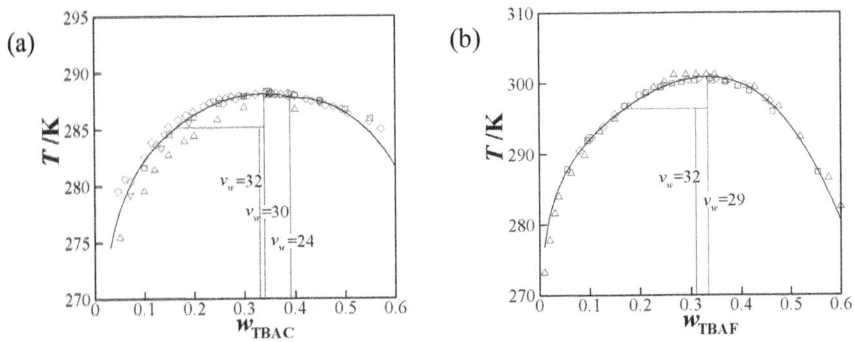

Figure 5.8. *Temperature composition diagrams of the TBAC + H_2O mixture a), calculated by considering the hydrate phases H_{32}, H_{30} and H_{24}, and phase diagram of the TBAF + H_2O mixture b), calculated by considering the hydrates phases H_{32} and H_{29}. The composition is expressed in terms of the salt weight fraction. The symbols denote the experimental data: a) squares [SAT 13]; triangles [SUN 11]; diamonds [ALA 96]; circles [NAK 87]; b) triangles [NAK 81]; squares [DYA 76]; diamonds [SAK 08]; circles [LEE 10]. The solid lines are the calculated SLE curves*

5.5. Thermodynamic model for gas semiclathrate hydrates

This chapter focuses on the specific case of gas semiclathrate hydrates. For a detailed explanation of the vdW-P theory and an extensive review of its various versions and applications to gas hydrates, the reader is directed to the original paper [VAN 59] and to the book of Sloan and Koh [SLO 08]. A semiclathrate hydrate is a solid phase composed of water molecules and ions. The anion is usually a halide while the cation is an alkylammonium or alkylphoshonium cation. Semiclathrate hydrates can form at atmospheric pressure without the presence of gas, as discussed in the previous section. They can encapsulate gas molecules and form at a lower pressure and higher temperature than gas hydrates. Then, tetraallkylammonium salts like TBAB are considered as thermodynamic promoters for gas hydrates. Two main thermodynamic approaches have been considered for the modeling of the dissociation conditions of gas semiclathrate hydrates: Paricaud [PAR 11] first proposed to extend of the thermodynamic approach for salt hydrates to gas semiclathrate hydrates by adding the vdW-P term to the Gibbs free energy of the salt hydrate phase. The main assumption of this approach is to consider that the crystalline structures of the TBAB gas semiclathrate hydrates are the same as those of the salt hydrate. However, it is possible to consider other types of crystalline structures in this approach, if needed. A very good description of gas–liquid–hydrate three phase lines as well as dissociation enthalpies could be obtained with few adjusted parameters. Besides the model for gas semiclathrate hydrates is consistent with the model for salt hydrates (without gas). The other approach was proposed by Eslamimanesh *et al.* [ESL 12]. This model is based on an empirical modification of the vdW-P model for gas hydrate: the gas semiclathrate hydrate is considered as a classical gas hydrate and a correlation of the Langmuir constant depending on the salt composition is used. This model can also describe accurately the H–L–V three phase lines, but more adjusted parameters are required. We now discuss both approaches in further details.

5.5.1. *Paricaud's approach*

We construct a multicomponent system at temperature T and pressure P composed of $C_{\nu_C}A_{\nu_A}$ salt (for example TBAB), water and N_g types of gas molecules (methane, carbon dioxide, hydrogen, etc.), which exhibits solid–liquid–vapor equilibrium with the formation of a gas semiclathrate hydrate phase (Figure 5.9).

The proposed approach is based on the combination of the thermodynamic approach for salt hydrates (section 5.4) with the vdW-P model [VAN 59, SLO 98, CAR 03] for gas hydrates. The key assumption of this approach is to consider that the metastable empty hydrate phase can have different structures that correspond to the same semiclathrate hydrate phases observed in the salt + H_2O binary system. For example, in the case of TBAB, we assume that the structures of gas semiclathrate hydrates of TBAB can be of types A and B, with hydrations numbers $\nu_w = 26$ and $\nu_w = 38$, respectively.

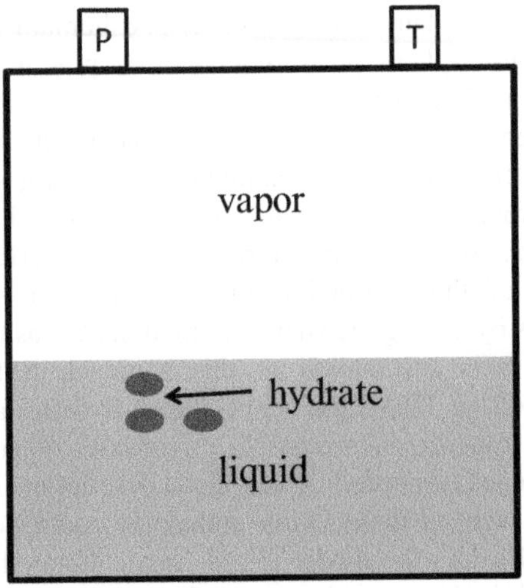

Figure 5.9. *Hydrate–liquid–vapor (HLV) three phase equilibrium inside the experimental cell of volume V. The initial volume of the introduced salt solution is V_{sol}. The global composition of the systems is expressed in terms of the global mole fractions z_i of chemical species i (water, gases or ions)*

The semigrand partition function Ξ of the gas semiclathrate hydrate phase can be derived by using the same hypothesizes as those suggested by van der Waals and Platteeuw [VAN 59], i.e. the positions of the host molecules (H_2O, cations and anions) are fixed and the cavities are not distorted by the gas molecules; the cavities are assumed to be spherical and can only contain one gas molecule; the guest–guest interactions are neglected, as well as quantum effects. The reader is directed to the original paper of Waals and Platteeuw [VAN 59] and to an alternative and rigorous derivation of the vdW-P model proposed by Wierzchowski and Monson [WIE 07]. We assume that the numbers N_w, N_C and N_A of water molecules, cations and anions are fixed, while the numbers N_j of gas molecules j in the hydrate phase vary. Following the derivation of van der Waals and Platteeuw [VAN 59, SLO 98], we can express Ξ as

$$\Xi = \exp\left(-\frac{A^{H-\beta}}{kT}\right)\sum_{N_{ij}}\prod_{i=1}^{N_{cav}}\left\{\frac{(n_i N_{salt})!}{\left(n_i N_{salt} - \sum_{j=1}^{N_g} N_{ij}\right)!\prod_{j=1}^{N_g} N_{ij}!}\prod_{j=1}^{N_g} q_{ij}^{N_{ij}}\lambda_j^{N_{ij}}\right\}, \qquad [5.47]$$

where $A^{H-\beta}$ is the Helmholtz free energy of the empty hydrate (semiclathrate hydrate phase without gas), which is denoted as the β phase [SLO 08]; k is the Boltzmann constant; N_{ij} is the number of gas molecules j that occupy the cavities of type i; N_{salt} is the number of salt molecules in the hydrate ($N_{salt} = N_C/\nu_C = N_A/\nu_A = N_w/\nu_w$); n_i is the number of cavities of type i per salt molecule in the hydrate; N_{cav} is the number of cavity types that encage gas molecules; q_{ij} is the partition function of molecule j in cavity i; and λ_j is the absolute activity of molecule j. The sum in equation [5.47] is a multiple sum over all values of the numbers N_{ij} under the constraints $0 \le \sum_{j=1}^{N_g} N_{ij} \le n_i N_{salt}$. Note that $N_j = \sum_{i=1}^{N_{cage}} N_{ij}$. By using the multinomial theorem, we can simplify equation [5.47] as [VAN 59]

$$\ln \Xi = -\frac{A^{H-\beta}}{kT} + N_{salt} \sum_{i=1}^{N_{cav}} n_i \ln\left(1 + \sum_{j=1}^{N_g} q_{ij}\lambda_j\right). \qquad [5.48]$$

The fraction Y_{ij} of cavities of type i occupied by gas molecules of type j is given by [VAN 59]

$$Y_{ij} = \frac{q_{ij}\lambda_j}{1 + \sum_{k=1}^{N_g} q_{ik}\lambda_k}. \qquad [5.49]$$

The molar Gibbs free energy of the gas semiclathrate phase (g^H) is given by [PAR 11]

$$g^H = g^{H-\beta} + RT \sum_{i=1}^{N_{cav}} n_i \ln\left(1 - \sum_{j=1}^{N_g} Y_{ij}\right). \qquad [5.50]$$

where $g^{H-\beta}$ is the molar Gibbs free energy of the empty semiclathrate hydrate phase (β phase). While the β phase does not exist for classical hydrates, it does exist in the case of semiclathrates as it corresponds to the salt hydrate. The vdW-P term on the right-hand side of equation [5.50] is negative, showing that the presence of encapsulated gas molecules stabilizes the hydrate phase by decreasing its molar Gibbs free energy. We can define the Langmuir constants C_{ij} as [PAR 72]

$C_{ij} = q_{ij}\lambda_j / f_j$, where f_j is the fugacity of molecule j. It can be shown that C_{ij} only depends on T and on the cell potential. The fractions Y_{ij} can be expressed as

$$Y_{ij} = \frac{C_{ij} f_j}{1 + \sum_{k=1}^{N_g} C_{ik} f_k} \quad [5.51]$$

The fugacities f_j are determined with an EoS. Paricaud and co-workers [PAR 11, FUK 14, BAB 16, FUK 15a, FUK 15b] used the SAFT-VRE EoS for both the liquid and vapor phases, while other authors used a cubic EoS for the gas phase and an activity coefficient model for the liquid phase [KWA 11, HER 14, NAJ 15]. By assuming that the cavities are spherical, we can express C_{ij} as

$$C_{ij} = \frac{4\pi}{kT} \int_0^{R_{cav,i}} \exp\left(\frac{-w_{ij}(r)}{kT} \right) r^2 dr, \quad [5.52]$$

where $R_{cav,i}$ is the effective radius of cavity i, r is the distance between the gas molecule center of mass and the center of the cavity and w_{ij} is the cell potential that characterizes the interactions between gas molecule j and host molecules forming the cavity. As proposed by Barrer and Edge [BAR 67] and Parrish and Prausnitz [PAR 72], we can use an SW potential for w_{ij}. In this case, the Langmuir constants can be expressed as [PAR 11, DUF 12]

$$C_{ij} = \frac{4\pi}{kT} V_{ij}^{cell} \exp\left(\frac{\varepsilon_{ij}^{cell}}{kT} \right), \quad [5.53]$$

where V_{ij}^{cell} is the free volume of molecule j inside cavity i and ε_{ij}^{cell} is the depth of the SW cell potential. The cell potential can also be described by the Kihara potential [HER 14].

The equilibrium condition [5.38] still holds due the stoichiometry of semiclathrate hydrate phase composition (fixed relative compositions of the ions and water) and can be written as

$$\Delta g_{dis} = \Delta g^{H-\beta} - RT \sum_{i=1}^{N_{cav}} n_i \ln\left(1 - \sum_{j=1}^{N_g} Y_{ij} \right) + RT\left(\nu_C \ln\left(x_C \gamma_C \right) + \nu_A \ln\left(x_A \gamma_A \right) + \nu_w \ln\left(x_w \gamma_w \right) \right) = 0, \quad [5.54]$$

where $\Delta g^{H-\beta} = \nu_w \mu_w^{ref(x)} + \nu_C \mu_C^{ref(x)} + \nu_A \mu_A^{ref(x)} - g^{H-\beta}$ is only a function of T and P and can be expressed as (equation [5.43]),

$$\frac{\Delta g^{H-\beta}}{RT} = \frac{\Delta h^{H-\beta}}{RT}\left(1-\frac{T}{T_{ref}}\right) - \frac{\Delta c_p^{H-\beta}}{R}\left(\frac{T_{ref}-T}{T}\right) + \frac{\Delta c_p^{H-\beta}}{R}\ln\left(\frac{T_{ref}}{T}\right) + \frac{\Delta v^{H-\beta}}{RT}\left(P-P_{ref}\right) + \frac{\Delta g^{H-\beta}\left(T_{ref},P_{ref}\right)}{RT}. \quad [5.55]$$

The parameters $\Delta h^{H-\beta}$, $\Delta c_p^{H-\beta}$ $\Delta v^{H-\beta}$ and T_{ref} can be determined on the available experimental data (SLE in salt + water binary systems and SLE in gas + salt + water ternary systems). Paricaud *et al.* proposed to assume that $\Delta c_p^{H-\beta}$=0, and fitted $\Delta h^{H-\beta}$, $\Delta v^{H-\beta}$ and T_{ref} on experimental SLE data. Each type of hydrate phase (for example, types A and B for TBAB hydrates) is characterized by a set of these parameters. Moreover, the reference temperature T_{ref} is chosen to be the congruent melting temperature of the salt hydrate at $P_{ref} = 1$ atm and $\Delta g^{H-\beta}(T_{ref}, P_{ref})$ is calculated with the thermodynamic model for the liquid phase, by applying the solid-liquid equilibrium condition at the reference point (equation [5.44]).

Paricaud [PAR 11] proposed an algorithm for the determination of the melting points of semiclathrate hydrates in the presence of one gas (CO_2): at the limit of appearance of the semiclathrate hydrate phase, the mass of the semiclathrate hydrate phase is negligible. The conditions for vapor–liquid equilibrium (VLE) are the equality of the pressure, temperature and chemical potentials of the different species (apart from the ions) in the vapor and liquid phases. For a given temperature, pressure and global composition of the mixture, the VLE can be solved by performing a vapor–liquid TP flash calculation. The condition of dissociation of the semiclathrate hydrate consists of finding the temperature T at which the gas semiclathrate phase disappears at a fixed pressure and fixed global composition.

Fukumoto *et al.* [FUK 15b] later proposed a rigorous algorithm to predict the hydrate–liquid–vapor three-phase equilibrium in the presence of gas mixtures. They first showed that the calculated HLV three phase line (melting temperature vs. pressure) does not depend much on the initial global composition of the system if only gas is present in the system. This is due to the fact that the total pressure is almost equal to the partial pressure of the gas (the partial pressure of water isnegligible). Besides, many authors have computed HLV three phase lines of gas hydrates by ignoring the VPE and assuming that the total pressure is equal to the partial pressure of the gas. Fukumoto *et al.* [FUK 15b] showed that this approximation cannot be made when two or more gases are present in the system. The position of the HLV three phase lines dramatically depends on the relative composition of the gas mixture in the vapor phase at equilibrium. This relative gas composition at equilibrium can be very different from the initial gas mixture composition, if the initial amount of salt solution in the cell is large, and if the solubility of one gas in the solution is significantly higher than the solubilities of the other gases. As a result, the position of the HLV curve may dramatically depend on the initial global composition. Fukumoto *et al.* [FUK 15b] proved that one should specify the initial amount of aqueous solution put into the cell to fully characterize the studied system.

The algorithm proposed by Fukumoto *et al.* [FUK 15b] is related to the experimental conditions: the known variables of the systems are the total volume of the experimental cell (V), the initial composition and volume of the aqueous salt solution (w_{salt}, V_{sol}) and the initial global composition of the gas mixture introduced into the experimental cell (mole fractions $y_{i,ini}$). The ratio r_{salt} between the number of moles of salt n_{salt} and the number of moles of water n_w is also known and is given by

$$r_{salt} = \frac{n_{salt}}{n_w} = \frac{w_{salt}/M_{salt}}{(1-w_{salt})/M_w}, \qquad [5.56]$$

where M_{salt} and Mw are the molecular weight of the salt and water, respectively. Let r_{gas} be the ratio between the initial total number of moles of gas molecules ($n_{g,ini}$) and the total number of moles of water: $r_{gas} = n_{g,ini}/n_w$. The ratio r_{gas} is unknown experimentally but it can be determined with an iterative procedure as shown later in this section. The total number of moles in the system is $n = n_w + n_C + n_A + n_{g,ini}$, and the global mole fraction of water is given by

$$z_w = \frac{n_w}{n} = \frac{1}{1+(\nu_C+\nu_A)r_{salt}+r_{gas}}. \qquad [5.57]$$

The global mole fractions of cations and anions (z_C and z_A, respectively) in the system are $z_C = \nu_C n_{salt}/n = \nu_C r_{salt}\, z_w$ and $z_A = \nu_A n_{salt}/n = \nu_A r_{salt}\, z_w$. The global mole fraction of gas i is $z_i = y_{i,ini}(1 - z_w - z_C - z_A)$. The initial total number of moles of water introduced into the cell at the beginning of the experiment (n_w) is given by $n_w = V_{sol}/[v_{sol}(1+(\nu_C + \nu_A)r_{salt})]$, where v_{sol} is the molar volume of the salt solution at ambient conditions and v_{sol} can be predicted with an EoS for the liquid phase or from a correlation. We can then define the volume ratio $r_V = V_{sol} / V$, as the initial fraction of volume occupied by the salt solution in the experimental cell at ambient conditions. The ratio r_V is usually known experimentally ($r_V = r_{V,exp}$) as both V_{sol} and V are known. At the limit of appearance of the hydrate phase (melting point), the system is at VPE at T and P, and the mass of the hydrate phase is negligible. T can be determined at fixed P and global mole fractions z_i by satisfying the SLE equilibrium condition (equation [5.54]). The VLE flash calculation at fixed (T, P, z_i) provides the compositions of the liquid (x_i) and vapor (y_i) phases, as well as the vaporization ratio α defined as $\alpha = n_{vap} / n$, where n_{vap} is the total number of moles of volatile chemical species in the vapor phase (gases and water). The vapor fraction α is related to the total volume V of the system as $V = n(\alpha\, v_{vap} + (1-\alpha)v_{liq})$, where v_{vap} and v_{liq} are molar volumes of the vapor and liquid phases at equilibrium, which can be predicted with an EoS (SAFT-VE). Since $n_w = z_w\, n$, we can relate the ratio r_V and α as

$$r_V = z_w \frac{v_{sol}(1+(\nu_C + \nu_A)r_{salt})}{\alpha\, v_{vap} + (1-\alpha)v_{liq}}. \qquad [5.58]$$

The melting temperature of the hydrate at a fixed pressure is determined with an iterative procedure according to the following steps:

1) fix the total pressure P of the system;

2) fix the initial weight fraction of the salt in the initial aqueous solution (w_{salt}), which is known experimentally; calculate r_{salt} (equation [5.56]);

3) use an initial value for r_{gas}: Fukumoto *et al.* proposed the following correlation: $r_{gas} \approx 0.006P / r_V$;

4) calculate the global mole fractions of all species (water, gases and ions) (equation [5 57]). Use an initial value for T: for instance, $T_{ini} = 300$ K;

5) perform the vapor–liquid T-P flash to determine the moles fractions x_i, y_i of the liquid and vapor phases, respectively, as well as the vapor fraction α. Change T (iterations over step 5) until the SLE condition (equations [5.54] and [5.55]) is

satisfied. The calculation is performed for all possible hydrate phases, and the most stable hydrate phase is kept;

6) calculate r_V from equation [5.58], and v_{vap} and v_{liq} from the EoS; change r_{gas} and come back to step 4, until $r_V = r_{V,\exp}$.

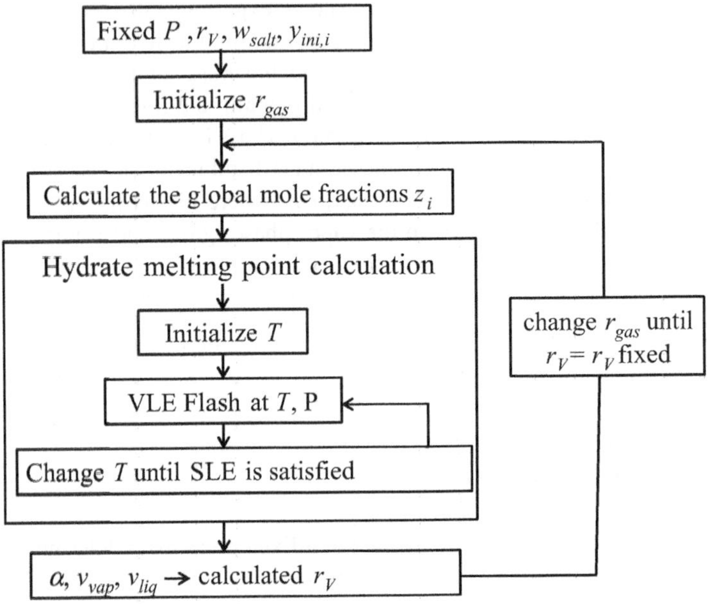

Figure 5.10. *Algorithm to determine the melting point of semiclathrate hydrates formed in the presence of a gas mixture [FUK 15b]. The fixed conditions are the pressure P, the initial volume fraction r_V occupied by the electrolyte solution of concentration w_{salt} in the experimental cell and the initial composition ($y_{ini,i}$) of the gas mixture*

The Paricaud model has been applied to model semiclathrate hydrates of carbon dioxide + dihydrogen, and an excellent description of the dissociation temperatures was obtained, over wide salt composition and pressure ranges, as shown in Figures 5.11–5.13. We must specify the possible hydrate phases that can be encountered (i.e. the possible hydration numbers), and the model is able to predict the change in phase from one crystalline structure to the other, by considering the most stable hydrate phase (lowest Helmholtz Gibbs free energy or highest melting point). We can observe such a change in hydrate phase in the pressure–temperature diagram (Figure 5.12) for the TBAB + CO_2 + H_2 system at the TBAB weight fraction w_{TBAB} = 0.329. The authors could predict the separation factors of these hydrates (capability of separating two gases) and obtained higher separation factors at moderate salt concentrations when the hydrate phase has a larger hydration number.

Herri *et al.* proposed a similar approach based on the NRTL-electrolyte model to describe the HLV three lines of TBAB systems in the presence of CO_2, N_2 and CH_4 [KWA 11, HER 14]. Garcia *et al.* [GAR 16] applied Paricaud's approach to the modeling of TBAB and TBAC semiclathrate in the presence of various gases. They obtained a very good description of the experimental data by considering a Kihara potential for the cell potential and the PSRK EoS [HOL 91] for calculating the properties of the liquid and vapor phases.

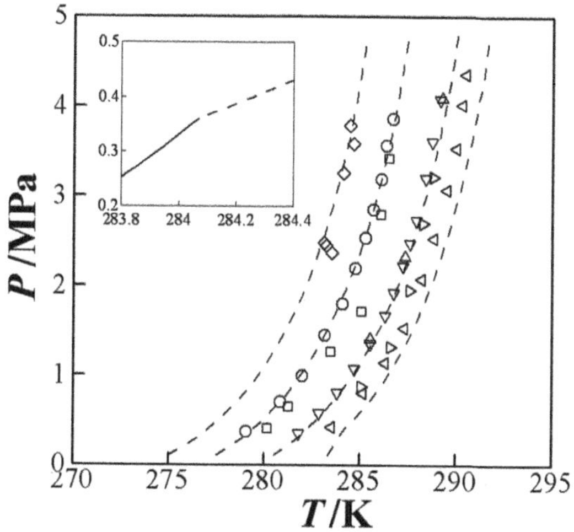

Figure 5.11. *Liquid–vapor–hydrate (L–V–H) three-phase lines for the CO_2 + TBAB + H_2O system calculated by considering the hydrate phases H_{38} and H_{26}. The symbols denote the experimental data at different TBAB weight fractions: (diamonds) w_{TBAB} = 0.03 from [OYA 08]; (squares) w_{TBAB} = 0.05 from [LI 10]; (circles) w_{TBAB} = 0.05 from [YE 12], respectively; (right triangles), (down triangles) and (up triangles) are at w_{TBAB} = 0.1 from [LI 10], [YE 12] and [ARJ 07], respectively; (left triangles) w_{TBAB} = 0.19 from [YE 12]. The lines, from the left to right, are the predictions of the model at w_{TBAB} = 0.03, 0.05, 0.1 and 0.19. The second figure is an enlargement of the region around the phase change from H_{26} (solid line) to H_{38} (dashed line) at w_{TBAB} = 0.19*

Paricaud's model and the algorithm proposed by Fukumoto *et al.* [FUK 15b] can also be applied to non-electrolyte systems such as mixed hydrates of tetrahydrofurane (THF), which have also been considered as promising materials for the separation and storage of gases [RIC 13]. The cages of THF hydrates are almost fully occupied by THF even at low pressures [MAR 08] so we can assume that TFH hydrates behave as solid compounds with a fixed composition. The algorithm could also be extended to three-phase equilibria with a non-negligible amount of hydrate phase. In this case, we must consider the proportion of hydrate phase as an additional variable, and the mass balance as well as the equilibrium conditions must be solved. Bouillot and Herri [BOU 16] proposed two algorithms of flash calculation for classical gas hydrates at fixed total volume (see the following chapter in this book). The first algorithm is based on the fact that the hydrate phase forms as layers that are in equilibrium with the instantaneous pressure. In this case, the hydrate is not homogenous and cannot reorganize itself during hydrate growth. The second algorithm assumes that the composition of the hydrate phase is homogenous and in equilibrium with the final pressure of the system. This means that the hydrate phase can reorganize itself during growth. Both algorithms provide good predictions of the final pressure and mass of crystallized water. The reader is directed to Chapter 6 of this book for a review by Bouillot and Herri on flash algorithms for hydrate systems.

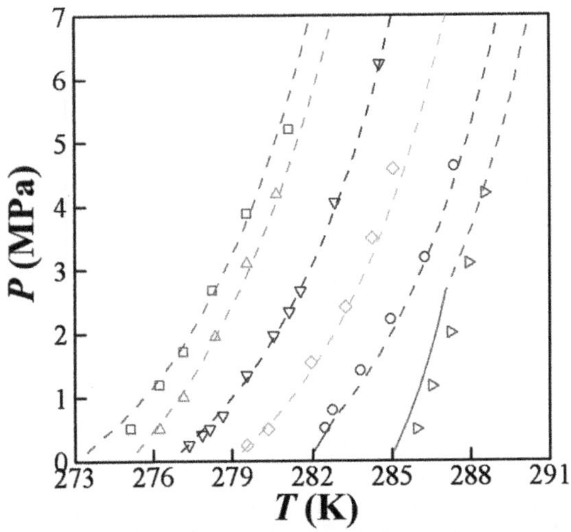

Figure 5.12. *Liquid–vapor–hydrate (L–V–H) three-phase lines for the CO_2 + H_2 + H_2O + TBAB system calculated by considering the hydrate phases H_{38} and H_{24}. Symbols are the experimental data obtained with a gas mixture of CO_2 39.2% and H_2 60.8% by Li et al. [LI 10]: (squares) w_{TBAB} = 0.0245, (up triangles) w_{TBAB} = 0.0363, (down triangles) w_{TBAB} = 0.0495, (diamonds) w_{TBAB} = 0.0826, (circles) w_{TBAB} = 0.153, (right-triangles) w_{TBAB} = 0.329. The lines are the corresponding calculated curves. The solid and dashed lines correspond to the H_{24} phase and the H_{38} phase, respectively. For a color version of this figure, see www.iste.co.uk/broseta/ hydrates1.zip*

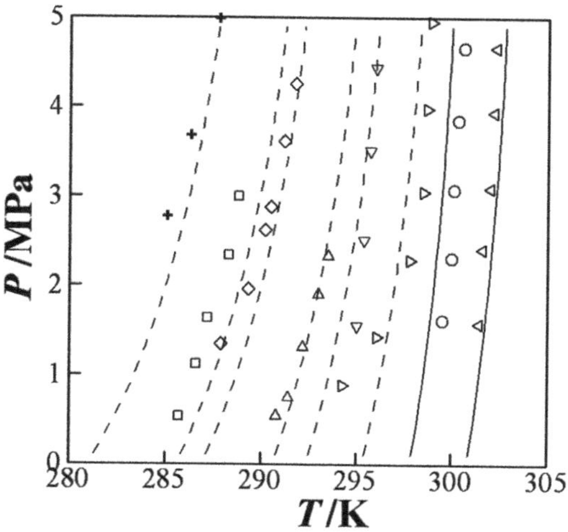

Figure 5.13. *Liquid–vapor–hydrate (L–V–H) three-phase lines for the CO_2 + TBAF + H_2O system calculated by considering the hydrate phases H_{32} and H_{29}. The symbols denote the experimental data at different TBAF weight fractions: (pluses) w_{TBAF} = 0.02 [MOH 13], (diamonds) w_{TBAF} = 0.05 [MOH 13], (squares) w_{TBAF} = 0.041 [LI 10]; (up triangles) w_{TBAF} = 0.083 [LI 10], (down triangles) w_{TBAF} = 0.104 [LEE 12], (right triangles) w_{TBAF} = 0.15 [MOH 13], (left triangles) w_{TBAF} = 0.31 [LEE 12], (circles) w_{TBAF} = 0.448 [LEE 12]. The lines are the predictions of the model. The solid lines correspond to the H_{29} phase, and the dashed lines indicate the H_{32} phase*

5.5.2. *The Eslamimanesh et al. model*

Eslamimanesh *et al.* [ESL 12] proposed an alternative model to describe the dissociation temperatures of gas semiclathrate hydrates. Their approach is based on the original vdw-P model for classical gas hydrates and the use of correlations for the molar Gibbs free energy of the empty hydrate phase. Moreover, the Langmuir constants are considered as empirical functions of the salt concentration. In this approach, the SLE condition is expressed as the equality of the fugacity of water in the liquid (f_w^L) and hydrate phase (f_w^H), and the salt (TBAB) is considered as a guest of the hydrate phase. The fugacity of water in the liquid phase is expressed as

$$f_w^L = x_w \gamma_w P_{sat,w} \exp\left(\frac{v_w^L (P - P_{sat,w})}{RT} \right), \qquad [5.59]$$

where $P_{sat,w}$ is the vapor pressure of pure water (correlation depending on T). The activity coefficient of water is calculated with the NRTL model [REN 68] with binary parameters adjusted to experimental data. The fugacity of water in the gas semiclathrate hydrate phase is given by

$$f_w^H f_w^{H-\beta} \exp\left(\frac{-\Delta\mu_w^{H-\beta}}{RT}\right), \qquad [5.60]$$

where $f_w^{H-\beta}$ is the fugacity of water in the semiclathrate hydrate phase without gas, and $\Delta\mu_w^{H-\beta}$ is the vdw-P term. The fugacity $f_w^{H-\beta}$ is calculated as

$$f_w^{H-\beta} = P_w^{H-\beta} \exp\left(\frac{v_w^{H-\beta}\left(P - P_w^{H-\beta}\right)}{RT}\right), \qquad [5.61]$$

where $P_w^{H-\beta}$ is a correlation of temperature and salt concentration, which can be assimilated to a sublimation pressure by analogy with equation [5.27]. $P_w^{H-\beta}$ is given by

$$P_w^{H-\beta} = 0.1\exp(17.44 - 6003.9/T + h \times w_{salt}), \qquad [5.62]$$

where h was adjusted to SLE experimental data w_{salt} is the initial weight fraction of salt in the aqueous solution. The salt is assumed to be the guest of only large cages while the gas molecules go into the small cages. The Langmuir constants of the salt and gas molecules are both empirical functions of the salt concentration. Two types of hydrate phases (types A and B) were considered for TBAB hydrates. The authors proposed an empirical function calculating the fugacity of the salt in order to compute the occupancy fraction for the salt in the large cages of the hydrate.

A good description of the HLV three phase lines of gas semiclathrate hydrates of TBAB could be obtained [ESL 12]. However, the model of Eslamimanesh *et al.* contains a large number of adjusted parameters, as discussed by Garcia *et al.* [GAR 16], and the vapor pressure of the TBAB salt needed to compute the fugacity of TBAB is not clearly defined. It is not clear either how the model can predict the SLE of the salt + water binary mixture and how good are the predictions for the dissociation enthalpies.

5.6. Conclusion

The methodologies used to compute SLE for engineering applications are now well established and most of them are based on thermodynamic cycles and well-chosen reference temperatures. These methods consist of expressing the chemical potentials of the species present in the solid phase at T, P by integrating the Gibbs–Helmholtz equation and considering the pressure dependence with Poynting correction factors. In the case of pure solid phase in equilibrium with a fluid mixture, two main approaches can be used: we can consider either the solid–liquid equilibrium of the pure substance (melting or triple point) or a sublimation point as the reference state. Similar ideas are used to compute SLE involving solid solutions, salt hydrates and gas hydrates. By combining the thermodynamic approach for salt hydrates and the vdw-P theory for gas hydrates, a new model for gas semiclathrate hydrates was proposed [PAR 11]. This model requires a few adjustable parameters and can predict the hydrate–liquid–vapor three phase lines in the presence of a gas mixture with a good accuracy.

Although this chapter focuses on engineering and practical approaches for computing SLE, it is worth mentioning that several equations of state based on statistical thermodynamics and perturbation theories have been developed for solid phases [COT 96a]. Solids are rather different in nature from gases and liquids, and these equations of state are specific to solid phases. A combination of theoretical equations of state for fluids and solids can lead to excellent predictions of molecular simulations of prototype models of molecules [VEG 02, BLA 03, VEG 03, BLA 03, VEG 04] (flexible, semiflexible and rigid chains including liquid crystals). Such approaches have mainly been used for pure substances, and the applications to mixtures are still limited to simple molecular models [COT 95, COT 96a, COT 96b, COT 97, MON 00] (hard spheres, Lennard-Jones particles). One main advantage of these approaches is the fact that they do not require the use of reference points for computing SLE and have strong physical bases. However, they are usually specific to only one possible crystalline structure, so they are not easy to use for real substances. Moreover, the extension of such equations of state to gas hydrates and solid compounds seems rather complex. The development of theoretical equations of state for solids is still a challenge but would be of great benefit for the chemical industry.

5.7. Bibliography

[ALA 03] ALADKO L.S., DYADIN Y.A., RODIONOVA T.V. *et al.*, "Effect of size and shape of cations and anions on clathrate formation in the system: halogenides of quaternary ammonium bases and water", *Journal of Molecular Liquids*, vol. 106, pp. 229–238, 2003.

[ALA 96] ALADKO L.S., DYADIN Y.A., "Phase equilibria in the system $(C_4H_9)_4NF-NH_4F-H_2O$", *Zhurnal Neorganicheskoi Khimii*, vol. 41, pp. 828–834, 1996.

[AMA 05] AMADO A.G., BLANCO L.H., "Isopiestic determination of the osmotic and activity coefficients of dilute aqueous solutions of symmetrical and unsymmetrical quaternary ammonium bromides with a new isopiestic cell at 298.15K", *Fluid Phase Equilibria*, vol. 233, pp. 230–233, 2005.

[ARJ 07] ARJMANDI M., CHAPOY A., TOHIDI B., "Equilibrium data of hydrogen, methane, nitrogen, carbon dioxide, and natural gas in semi clathrate hydrates of tetrabutyl ammonium bromide", *Journal of Chemical & Engineering Data*, vol. 52, pp. 2153–2158, 2007.

[BAB 16] BABU P., PARICAUD P., LINGA P., "Experimental measurements and modeling of the dissociation conditions of semiclathrate hydrates of tetrabutyl ammonium nitrate and carbon dioxide", *Fluid Phase Equilibria*, vol. 413, pp. 80–85, 2016.

[BAL 03] BLAS F.J., SANZ E., VEGA C. *et al.*, "Fluid-solid equilibria of flexible and linear rigid tangent chains from Wertheim's thermodynamic perturbation theory", *Journal of Chemical Physics*, vol. 119, pp. 10958–10971, 2003.

[BAR 67] BARRER R.M., EDGE A.V.J., "Gas hydrates containing argon, krypton and xenon: kinetics and energetics of formation and equilibria", *Proceedings of Royal Society of London A*, vol. 300, pp. 1–24, 1967.

[BEL 04] BELVEZE L.S., BRENNECKE J.F., STADTHERR M.A., "Modeling of activity coefficients of aqueous solutions of quaternary ammonium salts with the electrolyte-NRTL equation", *Industrial & Engineering Chemistry Research*, vol. 43, pp. 815–825, 2004.

[BLA 03] BLAS F.J., GALINDO A., VEGA C., "Study of the solid-liquid-vapour phase equilibria of flexible chain molecules using Wertheim's thermodynamic perturbation theory", *Molecular Physics*, vol. 101, pp. 449–458, 2003.

[BLU 75] BLUM L., "Mean spherical model for asymmetric electrolytes. 1. Method of solution", *Molecular Physics*, vol. 30, pp. 1529–1535, 1975.

[BLU 80] BLUM L., "Theoretical chemistry: advances and perspectives", in EYRING H. (ed.), *Theoretical Chemistry*, Henderson, New York, 1980.

[BOU 16] BOUILLOT B., HERRI J.-M., "Framework for clathrate hydrate flash calculations and implications on the crystal structure and final equilibrium of mixed hydrates", *Fluid Phase Equilibria*, vol. 413, pp. 184–195, 2016.

[CAM 14] CAMPESTRINI M., STRINGARI P., ARPENTINIER P., "Solid-liquid equilibrium prediction for binary mixtures of Ar, O-2, N-2, Kr, Xe, and CH_4 using the LJ-SLV-EoS", *Fluid Phase Equilibria*, vol. 379, pp. 139–147, 2014.

[CAR 03] CARROLL J., *Natural Gase Hydrates: A Guide for Engineers*, GPP, Amsterdam, 2003.

[CAR 09] CAROLL J., *Natural Gas Hydrates. A Guide for Engineers*, 2nd ed., Elsevier, Amsterdam, 2009.

[CHE 04] CHEN C.C., SONG Y., "Generalized electrolyte-NRTL model for mixed-solvent electrolyte systems", *AIChE Journal*, vol. 50, pp. 1928–1941, 2004.

[CHE 82] CHEN C.C., BRITT H.I., BOSTON J.F. *et al.*, "Local composition model for excess Gibbs energy of electrolyte systems. Part I: single solvent, single completely dissociated electrolyte systems", *AIChE Journal*, vol. 28, pp. 588–596, 1982.

[CON 59] CONDE M.M., TORRE J.P., MIQUEU C., "Revisiting the thermodynamic modelling of type I gas-hydroquinone clathrates", *Physical Chemistry Chemical Physics*, vol. 18, pp. 10018–10027, 2016.

[COT 95] COTTIN X., MONSON P.A., "A theory of solid-solutions and solid-fluid equilibria for mixtures", *International Journal of Thermophysics*, vol. 16, pp. 733–741, 1995.

[COT 96a] COTTIN X., PARAS E.P.A., VEGA C. *et al.*, "Solid-fluid equilibrium: new perspectives from molecular theory", *Fluid Phase Equilibria*, vol. 117, pp. 114–125, 1996.

[COT 96b] COTTIN X., MONSON P.A., "Solid-fluid phase equilibrium for single component and binary Lennard-Jones systems: a cell theory approach", *Journal of Chemical Physics*, vol. 105, pp. 10022–10029, 1996.

[COT 97] COTTIN X., MONSON P.A., "An application of classical thermodynamics to solid-fluid equilibrium in hard sphere mixtures", *Journal of Chemical Physics*, vol. 107, pp. 6855–6858, 1997.

[COU 16] COUPAN R., CHABOD M., DICHARRY C. *et al.*, "Experimental determination of phase equilibria and occupancies for CO_2, CH_4, and N-2 hydroquinone clathrates", *Journal of Chemical & Engineering Data*, vol. 61, pp. 2565–2572, 2016.

[DAR 05] DARBOURET M., COURNIL M.M., HERRI J.-M., "Rheological study of a hydrate slurry for air conditioning application", *Fifth International Conference on Gas Hydrates*, Trondheim, Norway, 2005.

[DE 05] DE HEMPTINNE J.C., "Benzene crystallization risks in the LIQUEFIN liquefied natural gas process", *Process Safety Progress*, vol. 24, pp. 203–212, 2005.

[DES 09] DESCHAMPS J., DALMAZZONE D., "Dissociation enthalpies and phase equilibrium for TBAB semi–clathrate hydrates of N_2, CO_2, N_2 + CO_2 and CH_4 + CO_2", *Journal of Thermal Analysis and Calorimetry*, vol. 98, pp. 113–118, 2009.

[DUC 07] DUC N.H., CHAUVY F., HERRI J.-M., "CO_2 capture by hydrate crystallization – a potential solution for gas emission of steelmaking industry", *Energy Conversion and Management*, vol. 48, pp. 1313–1322, 2007.

[DUF 12] DUFAL S., GALINDO A., JACKSON G. *et al.*, "Modelling the effect of methanol, glycol inhibitors and electrolytes on the equilibrium stability of hydrates with the SAFT-VR approach", *Molecular Physics*, vol. 110, pp. 1223–1240, 2012.

[DYA 76] DYADIN Y.A., TEREKHOVA I.S., OLYANSKAYA T.M. *et al.*, "Clathrate hydrates of tetrabutylammonium fluoride and oxalate", *Journal of Structural Chemistry*, vol. 17, pp. 566–571, 1976.

[DYA 84] DYADIN Y.A., UDACHIN K.A., "Clathrate formation in water-peralkylonium salts systems", *Journal of Inclusion Phenomena*, vol. 2, pp. 61–72, 1984.

[DYA 87] DYADIN Y.A., UDACHIN K.A., "Clathrate polyhydrates of peralkylonium salts and their analogs", *Journal of Structural Chemistry*, vol. 28, pp. 394–432, 1987.

[ESL 12] ESLAMIMANESH A., MOHAMMADI A.H., RICHON D., "Thermodynamic modeling of phase equilibria of semi-clathrate hydrates of CO_2, CH_4, or N-2 + tetra-n-butylammonium bromide aqueous solution", *Chemical Engineering Science*, vol. 81, pp. 319–328, 2012.

[FAR 04] FARELO F., LOPES A.M.C., FERRA M.I., "Solubilities of sodium chloride and potassium chloride in water plus ethanol mixtures from (298 to 323) K", *Journal of Chemical & Engineering Data*, vol. 49, pp. 1782–1788, 2004.

[FAR 05] FARELO F., FERNANDES C., AVELINO A., "Solubilities for six ternary systems: NaCl + NH_4Cl + H_2O, KCl + NH_4Cl + H_2O, NaCl + LiCl + H_2O, KCl + LiCl + H_2O, NaCl + $AlCl_3$ + H_2O, and KCl + $AlCl_3$ + H_2O at T = (298 to 333) K", *Journal of Chemical & Engineering Data*, vol. 50, pp. 1470–1477, 2005.

[FAR 93] FARELO F., VONBRACHEL G., OFFERMANN H., "Solid liquid equilibria in the ternary-system NACL-KCL-H_2O", *Canadian Journal of Chemical Engineering*, vol. 71, pp. 141–146, 1993.

[FOW 40] FOWLER D.L., LOEBENSTEIN W.V., PALL D.B. *et al.*, "Some unusual hydrates of quaternary ammonium salts", *Journal of the American Chemical Society*, vol. 62, pp. 1140–1142, 1940.

[FRE 75] FREDENSLUND A., JONES R.L., PRAUSNITZ J.M., "Group contribution estimation of activity coefficient in nonideal mixtures mixtures", *AIChE Journal*, vol. 21, pp. 1086–1099, 1975.

[FUK 15a] FUKUMOTO A., PARICAUD P., DALMAZZONE D. *et al.*, "Experimental measurements and modeling of the dissociation conditions of tetrabutylammonium chloride semi-clathrate hydrates in the presence of hydrogen", *Journal of Chemical & Engineering Data*, vol. 60, pp 343–350, 2015.

[FUK 15b] FUKUMOTO A., SALES SILVA L.P., PARICAUD P. *et al.*, "Modeling of the dissociation conditions of H_2 + CO_2 semiclathrate hydrate formed with TBAB, TBAC, TBAF, TBPB, and TBNO3 salts. Application to CO_2 capture from syngas", *International Journal of Hydrogen Energy*, vol. 40, pp. 9254–9266, 2015.

[FUK 14] FUKUMOTO A., PARICAUD P., DALMAZZONE D., *et al.*, "Modeling the dissociation conditions of carbon dioxide + TBAB, TBAC, TBAF, and TBPB semiclathrate hydrates", *Journal of Chemical & Engineering Data*, vol. 59, pp. 3193–3204, 2014.

[FUR 82] FURST W., RENON H., "Effect of the various parameters in the application of pitzers model to solid-liquid equilibrium – preliminary-study for strong 1-1 electrolytes", *Industrial & Engineering Chemistry Process Design and Development*, vol. 21, pp. 396–400, 1982.

[GAL 98] GALINDO A., DAVIES L.A., GIL-VILLEGAS A. *et al.*, "The thermodynamics of mixtures and the corresponding mixing rules in the SAFT-VRE approach for potentials of variable range", *Molecular Physics*, vol. 93, pp. 241–252, 1998.

[GAL 99] GALINDO A., GIL-VILLEGAS A., JACKSON G. *et al.*, "SAFT-VRE: phase behavior of electrolyte solutions with the statistical associating fluid theory for potentials of variable range", *Journal of Physical Chemistry B*, vol. 103, pp. 10272–10281, 1999.

[GAR 16] GARCIA M., MARRIOTT R., CLARKE M.A., "Modeling of the thermodynamic equilibrium conditions for the formation of TBAB and TBAC semiclathrates formed in the presence of Xe, Ar, CH_4, CO_2, N-2, and H-2", *Industrial & Engineering Chemistry Research*, vol. 55, pp. 777–787, 2016.

[HAA 63] HAASE R., NAAS H., THUMM H., "Experimentelle untersuchungen über das thermodynamische verhalten konzentrierter halogenwasserstoffsauren", *Zeitschrift für physikalische Chemie*, vol. 37 pp. 210–229, 1963.

[HAM 72] HAMER W.J., WU Y.-C., "Osmotic coefficient and mean activity coefficients of uni-univalent electrolytes in water at 25°C", *Journal of Physical and Chemical Reference Data*, vol. 1, pp. 1047–1099, 1972.

[HER 14] HERRI J.-M., BOUCHEMOUA A., KWATERSKI M. *et al.*, "Enhanced selectivity of the separation of CO_2 from N_2 during crystallization of semi-clathrates from quaternary ammonium solutions", *Oil & Gas Science and Technology – Rev. IFP Energies nouvelles*, vol. 69, pp. 947–968, 2014.

[HOL 91] HOLDERBAUM T., GMEHLING J., "PSRK. A group contribution equation of state based on UNIFAC", *Fluid Phase Equilibria*, vol. 70, pp. 251–265, 1991.

[HU 15] HU J.X., SANG S.H., LIU Q.Z., "Solid-liquid equilibria in the ternary systems KBr-$CaBr_2$-H_2O and NaBr-$CaBr_2$-H_2O at 348 K", *Journal of Chemical & Engineering Data*, vol. 60, pp. 993–998, 2015.

[JON 04] JONES H.C., GETMAN F.H., "Über das Vorhandensein von Hydraten in konzentrierten wässerigen Lösungen von Elektrolyten", *Zeitschrift für Physikalische Chemie*, vol. 49, pp. 385–455, 1904.

[KLE 09] KLEINER M., TURNAKAKA F., SADOWSKI G., "Thermodynamic modeling of complex systems", in LU X., HU Y. (eds), *Molecular Thermodynamics of Complex Systems*, Springer, Berlin, 2009.

[KON 10] KONTOGEORGIS G., FOLAS G.K., *Thermodynamic Models for Industrial Applications: From Classical and Advanced Mixing Rules to Association Theories*, Wiley, 2010.

[KWA 11] KWATERSKI M., HERRI J.-M., "Thermodynamic modelling of gas semi-clathrate hydrates using the electrolyte NRTL model", *ICGH 2011 Proceeeding*, Edinburgh, Scotland, 2011.

[LAC 15] LACH A., ANDRÉ L., LASSIN A. *et al.*, "A new pitzer parameterization for the binary NaOH–H_2O and ternary NaOH–NaCl–H_2O and NaOH–LiOH–H_2O systems up to NaOH solid salt saturation, from 273.15 to 523.15 K and at saturated vapor pressure", *Journal of Solution Chemistry*, vol. 44, pp. 1424–1451, 2015.

[LEE 08] LEE L.L., *Molecular Thermodynamics of Electrolyte Solutions*, World Scientific, 2008.

[LEE 10] LEE S., LEE Y., PARK S. *et al.*, "Phase equilibria of semiclathrate hydrate for nitrogen in the presence of tetra-n-butylammonium bromide and fluoride", *Journal of Chemical & Engineering Data*, vol. 55, pp. 5883–5886, 2010.

[LEE 12] LEE S., LEE Y., PARK S. *et al.*, "Thermodynamic and spectroscopic identification of guest gas enclathration in the double tetra-n-butylammonium fluoride semiclathrates", *Journal of Physical Chemistry B*, vol. 116, pp. 9075–9081, 2012.

[LEE 15] LEE J.W., DOTEL P., PARK J. *et al.*, "Separation of CO_2 from flue gases using hydroquinone clathrate compounds", *Korean Journal of Chemical Engineering*, vol. 32, pp. 2507–2511, 2015.

[LI 07] LI J.T., WANG J.K., WANG Y.L., "Solubility of KCl and MgCl2 in binary solvents formed by acetone and water in the temperature range between (293.15 and 323.15) K", *Journal of Chemical & Engineering Data*, vol. 52, pp. 1069–1071, 2007.

[LI 09] LI S., FAN S., WANG J. *et al.*, "CO_2 capture from binary mixture via forming hydrate with the help of tetra-n-butyl ammonium bromide", *Journal of Natural Gas Chemistry*, vol. 18, pp. 15–20, 2009.

[LI 10] LI S.F., FAN S.S., WANG J.Q. *et al.*, "Semiclathrate hydrate phase equilibria for CO_2 in the presence of tetra-n-butyl ammonium halide (bromide, chloride, or fluoride)", *Journal of Chemical & Engineering Data*, vol. 55, pp. 3212–3215, 2010.

[LI 10] LI X.-S., XIA Z.-M., CHEN Z.-Y., *et al.*, "Equilibrium hydrate formation conditions for the mixtures of CO_2 + H_2 + tetrabutyl ammonium bromide", *Journal of Chemical & Engineering Data*, vol. 55, pp. 2180–2184, 2010.

[LI 11] LI M.-Y., WANG L.-S., GMEHLING J., "Thermodynamics of phase equilibria in aqueous strong electrolyte systems", *Industrial & Engineering Chemistry Research*, vol. 50, pp. 3621–3631, 2011.

[LIN 13] LIN W., DALMAZZONE D., FURST W. *et al.*, "Accurate DSC measurement of the phase transition temperature in the TBPB-water system", *Journal of Chemical Thermodynamics*, vol. 61, pp. 132–137, 2013.

[LIN 64] LINDENBAUM S., BOYD G.E., "Osmotic and activity coefficients for the symmetrical tetraalkyl ammonium halides and aqueous solution at 25°C", *Journal of Physical Chemistry*, vol. 68, pp. 911–917, 1964.

[LIP 02] LIPKOWSKI J., KOMAROV V.Y., RODIONOVA T.V. *et al.*, "The structure of tetrabutylammonium bromide hydrate $(C_4H_9)4NBr{\cdot}21/3H_2O$", *Journal of Supramolecular Chemistry*, vol. 2, pp. 435–439, 2002.

[LIP 02] LIPKOWSKI J., KOMAROV V.Y., RODIONOVA T.V. *et al.*, "The structure of tetrabutylammonium bromide hydrate", *Journal Supramolecular Chemistry*, vol. 2, pp. 435–439, 2002.

[MAR 08] MARTINEZ M.C., DALMAZZONE D., FÜRST W. *et al.*, "Thermodynamic properties of THF +CO2 hydrates in relation with refrigeration applications", *AiChE Journal*, vol. 54, pp. 1088–1095, 2008.

[MAS 04] MASOUDI R., TOHIDI B., DANESH A. *et al.*, "A new approach in modelling phase equilibria and gas solubility in electrolyte solutions and its applications to gas hydrates", *Fluid Phase Equilibria*, vol. 215, pp. 163–174, 2004.

[MAY 10] MAYOUFI N., DALMAZZONE D., FÜRST W. *et al.*, "CO_2 enclathration in hydrates of peralkyl-(ammonium/phosphonium) salts: stability conditions and dissociation enthalpies", *Journal of Chemical & Engineering Data*, vol. 55, pp. 1271–1275, 2010.

[MGA 91] MGAIDI A., FURST W., RENON H., "Representation of the solubility of lead chloride in various chloride solutions with pitzer model", *Metallurgical Transactions B-Process Metallurgy*, vol. 22, pp. 491–498, 1991.

[MOH 10] MOHAMMADI A.H., BELANDRIA V., RICHON D., "Use of an artificial neural network algorithm to predict hydrate dissociation conditions for hydrogen + water and hydrogen + tetra-n-butyl ammonium bromide + water systems", *Chemical Engineering Science*, vol. 65, pp. 4302–4305, 2010.

[MOH 13] MOHAMMADI A., MANTEGHIAN M., MOHAMMADI A.H., "Dissociation data of semiclathrate hydrates for the systems of tetra-n-butylammonium fluoride (TBAF) + methane + water, TBAF + carbon dioxide + water, and TBAF + nitrogen + water", *Journal of Chemical & Engineering Data*, vol. 58, pp. 3545–3550, 2013.

[MON 00] MONSON P.A., KOFKE D.A., "Solid-fluid equilibrium: insights from simple molecular models", *Advances in Chemical Physics*, vol. 115, pp. 113–179, 2000.

[MON 02] MONNIN C., DUBOIS M., PAPAICONOMOU N. *et al.*, "Thermodynamics of the LiCl + H_2O system", *Journal of Chemical & Engineering Data*, vol. 47, pp. 1331–1336, 2002.

[NAJ 15] NAJIBI H., MOMENI K., SADEGHI M.T. *et al.*, "Experimental measurement and thermodynamic modelling of phase equilibria of semi-clathrate hydrates of (CO_2 + tetra-n-butyl-ammonium bromide) aqueous solution", *Journal of Chemical Thermodynamics*, vol. 87, pp. 122–128, 2015.

[NAK 81] NAKAYAMA H., "Solid-liquid and liquid-liquid phase equilibria in the symmetrical tetraalkylammonium halide–water systems", *Bull. Chem. Soc. Jpn.*, vol. 54, pp. 3717–3722, 1981.

[NAK 87] NAKAYAMA H., "Hydrates of organic-compounds 11. Determination of the melting-point and hydration numbers of the clathrate-like hydrate of tetrabutylammonium chloride by differential scanning calorimetry", *Bulletin of the Chemical Society of Japan*, vol. 60, pp. 839–843, 1987.

[NIK 16] NIKOLAIDIS I.K., BOULOUGOURIS G.C., PERISTERAS L.D. *et al.*, "Equation-of-state modeling of solid liquid gas equilibrium of CO_2 binary mixtures", *Industrial & Engineering Chemistry Research*, vol. 55, pp. 6213–6226, 2016.

[OKU 16] OKUNIEWSKI M., PADUSZYNSKI K., DOMANSKA U., "(Solid plus liquid) equilibrium phase diagrams in binary mixtures containing terpenes: new experimental data and analysis of several modelling strategies with modified UNIFAC (Dortmund) and PC-SAFT equation of state", *Fluid Phase Equilibria*, vol. 422, pp. 66–77, 2016.

[OYA 05] OYAMA H., SHIMADA W., EBINUMA T. *et al.*, "Phase diagram, latent heat, and specific heat of TBAB semiclathrate hydrate crystals", *Fluid Phase Equilibria*, vol. 234, pp. 131–135, 2005.

[OYA 08] OYAMA H., EBINUMA T., NAGAO J. *et al.*, "Phase behaviour of TBAB semiclathrate hydrate crystal under several vapor components", *6th International Conference on Gas Hydrates*, Vancouver, 2008.

[PAP15] PAPAICONOMOUA N., SIMONIN J.P., BERNARD O., "Aqueous solutions of ionic liquids. Description of osmotic coefficients within the binding mean spherical approximation", *Journal of Chemical Thermodynamics*, vol. 91, pp. 445–451, 2015.

[PAR 10] PARICAUD P., TAZI L., BORGARD J.-M., "Modeling the phase equilibria of the HIx mixture using the SAFT-VRE equation of state: binary systems", *International Journal of Hydrogen Energy*, vol. 35, pp. 978–991, 2010.

[PAR 11] PARICAUD P., "Modeling the dissociation conditions of salt hydrates and gas semiclathrate hydrates: application to lithium bromide, hydrogen iodide, and tetra-n-butylammonium bromide + carbon dioxide systems", *Journal of Physical Chemistry B*, vol. 115, pp. 288–299, 2011.

[PAR 72] PARRISH W.R., PRAUSNITZ J.M., "Dissociation pressures of gas hydrates formed by gas mixtures", *Industrial & Engineering Chemistry Process Design and Development*, vol. 11, pp. 26–35, 1972.

[PAS 60] PASCAL P., *Nouveau Traité de Chimie Minérale, Tome XVI*, Masson et Cie, Paris, 1960.

[PÁT 06] PÁTEK J., KLOMFAR J., "Solid–liquid phase equilibrium in the systems of LiBr–H_2O and LiCl–H_2O", *Fluid Phase Equilibria*, vol. 250, pp. 138–149, 2006.

[PIC 93] PICKERING S.U., "Die hydrate der jodwasserstoffsäure", *Berichte der Deutschen Chemischen Gesellschaft*, vol. 26, pp. 2307–2310, 1893.

[PIN 05] PINHO S.P., MACEDO E.A., "Solubility of NaCl, NaBr, and KCI in water, methanol, ethanol, and their mixed solvents", *Journal of Chemical & Engineering Data*, vol. 50, pp. 29–32, 2005.

[PIT 73] PITZER K.S., MAYORGA G., "Thermodynamics of electrolytes: II activity and osmotic coefficients for strong electrolytes with one or both ions univalent", *Journal of Physical Chemistry*, vol. 77, pp. 2300–2308, 1973.

[PRA 00] PRAUSNITZ J.M., POLING B.E., O'CONNELL J.P., *The Properties of Gases and Liquids*, 5th ed., McGraw-Hill, New York, 2000.

[PRA 99] PRAUSNITZ J.M., LICHTENTHALER R.N., GOMES DE AZEVEDO E., *Molecular Thermodynamics of Fluid-Phase Equilibria*, 2nd ed., Prentice Hall, Englewood Cliffs, 1999.

[PRI 12] PRIVAT R., CONTE E., JAUBERT J.-N. *et al.*, "Are safe results obtained when SAFT equations are applied to ordinary chemicals? Part 2: study of solid-liquid equilibria in binary systems", *Fluid Phase Equilibria,* vol. 318, pp. 61–76, 2012.

[PRO 13] PROSIM, Simulis Thermodynamics®, Software, 2013.

[REN 68] RENON H., PRAUSNITZ J.M., "Local compositions in thermodynamic excess functions for liquid mixtures", *AIChE Journal*, vol. 14, pp. 135–144, 1968.

[RIC 13] RICAURTE M., DICHARRY C., BROSETA D. *et al.*, "CO_2 removal from a CO_2-CH_4 gas mixture by clathrate hydrate formation using THF and SDS as water-soluble hydrate promoters", *Industrial & Engineering Chemistry Research*, vol. 52, pp. 899–910, 2013.

[RIV 14] RIVA M., CAMPESTRINI M., TOUBASSY J. *et al.*, "Solid-liquid-vapor equilibrium models for cryogenic biogas upgrading", *Industrial & Engineering Chemistry Research*, vol. 53, pp. 17506–17514, 2014.

[ROB 02] ROBINSON R.A., STOKES R.H., *Electrolyte Solutions*, 2nd revised ed., Dover Publications, New York, 2002.

[RUT 11] RUTHER F., SADOWSKI G., "Thermodynamic modeling of solubility", *Chemie Ingenieur Technik*, vol. 83, pp. 496–502, 2011.

[SAK 08] SAKAMOTO J., HASHIMOTO S., TSUDA T. *et al.*, "Thermodynamic and Raman spectroscopic studies on hydrogen plus tetra-n-butyl ammonium fluoride semi-clathrate hydrates", *Chemical Engineering Science*, vol. 63, pp. 5789–5794, 2008.

[SAP 29] SAPGIR VII S., "Application de l'analyse thermique à des mélanges binaires de composés organiques fondant à tres basse temperature", *Bulletin des Sociétés Chimiques Belges*, vol. 38, pp. 392–408, 1929.

[SAT 13] SATO K., TOKUTOMI H., OHMURA R., "Phase equilibrium of ionic semiclathrate hydrates formed with tetrabutylammonium bromide and tetrabutylammonium chloride", *Fluid Phase Equilibria*, vol. 337, pp. 115–118, 2013.

[SEI 01] SEILER M., GROSS J., BUNGERT B. *et al.*, "Modeling of solid/fluid phase equilibria in multicomponent systems at high pressure", *Chemical Engineering and Technology*, vol. 24, pp. 607–612, 2001.

[SHI 03] SHIMADA W., EBINUMA T., OYAMA H. *et al.*, "Separation of gas molecule using tetra-n-butyl ammonium bromide semi-clathrate hydrate crystals", *Japanese Journal of Applied Physics*, vol. 42, pp. L129–L131, 2003.

[SHI 05] SHIMADA W., SHIRO M., KONDO H. *et al.*, "Tetra-n-butylammonium bromide + water (1/38)", *Acta Crystallographica*, vol. C61, pp. o65–o66, 2005.

[SLO 98] SLOAN E.D., *Clathrate Hydrates of Natural Gases*, 2nd ed., Taylor & Francis, New York, 1998.

[SLO 08] SLOAN E.D., KOH C.A., *Clathrate Hydrates of Natural Gases*, 3rd ed., Taylor & Francis, New York, 2008.

[SLU 97] SLUSHER J.T., CUMMINGS P.T., "Molecular simulation study of tetraalkylammonium halides. 1. solvation structure and hydrogen bonding in aqueous solutions", *J. Phys. Chem. B*, vol. 101, pp. 3818–3826, 1997.

[STE 27] STEUDEMANN W., "Die thermische Analyse der Systeme desWassers mit den Lithiumhalogeniden", Jena, 1927.

[STR 13] STRINGARI P., CAMPESTRINI M., "Application of the SLV-EoS for representing phase equilibria of binary Lennard-Jones mixtures including solid phases", *Fluid Phase Equilibria*, vol. 358, pp. 68–77, 2013.

[STR 14] STRINGARI P., CAMPESTRINI M., COQUELET C. *et al.*, "An equation of state for solid-liquid-vapor equilibrium applied to gas processing and natural gas liquefaction", *Fluid Phase Equilibria*, vol. 362, pp. 258–267, 2014.

[SUG 12] SUGINAKA T., SAKAMOTO H., IINO K. *et al.*, "Thermodynamic properties of ionic semiclathrate hydrate formed with tetrabutylphosphonium bromide", *Fluid Phase Equilibria*, vol. 317, pp. 25–28, 2012.

[SUN 08] SUN Z.-G., JIANG C.-M., XIE N.-L., "Hydrate equilibrium conditions for tetra-n-butyl ammonium bromide", *Journal of Chemical & Engineering Data*, vol. 53, pp. 2375–2377, 2008.

[SUN 11] SUN Z.-G., LIU C.-G., ZHOU B. *et al.*, "Phase equilibrium and latent heat of tetra-n-butylammonium chloride semi-clathrate hydrate", *Journal of Chemical & Engineering Data*, vol. 56, pp. 3416–3418, 2011.

[TSI 09] TSIVINTZELIS I., ECONOMOU I.G., KONTOGEORGIS G.M., "Modeling the solid-liquid equilibrium in pharmaceutical-solvent mixtures: systems with complex hydrogen bonding behvaior", *Aiche Journal*, vol. 55, pp. 756–770, 2009.

[TUM 07] TUMAKAKA F., PRIKHODKO I.V., SADOWSKI G., "Modeling of solid–liquid equilibria for systems with solid-complex phase formation", *Fluid Phase Equilibria*, vol. 260, pp. 98–104, 2007.

[VAN 59] VAN DER WAALS J.H., PLATTEEUW J.C., "Clathrate solutions", *Advances in Chemical Physics*, vol. 2, pp. 1–57, 1959.

[VAN 04] VAN DER WAALS J.D., *On the Continuity of the Gaseous and Liquid States*, Dover Publications Inc, New York, 2004.

[VEG 02] VEGA C., BLAS F.J., GALINDO A., "Extending Wertheim's perturbation theory to the solid phase of Lennard-Jones chains: Determination of the global phase diagram", *Journal of Chemical Physics*, vol. 116, pp. 7645–7655, 2002.

[VEG 03] VEGA C., MCBRIDE C., DE MIGUEL E. *et al.*, "The phase diagram of the two center Lennard-Jones model as obtained from computer simulation and Wertheim's thermodynamic perturbation theory", *Journal of Chemical Physics*, vol. 118, pp. 10696–10706, 2003.

[VEG 04] VEGA C., MACDOWELL L.G., MCBRIDE C. *et al.*, "Molecular modeling of flexible molecules. Vapor-liquid and fluid-solid equilibria", *Journal of Molecular Liquids*, vol. 113, pp. 37–51, 2004.

[WAG 82] WAGMAN D.D., EVANS W.H., PARKER V.B. *et al.*, "The NBS tables of chemical thermodynamic properties: Selected values for inorganic and C21 and C2 organic substances in SI units", *Journal of Physical and Chemical Reference Data,* vol. 2, 1982.

[WIE 07] WIERZCHOWSKI S.J., MONSON P.A., "Calculation of free energies and chemical potentials for gas hydrates using Monte Carlo simulations", *Journal of Physical Chemistry B*, vol. 111, pp. 7274–7282, 2007.

[YE 12] YE N., ZHANG P., "Equilibrium data and morphology of tetra-n-butyl ammonium bromide semiclathrate hydrate with carbon dioxide", *Journal of Chemical & Engineering Data*, vol. 57, pp. 1557–1562, 2012.

[YOK 03] YOKOZEKI A., "Analytical equation of state for solid-liquid-vapor phases", *International Journal of Thermophysics*, vol. 24, pp. 589–620, 2003.

[YOK 04] YOKOZEKI A., "Solid-liquid-vapor phases of water and water-carbon dioxide mixtures using a simple analytical equation of state", *Fluid Phase Equilibria*, vol. 222, pp. 55–66, 2004.

[YOK 05] YOKOZEKI A., "Methane gas hydrates viewed through unified solid-liquid-vapor equations of state", *International Journal of Thermophysics*, vol. 26, pp. 743–765, 2005.

[ZHA 13] ZHANG P., YE N., ZHU H. *et al.*, "Hydrate equilibrium conditions of tetra-n-butylphosphonium bromide plus carbon dioxide and the crystal morphologies", *Journal of Chemical & Engineering Data*, vol. 58, pp. 1781–1786, 2013.

6

Volume and Non-Equilibrium Crystallization of Clathrate Hydrates

6.1. Introduction

Phase equilibria involving clathrate hydrates are usually modelled by using equations and tools from standard thermodynamics. For given temperature and pressure conditions (*T* and *P*), the equilibria between the different phases (hydrate, liquids, vapor, etc.) are obtained by minimizing the Gibbs free energy or, if there is no ambiguity, by equalizing the chemical potentials (or fugacities) of each component in all their different phases, as described in numerous books [PRA 99, DEH 13] (see also the book by Sloan and Koh [SLO 07] or Chapter 5 for hydrates).

Usually, to model liquid-hydrate equilibrium (LHE), it is necessary to determine the equilibrium curve (*PT*) for a given gas composition. In addition, the hydrate cage filling is commonly resolved with a Langmuir-like approach. Furthermore, the compositions of the fluid phases in equilibrium with the hydrate are also of interest. Gibbs' phase law states that the degree of freedom (intensive variables) is 2 plus the number of components minus the number of phases at equilibrium (n + 2 - φ). As a consequence, when considering the vapor-liquid-hydrate equilibrium, the hydrate composition x^H (fraction of guest molecules in occupied cavities) can be seen as a function of the temperature (or pressure), and the liquid (or gas) composition:

$$x^H = f(T, x^L) = f'(P, x^L) = g(T, x^V) = g'(P, x^V) \qquad [6.1]$$

First, there is little hydrate composition data available in the literature because this measurement remains a challenge. While this composition can often be obtained by mass balance calculations in instrumented high pressure cells [HER 11], or by the

Chapter written by Baptiste BOUILLOT and Jean-Michel HERRI.

use of analytical techniques such as Raman spectroscopy, nuclear magnetic resonance (NMR), X-ray, or neutron diffraction (see Chapters 1 and 2 in this book), these methods require a significant amount of hydrate, especially for mass balance calculations. In particular, it is difficult to determine the composition of nuclei that start crystallizing at liquid saturation of gas molecules. If the liquid composition changes during the crystallization process, as is the case for batch experiments, the hydrate composition changes accordingly. Hence, it becomes harder to obtain the hydrate composition with accuracy.

Second, for practical applications, pipelines for example, predicting the amount of hydrates to be formed in such crystallization processes is crucial. For instance, the expected hydrate volume in flow-assurance applications is used to determine the quantity of kinetic hydrate inhibitors (KHIs), or anti-agglomerants (AAs) to use [SLO 04]. Also, for gas capture and storage by means of hydrates, the amount of the solid phase (linked to the capture or storage capacities of the hydrates) must be simulated. In process simulations, thermodynamic flash calculations are conducted to combine phase equilibria and mass balance. Indeed, such calculations provide a complete description of both the intensive and extensive properties of the system, assuming that each phase is in equilibrium with the other. This assumption for species like hydrates may be presumptuous since their composition (cage filling) is not stoichiometric. In fact, the guest molecule composition is supposed to depend on the local fluid composition which can change during crystallization. From this perspective, the final state can be dependent on the pathway of crystallization.

Third, for crystallization to occur, with all transfer phenomena involved, a non-equilibrium state is needed. When the system is not yet at thermodynamic equilibrium, there is a driving force toward stability. Equation [6.1] is interesting for this purpose. The crystallization conditions are indeed a function of the distance from thermodynamic equilibrium. In crystallization science of a solute dissolved in a solvent, the supersaturation ratio is usually defined as the solute concentration relative to the saturation (equilibrium) concentration: $S = c/c^*$ [MUL 01]. The larger S, the larger the driving force for crystallization. Other expressions for the driving force are $\Delta c = c-c^*$, or $S = a/a^*$, with a the activity of the solute (more rigorous). Nevertheless, these expressions cannot be simply adapted to the case of clathrate hydrates. Indeed, there are several formers (water + guest molecules), and the crystalline form is not stoichiometric.

However, the driving force for crystallization remains of prime importance because it controls the kinetics of the clathrate formation process. Moreover, this force can correspond to different guest fillings, or polymorphic structures. This is why section 6.2 deals with the expression of the driving force for clathrate hydrates. Then, some evidence in the literature of non-equilibrium formation of clathrate hydrates is presented and discussed.

In sections 6.3, 6.4 and 6.5 thermodynamics at equilibrium and non-equilibrium conditions are used to model the hydrate composition and volume. Standard approaches (isobaric flash, semi-batch conditions) as well as non-stoichiometric frameworks are detailed briefly. Note that, "non-stoichiometric" stands for heterogeneous guest filling in the hydrate structure. Therefore, diverse compositions can be considered, along with different percentages of occupancy.

Finally, in section 6.6, various kinetics approaches are considered to simulate the non-equilibrium hydrate growth from mass transfer laws.

6.2 Driving force and evidence for non-equilibrium gas hydrate crystallization

6.2.1. *Driving force*

Often, the driving force of the hydrate crystallization is considered to be the temperature or pressure difference compared to the equilibrium state (ΔT or ΔP). As discussed previously, these expressions are different from those used for solute crystallization in a solvent [MUL 01]. Certainly, there is not a single former molecule, and the hydrate phase is non-stoichiometric. For the simple guest hydrate case, the reaction can be written as follows:

$$G + n_w H_2O \rightleftharpoons G.n_w H_2O \qquad [6.2]$$

In 2002, Kashchiev and Firoozabadi [KAS 02a] provided a more rigorous expression for the driving force of single guest clathrate:

$$\Delta\mu = \mu^H - \mu^L_{gas} - n_w \mu^L_w \qquad [6.3]$$

This equation is the difference between the chemical potential of the hydrate (μ^H), and the chemical potentials of the water (μ^L_w) and dissolved gas (μ^L_{gas}) in the aqueous phase. n_w is the hydration number, i.e. the number of moles of water per mole of gas molecules in the hydrate. Thus, this can be considered as the stoichiometry of the crystallization reaction, or the stoichiometry of the cage filling. Note that not all cages necessarily host a guest molecule. Thus, the hydration number includes both the guest composition and the number of empty cavities for a given hydrate structure. For several guest molecules, identifying the number of gas molecules, as well as the number of water molecules, is more suitable. If $\Delta\mu < 0$, the system is metastable. Crystallization should occur, if the critical nucleus size is reached, and existing crystals will grow. Otherwise, the hydrate crystals that could exist are metastable and should dissociate. When $\Delta\mu = 0$, the hydrate liquid phases are at thermodynamic equilibrium, and nothing happens. Since the main concern of hydrates is their formation or dissociation, the latter case is not relevant in dynamic systems. Thus, $\Delta\mu$ needs an appropriate expression.

The chemical potential of a single gas molecule (*j*) dissolved in the liquid phase, according to an excess approach, can be written [FIR 99]:

$$\mu_{gas,j}^{L} = \mu_{gas,j}^{L,0} + RTln(x_j\gamma_j) \quad [6.4]$$

$\mu_{gas,j}^{L,0}$ is the chemical potential in the reference state (pure gas or, more likely, infinite dilution according to Henry's approach). γ_j is the activity coefficient, and is a function of *T*, *P* and the mole fraction x_j. Considering that the number of moles for gas molecules is negligible compared to the number of water molecules ($N_j \ll N_w$), the mole fraction x_j is a function of the gas concentration and the water molar volume, $x_j = C_j V_{w,m}$.

From equation [6.4], the driving force becomes:

$$\Delta\mu(P,T,C_j) = \mu^H - \mu_{gas,j}^{L,0}(P,T) - RTln\left(C_j V_{w,m}\gamma_j(T,P,C_j)\right) - n_w\mu_w^L(P,T) \quad [6.5]$$

This means that the driving force is logarithmically dependent on the gas concentration in the liquid phase. In addition, the crystal stoichiometry, or cage filling, is usually assumed to be constant (full occupancy of cavities). However, this is not correct in defining actual conditions, especially if there are several guest molecules. This will be discussed in section 6.2.2.

The Gibbs-Duhem equation provides the expression for the chemical potential of molecule *j* in phase π [PRA 99, FIR 99]:

$$\forall_{j\in molecule,}\forall_{\pi\in phase,}\mu_{\pi,j}(P,T) = \mu_{\pi,j}^0(P^0,T^0) + \int_{P^0}^{P} V_{m,j}dP + \int_{T^0}^{T} S_m dT +$$

$$RTln\left(\frac{f_{\pi,j}(P,T,x_j)}{f_{\pi,j}^0(P,T,x_j)}\right) \quad [6.6]$$

with subscript *0* designating the reference state, and *f* the fugacity. The fugacity ratio may be expressed within the residual (φ) or excess (γ) approaches. This is valid for all the fugacities in equation [6.6], and thus on all chemical potentials in equation [6.5]. In addition, the water chemical potential μ_w^L can be considered as a function of (*T*, *P*) with the assumption ($N_j \ll N_w$). Note: the influence of the pressure is provided by a less significant water molar volume.

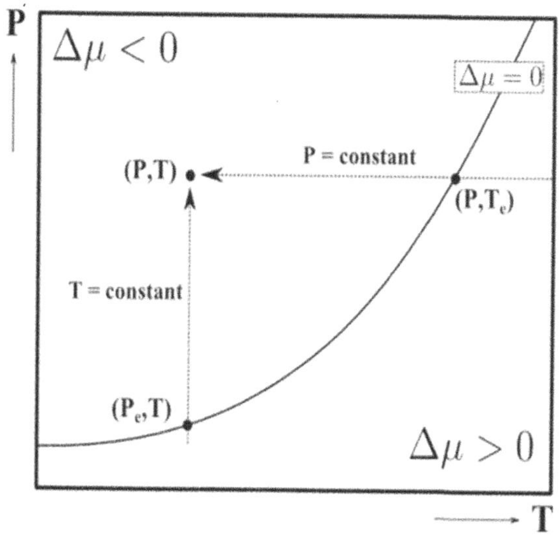

Figure 6.1. *P,T hydrate phase diagram Driving force at isobaric and isothermal regimes [KAS 02a]*

As for Kashchiev and Firoozabadi, two regimes can be studied: isobaric and isothermal (Figure 6.1). In addition, for both, the gas phase may or not be at thermodynamic equilibrium with the aqueous phase. In this last situation, the driving force is also affected by the dynamics of gas-to-liquid mass transfer phenomena. Of course, if crystallization is supposed to happen in bulk, the guest molecules need to move from the gas to the hydrate phase. The rates of mass transfer at the gas/liquid interface, in the bulk and at the liquid/hydrate interface influence the local gas concentration, and so the driving force. Hence, for transient regimes, $\Delta\mu$ can be significantly affected by the time *t*, and the operating conditions (geometry, stirring, etc.).

As an example, for an isothermal regime, when the gas and the liquid phases are at thermodynamic equilibrium, Kashchiev and Firoozabadi [KAS 02a] suggested the following expression for the driving force (or supersaturation), for single guest hydrates:

$$\Delta\mu = RTln\left(\frac{\varphi(P,T)P}{\varphi(P_e,T)P_e}\right) + \Delta v_e(P - P_e) \qquad [6.7]$$

where Δv_e is the difference between the volumes of n_w moles of water in the solution and one mole of a hydrate building unit. Note that, in this case, both vapor and liquid phases are at thermodynamic equilibrium. Thus, the chemical potential for both phases are the same, and its expression for the liquid phase can be written as a function of the fugacity coefficient only.

If more than one guest molecule is involved, equation [6.2] has to take into account all guest compounds. Equation [6.3] is changed and the terms need to be weighted by the guest fraction in the mixed hydrate. This expression was not established by Kashchiev and Firoozabadi. However, Gnanendran *et al.* [GNA 04] proposed an expression for hydrates with multicomponent guests (each guest molecule being labeled with subscript m):

$$\Delta\mu = kT\left(\sum_m x_m^h ln\left(\frac{\varphi(P,T)_m P}{\varphi(P_e,T)_m P_e}\right)\right) + \Delta v_e(P - P_e) \qquad [6.8]$$

where x_m^h is the water-free mole fraction of guest m in the hydrate phase.

Another approach was suggested by Arjmandi *et al.* [ARJ 05]. Since most of the experimental studies are performed at isothermal and isobaric conditions, to simulate field situations, they compared the rigorous driving force to subcooling ($\Delta T = T - T_e$, see also Figure 6.1). They concluded that, for single guest hydrates, the driving force is proportional to ΔT. In addition, at constant subcooling, the higher the pressure (> 20 MPa), the better ΔT is representative for $\Delta\mu$. For double hydrates, they observed the same tendency, although with less accuracy. Under 20 MPa, the use of ΔT underestimates the real driving force. Beyond this value, approximations are good, and a constant degree of subcooling can be considered as constant driving force.

A different idea was also put forward by Subramanian and Sloan [SUB 02], and Ohmura *et al.* [OHM 04]. Although the driving force can be correlated to pressure and thermal conditions, it can also be observed from the dissolved gas concentration. At given temperature and pressure, vapor-liquid equilibrium (VLE) provides the gas concentration in the aqueous phase at saturation. Otherwise, liquid-hydrate equilibrium limits this concentration. Figure 6.2 presents this idea at constant pressure.

If the three phase system is not at equilibrium, there should be a mass transfer in the aqueous phase, from bulk (at x^{VLE}) to the hydrate film surface. The driving force for the mass transfer is then dynamically changing, and the composition at the hydrate interface should not be x^{VLE} for a quick mass transfer process.

To study the hydrate crystallization driving force ($\Delta\mu$) exactly, more attention needs to be paid to the cage filling, n_w, which is somewhat connected to the hydrate phase, since this term represents the hydration number. An expression for the occupancy of the hydrate cavities is essential. The occupancy matrix θ (see section 6.2.2 for its definition) may be determined under conditions of thermodynamic equilibrium. If non-thermodynamic equilibrium is examined, note that the driving force may remain negative ($\Delta\mu < 0$, saturated liquid phase compared to a given hydrate structure and filling) while the occupancy of the cavities is not that which is expected at equilibrium. In other words, multiple hydrate guest occupancies may present a theoretically lower chemical potential than the hydrate formers. Thus, such crystals, not the most stable, could be formed. First, let us discuss cage occupancy from equilibrium thermodynamics, and non-equilibrium clathrate formation in further sections.

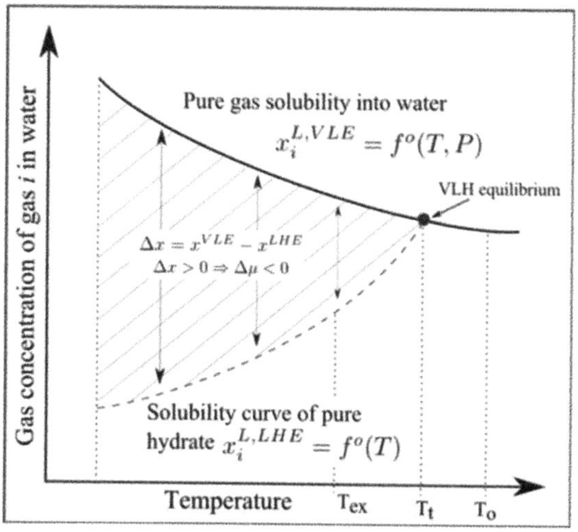

Figure 6.2. *Dependency of a dissolved gas molecule according to liquid-hydrate and vapor-liquid equilibria on the temperature (at a given vapor pressure). Δx > 0 represents a driving force of crystallization, showing an oversaturated gas concentration regarding LHE*

6.2.2. *Cage occupancy from equilibrium thermodynamics*

Cage occupancy is usually modeled within the theory of van der Waals and Platteeuw [VAN 59]. In this theory, the chemical potential of the hydrate phase is described as:

$$\mu^H = \mu^{H-\beta} + RT \sum_i \vartheta_i \ln\left(1 - \sum_j \theta_i^j\right) \quad [6.9]$$

where $\mu^{H-\beta}$ is the chemical potential of the hypothetic empty hydrate, R the gas constant, T the temperature and ϑ_i the number of cavities i per water molecule. This expression also involves the so-called occupancy factor of the cavities, θ_i^j, i.e. the fraction of cavities i hosting a molecule j (θ_i^j is the matrix θ mentioned above). This factor is usually determined from a Langmuir approach as follows:

$$\theta_i^j = \frac{C_i^j f^j(T,P)}{1+\sum_j C_i^j f^j(T,P)} \qquad [6.10]$$

In this model, C_i^j is the Langmuir constant for molecule j and cavity i, and f^j is the fugacity of gas molecule j. The Langmuir constants are intrinsic parameters of each molecule j for a given cavity i. They can be considered as constants under some assumptions (at a set temperature, cavities of given size). Using an interaction potential, usually the Kihara potential [MCK 63], C_i^j is calculated as follows:

$$C_i^j = \frac{4\pi}{k_B T}\int_0^{R-a} \exp\left(-\frac{w(r)}{k_B T}\right) r^2 dr \qquad [6.11]$$

This Langmuir constant is only a function of the interaction potential used, w, the temperature, T, the distance from the cavity, R–a, R being the cavity radius, and a, the hard core radius of molecule j.

At thermodynamic equilibrium, the concentration of the gas in the liquid phase only depends on the partial pressure of the gas and the temperature. The fugacities of the hydrate guest molecules are the same in the vapor phase and the liquid phase. However, these fugacities in the liquid phase are time dependent in any dynamic process. Therefore, θ_i^j is also a function of time and, by extension, the driving force for the thermodynamically stable hydrate as well.

In this first approach, the hydrate phase is considered to be an ideal solid mixture. In 2002, Ballard and Sloan [BAL 02b] provided a modification in equation [6.9] to account for the distortion of hydrates due to guest filling (Figure 6.3).

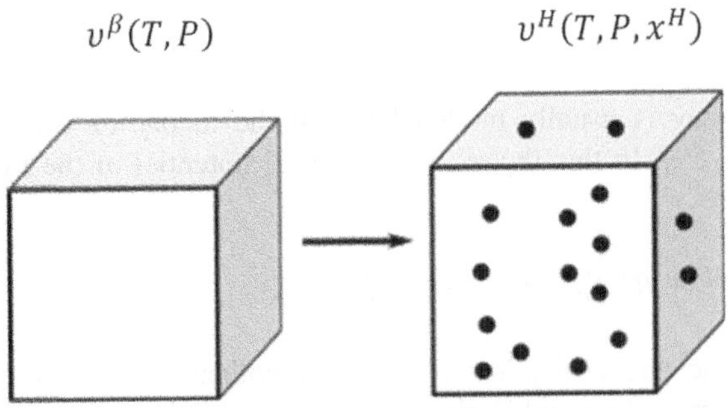

Figure 6.3. *Ballard and Sloan description of cavity distortion due to guest molecules [BAL 02b]*

They simply included an activity coefficient γ_w^H to account for the excess chemical potential of the hydrate phase, and model the excess volume of mixing $\Delta v^{H,Ex}$. Hence, equation [6.9] becomes:

$$\mu^H = \mu^{H-\beta} + RT\sum_i \vartheta_i \ln\left(1 - \sum_j \theta_i^j\right) + RTln(\gamma_w^H) \qquad [6.12]$$

Due to this excess volume, the cavity radius is affected, and the integration in equation [6.11] changes too. This modification provides better results at high pressure (P > 200 bars). It is also capable of capturing the maximum temperature equilibrium for SI methane hydrates.

Finally, the driving force mostly depends on pressure, temperature, gas concentration and cage filling. Most of all, it can remain negative in many situations, so that a crystal can be formed at a metastable state. Hence, different hydrates (structure and filling) can be formed in theory, just looking at their corresponding driving force. Some of them will be metastable, and will eventually change into a stable form over time. Evidence of this metastable hydrate formation is observed from experiments and molecular simulations. Some examples to support this "new" approach of non-equilibrium thermodynamics of mixed clathrate hydrates will be provided next.

6.3. Non-equilibrium hydrate formation?

In section 6.2.2, the driving force for the clathrate hydrate crystallization was discussed, and the non-equilibrium formation suggested. Before going any deeper in the literature, further observations can be made theoretically. Looking at hydrate formation thermodynamically, various issues arise in the evaluation of the stability, as highlighted by Kvamme *et al.* [KVA 14]. First, crystallization can occur in places other than the liquid bulk, such as interfaces in the system (minerals in sediments for instance). Thus, guest molecules that form the new hydrate phase can be in the bulk, gas phase, or adsorbed. Each time, they differ in chemical potential, and hence the driving force for crystallization is different. Then, if the cage filling is distinct for each new hydrate that is formed, each of these should be considered as separate phases. As a consequence, Gibbs' phase rule predicts an over-determined thermodynamic system, without any possibility of reaching thermodynamic equilibrium without forming new phases, which further overdetermine the system. This is why Kvamme *et al.* [KVA 14] suggested that thermodynamic equilibrium could not be reached.

6.3.1. *Evidence from experimental studies*

Section 6.2 provided theoretical clues to understand hydrate crystallization and its potential non-equilibrium behavior. Although equilibrium thermodynamics is capable of calculating the most stable state (structure and filling), this does not mean that a metastable state will not occur. This metastable state can be a difference in cage occupancy, and/or in hydrate structure (polymorphism). Hydrate structure is sometimes not expected, especially if the gas mixture composition changes during the crystallization process. Subramanian *et al.* [SUB 00] observed that, even if methane/ethane mixture forms SI hydrate, there is a transition from SI to SII under a certain range of compositions. Therefore, when hydrate from methane-ethane gas mixture is crystallized, there is a possibility that this transition zone is attained. Will the final solid phase be at thermodynamic equilibrium? Will the solid phase be transformed to reach the minimum Gibbs energy? This should be related to mass transfer phenomena, as well as kinetics of crystallization.

In addition, metastable states have been observed experimentally. Different techniques can be utilized to study phase equilibrium: high pressure cells, Raman and X-ray spectroscopy, nuclear magnetic resonance (NMR), or neutron diffraction [SLO 07, ESL 12]. Modeling can also be used as a tool to detect discrepancies between experimental results and thermodynamics. By this means, erroneous data or measurements resulting from metastable crystallization can be exposed [ESL 13, LEQ 16].

In this spirit, Eslimanesh *et al.* [ESL 13] suggested a statistical approach to discuss the degree of confidence of hydrate dissociation data. They compared a significant number of dissociation data at given vapor compositions to standard thermodynamic models and a correlation [ADI 91]. The use of dissociation data is relevant since it does not require any information on the hydrate phase (volume or composition). Among the large amount of data observed, Eslimanesh *et al.* found five probably doubtful experimental data.

Experimentally, many authors pointed out the non-equilibrium behavior of gas hydrate crystallization. Schicks and Ripmeester [SCH 04] witnessed the formation of both structure I and II from pure methane under moderatre conditions ($P <$ 100MPa, $T < 20$°C), while structure II is unlikely to occur. Raman spectroscopy was used to detect structure II in a small pressurized cell in pressure range of 3–9MPa, and temperature range +1.5–12°C.

Shin *et al.* [SHI 12] investigated the hydrate formation from water-methane-hexamethyleneimine (HMI) mixture. HMI is a highly water-soluble large guest molecule (LGM) capable of forming SH hydrates. Experiments were performed from ground frozen aqueous solution of 2.9 mol% HMI, then pressurized with CH_4 at 120 bars, and finally matured at 273.15 K. The authors investigated the crystal structure as function of the elapsed time by ^{13}C NMR and Raman spectroscopy. They witnessed the formation of both SII and SH structures in roughly equivalent amounts at the beginning of the crystallization, during the first 30 minutes. Then, in the second period, they noted a disappearance of the SII signal in favor of the SH structure (Figure 6.4(a)). Therefore, they concluded that, while SH is the most thermodynamically stable structure, SII is kinetically preferred over SH, and is formed as a metastable phase in the first step of the hydrate formation. They also did the same experiments starting with a different HMI content in water (1.23 mol%) and discovered another phase partition and time evolution (Figure 6.4(b)): the SII structure was no longer dominant at the very beginning and both SI and SH structures were present. Over time, the SII crystals disappeared, while SI increased to three times greater, as SH became the dominant solid phase (Figure 6.4(b)).

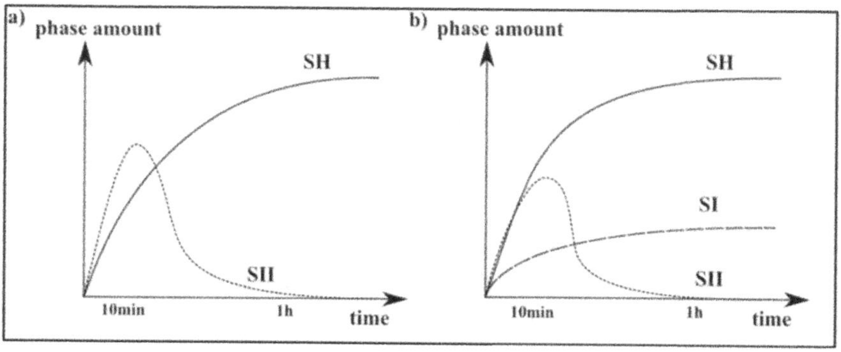

Figure 6.4. *Simplified results from Shin et al. [SHI 12] for HMI+CH4 system in water for: a) 2.9 mol% HMI and b) 1.23 mol% HMI*

According to the authors, the 5^{12} (pentagonal dodecahedra) face sharing of structure II, with the presence of small guest molecules such as CH_4, facilitate, kinetically, the formation of structure II hydrates. This explanation is supported by Molecular Dynamics (MD) simulations [WAL 09]. Put simply the conclusion is the following: "the kinetic preference for a specific structure does not coincide with thermodynamic preferences".

The same observation was made by Zhu *et al.* [ZHU 14] concerning CO hydrates. In their study, they witnessed an SII hydrate formation after the first SI crystallization under moderate pressure/ low temperature conditions (about 250 K and 170 bars). During the first two weeks, only SI hydrates were formed. Then, after

17 weeks, both SI and SII structures were found. However, although it was demonstrated that SII is thermodynamically more stable [MIL 85], other experimental work showed that CO hydrate usually forms structure I [DAV 87, DES 90, MOH 05, MOH 09, MOH 10, DAR 11]. According to Zhu *et al.*, structure I is kinetically favored, while SII structure is thermodynamically more stable. As expected, when it comes to thermodynamic studies and kinetic phenomena, there is a discrepancy between the model and the experimental results.

Le Quang *et al.* [LEQ 16] compared the results of hydrate crystallization at quick and slow rates from hydrocarbon gas mixtures in a pressurized batch reactor. These results were compared to the predictions of the standard thermodynamic approach for liquid hydrate equilibrium (LHE). In a CO_2-CH_4-C_2H_6 gas mixture, for a given initial gas composition, this approach was able to reliably predict the pressure and hydrate composition of the experimental data obtained at low crystallization rate. In contrast, the results provided by a quick crystallization process were not well predicted. Since these batch experiments, at constant volume, produce a significant amount of hydrates, the collected results can only be mean values. They could not take into account the probable heterogeneous hydrate composition. The results obtained from mass balance calculations consider the hydrate phase as a homogeneous solid. Therefore, the path taken of the local crystallization process cannot be analyzed by this means. However, the fact that the simulated results are closer to experimental observations for the slow crystallization process gives a clue on the mechanism. Quick and slow processes do not produce the same hydrate phase. Moreover, a slow crystallization process is better predicted by the standard thermodynamic approach. Thus, this procedure forms a more homogeneous hydrate phase, contrary to quick crystallization. In the end, the question remains: how can we predict the final state after a non-stoichiometric crystallization process of mixed hydrates? In other words: How can we simulate the path taken in mixed hydrate crystallization?

6.3.2. *Clathrate hydrates in fluid inclusions*

Similar non-equilibrium observations were obtained by means of microthermometric measurements combined with micro-Raman spectrometry by Murphy and Roberts [MUR 97] for a different chemical system: aqueo-carbonic fluid inclusions (isochoric processes). They had detected two years before [MUR 95] that, in a given fluid inclusion, two or more clathrate phases with diverse melting temperatures, could be present. Their conclusion was that the clathrate phase does not rapidly change in composition once formed, so that thermodynamic equilibrium cannot be reached. Working with CO_2-rich gases (such as mixtures of

CO_2, N_2 and CH_4), they noticed in particular that the composition of gas hydrates present in the bulk of the aqueous phase, i.e. far from the water/gas interface, were richer in CO_2 than the hydrates formed near the interface. Many clathrates of different compositions can coexist within an inclusion. They pointed out that standard thermodynamic approaches are not suited to model these inclusions. However, a local approach could be used (see section 6.5.4.3 for isochoric non-equilibrium approach).

6.3.3. *Evidence from molecular dynamics*

Non-equilibrium behavior has also been investigated and examined through Molecular Dynamics (MD). In 2011, Sum *et al.* [SUM 11] published a review on this subject. Among the perspectives and issues of modeling clathrate hydrates, the nucleation and growth steps are particularly interesting. Indeed, molecular simulation reveals key aspects that cannot -or hardly- be detected experimentally. Along this idea, the works of Vatamanu and Kusalik [VAT 06], Walsh *et al.* [WAL 09], or Jacobson *et al.* [JAC 10] are remarkable. Vatamanu and Kusalik studied the heterogeneous crystallization of methane hydrates. They discovered that the [001] crystallographic face of SI hydrate could support the crystal growth of face [001] of structure II. Walsh *et al.* observed the formation of 5^{12}, $5^{12}6^2$, $5^{12}6^3$ $5^{12}6^4$ cages during nucleation and initial growth. Cages 5^{12} and $5^{12}6^2$ lead to SI while 5^{12} and $5^{12}6^4$ to SII. Cages $5^{12}6^3$ are uncommon and do not form any standard structure. According to Walsh *et al.*, this allows the coexistence of both SI and SII. At the end of the simulation, both structures coexist. As seen in paragraph 6.3.1, this has later been experimentally confirmed by Shin *et al.* [SHI 12] for hydrates hosting methane and a large guest molecule (such as HMI). This conclusion is the basis for the question of which is the kinetically favored structure? Jacobson *et al.* [JAC 10] also noticed this competition between SI and SII methane hydrates by MD. In addition, they observed an amorphous phase containing $5^{12}6^n$ cages prior to the crystalline structure. According to them, the crystallization process of clathrate hydrate can be described as: *solution* $\leftrightarrow$ *blob* $\rightarrow$ *amorphous clathrate* $\rightarrow$ *crystalline clathrate* (a blob is "an amorphous cluster involving multiple guest molecules in water-mediated configurations", according to Jacobson *et al.*). In this spirit, Subramanian and Sloan [SUB 99], as well as Schicks *et al.* [SCH 13, SCH 15], detected the formation of 5^{12} cages, with Raman spectroscopy, before the formation of the large cages, from methane, and thus at the early stage before crystallization.

Following the works of Jacobson *et al.*, Lauricella *et al.* investigated, by MD, the crystallization of methane [LAU 14] and hydrogen sulfure (H_2S) hydrates [LAU 15], both known to be SI hydrates [SLO 07, LIA 11]. In the first work, they obtained only SII hydrates at 273 K and 500 *atmospheres*. The local composition in water at the nanoscale could be responsible for this result. Moreover, in the case of H_2S hydrates, structure II could be enhanced kinetically, whereas the most stable form is known to be Structure I.

In the same vein, Nguyen *et al.* [NGU 12, NGU 13, NGU 14] investigated cross-nucleation between structure I and II of some clathrates (methane hydrates in 2012, bromine clathrates in 2013 and 2014). They observed that the structure that is formed is not necessarily the most stable thermodynamically, but the one that shows the highest growth rate. Like the works of Walsh *et al.* [WAL 09] and Jacobson *et al.* [JAC 10], uncommon $5^{12}6^3$ cages are necessary at the interface of SI and SII phases.

This non-exhaustive review of clathrate hydrate formation simulation through MD, while interesting in its own right will not be developed any further since it is not within the scope of this chapter. Note to our readers: such new developments open original perspectives on the understanding and control of hydrate structures.

6.3.4. *Experimental and modeling issues*

Summing up the previous paragraph, the structure and content of the clathrate hydrates can be driven by kinetics. At the onset, the initially formed, kinetically preferred, crystalline structure may remain and coexist with other structures. Based on the speed of crystallization, it is also possible to form a more stable and homogeneous hydrate phase, if done slowly.

In order to investigate the non-equilibrium behavior, experimental data are needed, as well as modeling. Unfortunately, while many experiments provide the equilibrium pressure of gas hydrate for a given temperature and gas phase composition, few of them also give indications on the hydrate volume, water conversion, or hydrate composition as well (note that a databank exists online [KRO 09]).

Moreover, there is a need for accurate models to predict both hydrate composition and hydrate volume. Some researchers suggested thermodynamic flash methods, or kinetic approaches. However, because of the lack of information on the hydrate phase and volume, they could not be verified against a wide range of experimental data. Crucial information is often missing in the literature, like the initial quantity of compounds in the system, gas, additives, etc. These data are sometimes provided at equilibrium only, i.e. at total dissociation point (end of the experiment), or with the presence of the hydrate phase.

Henceforth, we will discuss the modeling of the crystallization and evolution of the hydrate phase from the perspectives of equilibrium and non-equilibrium thermodynamics, and then kinetics.

6.4. Modeling gas to hydrate transfer: equilibrium thermodynamics versus kinetics

Previously, questions about the way to model the hydrate formation have been posed. If thermodynamic equilibrium conditions are expected, a standard flash calculation based on thermodynamic equations can be used. Therefore, all phases are thermodynamically equilibrated, homogeneous (top of Figure 6.5), and the Gibbs Energy is at its minimum. Of course, this is a simplified approach, and more phases, several solid phases for instance, could be taken into account. In addition, metastable states may exist in hydrate systems, as mentioned previously. As a consequence, an approach under non-equilibrium conditions should be considered. Hence, the system change pathway of crystallization and mass transfer have to be studied.

In Figure 6.5 (bottom) a kinetic approach illustrated by the two-film theory is also presented. This representation has been suggested by Vysniauskas and Bishnoi [VYS 83] and Jones *et al.* [JON 92]. In this concept, mass transfer phenomena are taken into account in the whole volume. Therefore, local composition is considered. The first zone is the gas phase. Then, there is the first film at the gas-to-liquid interface, on the liquid side, whose thickness is about a few tens of micrometers. The insignificant film at the gas side is neglected. Due to the thickness of the interface between the gas and the liquid phase, there is a linear concentration gradient. The associated mass transfer for molecule j is here written ϕ_j^{VL}. According to the high level of dissolved gas concentration in this area, primary nucleation is likely to occur on and around this film. In the homogeneous liquid phase (bulk), primary and secondary nucleations as well as growth and agglomeration occur. In the bulk is the "diffusion layer".

It is obvious that the kinetic approach requires a large amount and variety of data, mainly based on the experimental conditions (kinetic constants for mass transfer, surface area, etc.). In this modeling, some assumptions are needed or a more precise and complex representation of the geometry of the interfaces.

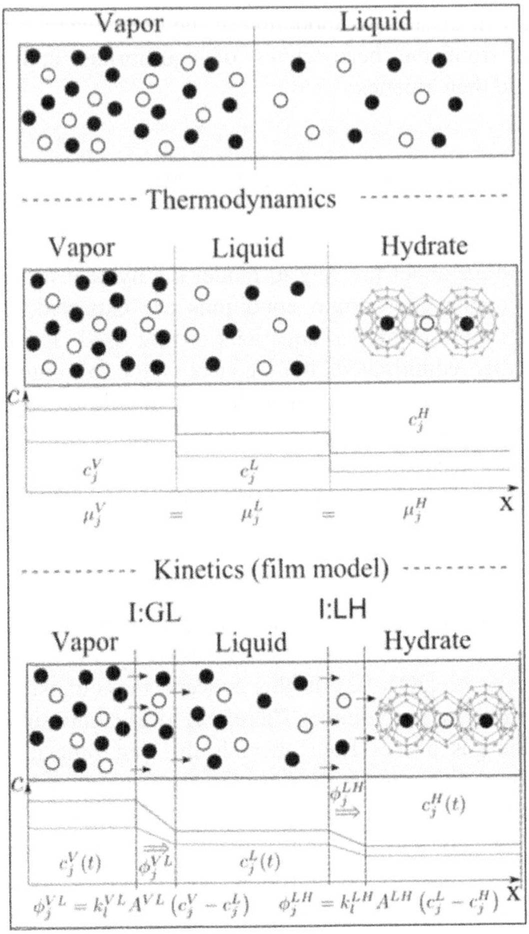

Figure 6.5. *Simplified three-phase approach for hydrate formation from a vapor-liquid system*

6.5. Non-equilibrium flash calculations

6.5.1. *Basics of flash calculations*

The most common situation is the vapor-liquid flash at given temperature and pressure, also known as isothermal flash [MIC 82a, MIC 82b], as illustrated in Figure 6.6.

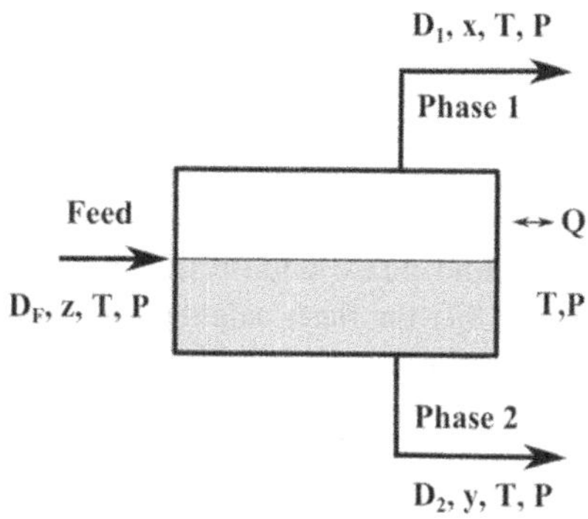

Figure 6.6. *Flash PT (D is the flow rates, and Q the thermal energy exchange, (x, y, z) the flow compositions)*

In this simple case, all phases are homogeneous and the standard approach is to combine the material balance with partition coefficients:

$$\begin{cases} \forall_{j \in molecule}, D_F z_i = D_1 x_j + D_2 y_j \\ y_j = K_j x_j \end{cases} \quad [6.13]$$

where symbol D stands for the flows, and z, x and y the composition according to Figure 6.6.

The partition coefficients K_j (= ratio of the fugacity coefficients, see equation [6.14] below) can be calculated by means of thermodynamic models, like an equation of state or an activity coefficient model, according to the chosen approach (residual or excess energy).

In a generalized case, the Gibbs energy minimization, through the tangent plane criterion [BAK 82], is widely used. This method is usually more effective than those based on the equality of chemical potentials, or fugacities, since the minimization of the Gibbs energy implies the latter equality (whereas the opposite may not be true).

6.5.2. *Conventional flash approach for clathrate hydrates*

6.5.2.1. *Review of standard approaches (Gupta-based models)*

Hydrate flash calculation approaches are mostly based on the van der Waals and Platteeuw model [VAN 59]. Some authors have suggested different improvements from this model, like Chen and Guo [CHE 96, CHE 98] or Ballard and Sloan

[BAL 02a] (see also section 6.2.2 for the impact of pressure on hydrate volume). These modifications will not be deeply discussed.

The first work on clathrates multiphase flash calculations was carried out by Bishnoi *et al.* [BIS 89]. They presented a methodology for isobaric-isothermal (PT) flash calculations for hydrate-fluid phase equilibria. This method, which relies on the PhD thesis of Gupta [GUP 90] for flash calculation stability, minimizes Gibbs free energy. It also calculates equilibrium distribution ratios:

$$\forall_{j \in molecules} \forall_{\pi \in phase}, K_j^{\pi} = \frac{\varphi_j^0}{\varphi_j^{\pi}} \quad [6.14]$$

where φ is the fugacity coefficient, with *0* the reference state (vapor or liquid) and π another phase. When *0* and π are fluid phases, $K_{i\pi}$ are determined from an equation of state. The van der Waals and Platteeuw model [VAN 59] is used when π is the hydrate phase, whatever the structure. Therefore, the fugacity coefficient of the hydrate phase *H* is written as:

$$\varphi_j^H = \frac{f_j^H}{x_j^H P} \quad [6.15]$$

and

$$x_j^H = \sum_i \frac{\vartheta_i \theta_i^j}{(1+\sum_m \vartheta_m \sum_k \theta_m^k)} \quad [6.16]$$

The mole fractions are thus calculated from the occupancy factors (equation [6.10]).

Hence, this is a standard flash calculation considering a solid phase at given composition, in equilibrium with the unlimited gas phase. In order to improve the algorithm, Gupta [GUP 90] and Gupta *et al.* [GUP 91] introduced a stability variable, λ_π, that was found to be the same for all components, defined as:

$$\lambda_\pi = \ln\left(\frac{f_j^{\pi}}{f_j^0}\right) \quad [6.17]$$

Given this variable, the phase stability and equilibrium for Π phases can be described as follows:

$$\begin{cases} x_j^{\pi} = K_j^{\pi} x_j^0 e^{\lambda_\pi} & (j = 1, \dots, N; \pi = 1, \dots, \Pi; \pi \neq 0) \\ \alpha_\pi \lambda_\pi = 0 & (\pi = 1, \dots, \Pi; \pi \neq 0) \end{cases} \quad [6.18]$$

α_π is the phase fraction, so that $\sum_\pi \alpha_\pi = 1$.

Unfortunately, the experimental data were scarce at that time, nevertheless, this work opened prospects for the future of clathrate phase modeling.

At the same time, Cole and Goodwin [COL 90] suggested a similar approach for isothermal gas hydrate flash calculations. The calculation of the hydrate stoichiometry is once again expressed from equation [6.16], and the Langmuir approach is utilized to calculate the occupancy factors (equation [6.10]). The computations are based on the Michelsen flash PT algorithm based on Gibbs free energy minimization. Still, the lack of experimental data on hydrate compositions and phase fractions has hindered the evaluation of this method.

In 2004, Ballard and Sloan [BAL 04a] presented their implementation of the earlier methods in the software CSMGem, taking into account hydrate cavity distortion at high pressure (see section 6.2.2) [BAL 02b, BAL 04a, BAL 04b]. This change in the hydrate fugacity model significantly improves the simulation of hydrate equilibria at high pressures (up to 800 MPa for methane hydrate). Figure 6.7 illustrates all the above algorithms.

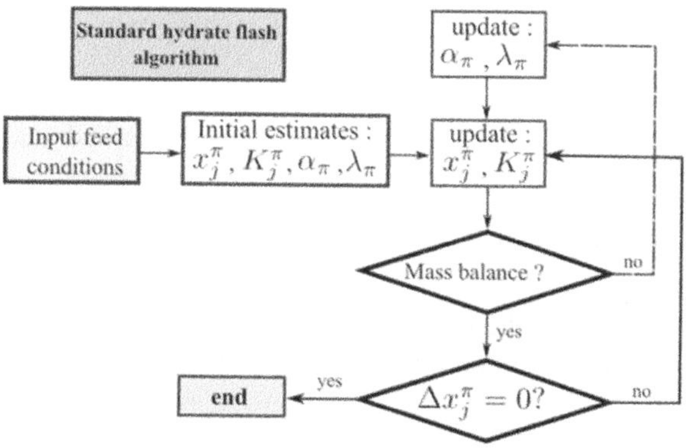

Figure 6.7. *Standard hydrate flash as suggested by Ballard and Sloan [BAL 04]*

Still based on the work of Gupta *et al.* [GUP 91], and Ballard and Sloan [BAL 02b, BAL 04a, BAL 04b], Segtovich *et al.* [SEG 16] suggested an algorithm to improve the robustness of the flash calculations and managed to obtain the generation of complex behavior P-T phase diagrams, even though they did not compare their simulations to extensive experimental data.

In addition, Mahabadian *et al.* [MAH 16] presented another Gibbs energy minimization approach based on the Michelsen algorithm for multiphase flash [MIC 94]. They coupled their model with the Cubic Plus Association – Peng Robinson equation of state (CPA-PR EoS). According to the authors, the algorithm is fast and robust. In addition, it takes into account the stability of more than one hydrate structure, and pseudo-retrograde behavior in hydrate formation. Unfortunately, the simulated results could not be corroborated with experimental phase fractions since these data were absent from the set of experimental results they considered.

6.5.2.2. *Chen and Guo based approach*

In 2013, Ma *et al.* [MA 13a, MA 13b] suggested an isofugacity approach with a four phase equilibrium algorithm based on the work of Cole and Goodwin [COL 90], and Chen and Guo [CHE 96, CHE 98], combining vapor-liquid-liquid flash calculations through a standard equation of state, with the vapor-hydrate equilibrium model by Chen and Guo [CHE 98]. Chen and Guo divided the hydrate formation into two mechanisms:

– formation of a stoichiometric basic hydrate (with fully occupied large cavities, n_L being the number of large cavities per water molecule, and empty small cavities): $H_2O + n_L G \rightarrow \left(G_{n_L} . H_2O\right)^{Hyd}$;

– adsorption of gas molecules into the small cavities (non-stoichiometric nature of hydrates).

From the first step, the chemical potential of the basic hydrate phase, $\mu^{B,0}$, is expressed as

$$\mu^{B,0} = \mu^{water} + n_L \mu^{gas} \qquad [6.19]$$

Since the adsorption of the gas molecules into the small cavities decreases the chemical potential, the expression for the filled basic hydrate becomes:

$$\mu^{B} = \mu^{B,0} + n_S RT \ln\left(1 - \sum_j \theta_S^j\right) \qquad [6.20]$$

For the standard calculation of the occupancy factor θ_S^j (see also equation [6.10]), Chen and Guo suggested the use of Lennard-Jones potential for the Langmuir constants, and an Antoine-type equation for the fugacity. Ma *et al.* chose to use a temperature dependent correlation for the calculation of the Langmuir constants.

In the end, this work differs from the previous one on two points: Kihara potential is not used, but could be, and the above basic hydrate, with the large cages fully occupied by the guest molecules, is adopted as the reference state. Thus, the reference state is not hypothetical anymore.

The algorithm is illustrated in Figure 6.8. The authors observed quite accurate results of the gas and hydrate compositions for the following systems: {H_2-CH_4-C_2H_4-C_2H_6-C_3H_8-$C_{10}H_{22}$-H_2O}, {H_2-CH_4-C_2H_4-$C_{10}H_{22}$-H_2O}, and {H_2-CH_4-C_2H_4-C_2H_6-C_3H_8-$C_{10}H_{22}$-H_2O}. Regrettably, they did not compare the simulations' results with the experimental phase fraction. Also that year, they did the same work with THF as additive, and other gas molecules ({H_2-Ar-N_2-CH_4} mixtures). They also chose to use the Cubic Plus Association (CPA) EoS for the calculation of fluid equilibria.

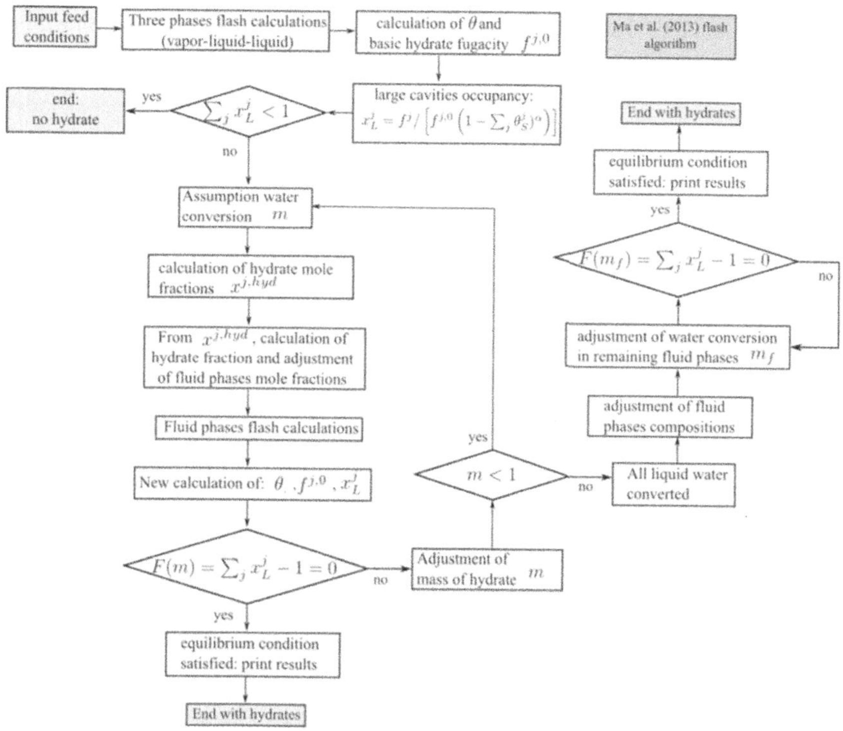

Figure 6.8. *Hydrate flash algorithm according to Ma et al. [MA 13a, MA 13b] (with α being here the ratio between the number of small cavities over the number of large cavities, and subscript L accounts for the large cavities)*

6.5.2.3. *Tavasoli and Feyzi approach*

Another creative work deserves to be mentioned. Tavasoli and Feyzi [TAV 15] suggested an algorithm for vapor-aqueous-hydrate systems in batch operations (constant volume) where pressure is not constant anymore. Hence, their framework takes into account this change of pressure, and its influence on the hydrate equilibrium conditions. Of course, if the pressure predicted by the fluid phase flash calculations does not match the hydrate equilibrium pressure, the assumed water conversion is wrong and has to be adjusted. The basic algorithm is in Figure 6.9 (note: the water conversion into hydrate can be 0% or 100% depending on the pressure).

Contrary to the conventional approaches, such as those based on the work by Gupta *et al.* [GUP 91], and similarly to the model of Chen and Guo [CHE 96, CHE 98], the approach of Tavasoli and Feyzi does not use Gibbs Energy minimization as a criterion for equilibrium, but the equality of fugacities.

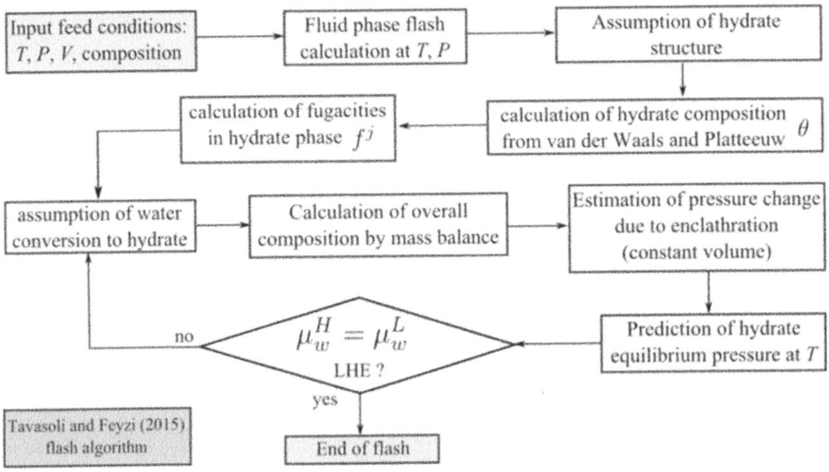

Figure 6.9. *Hydrate flash algorithm at T and V according to Tavasoli and Feyzi [TAV 15]*

6.5.3. *Conclusions on standard flash approaches*

Since the early work of Gupta [GUP 90], various flash approaches involving hydrates have been developed. Most of them are based on Gibbs energy minimization, and some others on the weaker isofugacity criterion. Hydrate phase composition is usually predicted from the van der Waals and Platteeuw model [VAN 59].

Some of these works are implemented into thermodynamic software. However, their accuracy is mainly assessed by comparison with dissociation curves. Because there is a lack of data in the literature for properties such as hydrate composition or phase fractions, the efficiency of these models is not warranted for predicting other features such as the hydrate volume. However, isobaric-isothermal flash calculations usually lead either to total water consumption (the aqueous phase disappears), or to total gas consumption (the gas phase disappears). The volume of the hydrate phase can then be fully determined.

The work of Ballard and Sloan [BAL 02a, BAL 02b, BAL 04a] has been embedded in the program called CSMGem (CSM: Colorado School of Mines). Another program developed at CSM, called CSMHyd, also performs thermodynamic equilibrium with hydrates. There are other commercially available programs, including DBRHydrate (Schlumberger), Multiflash (KBC), and PVTSim (Calsep). A comparison of these programs was suggested by Ballard and Sloan [BAL 04b]. Only dissociation points were considered. So, this is mostly a comparison of the thermodynamics inside the softwares, i.e. the accuracy of the predicted dissociation pressure for a given gas composition. Ballard and Sloan concluded that the programs are quite similar in terms of results. However, they noticed that the improvement in the hydrate phase modeling implemented in CSMGem gives better results at high pressures, and can predict phase transitions quite well. Another program, not mentioned in the comparison of Ballard, is HWHYD, based on the isofugacity criterion, which was developed by Heriot-Watt University (both in commercial and research versions). More recently, Hydrafact Ltd (hosted at Heriot-Watt University) developed HydraFLASH®.

Finally, one of the main issues of clathrate hydrate modeling is the non-stoichiometric behavior of the solid phase. The use of partition coefficients is not relevant if the final crystal is not formed at thermodynamic equilibrium, although it gives boundaries. To this end, non-equilibrium thermodynamic algorithms can be considered. Logically, Gibbs energy minimization is not then the right approach since the final state of the system will be metastable.

6.5.4. *Non-stoichiometric flash approaches*

6.5.4.1. *Introduction*

In these approaches, the hydrate phase is not homogeneous, nor the consequence of the crystallization process. The resulting hydrate phase is formed under non-equilibrium conditions, even if local thermodynamic equilibrium is assumed and the calculations are based on equilibrium thermodynamic equations. Two formalisms can be found in the literature. The first was suggested by Tsuji *et al.* [TSU 05], and the second by Kobayashi and Mori [KOB 07], and Bouillot and Herri [BOU 16]. The first formalism is adapted to hydrate formation at constant pressure (isobaric, semi-batch) conditions in a quasistatic process, while gas-limited supply at constant volume (isochoric, batch conditions) is considered in the other two studies. The constant pressure approach indicates a continuous feed of the system (semi-batch conditions). The other one corresponds to batch conditions, such as those encountered in closed reactor experiments, in closed reservoirs [MAK 10] or in fluid inclusions in minerals [COL 79] (see evidence in section 6.3.2).

In both formalisms, the vapor composition changes during crystallization as a consequence of gas consumption. Subsequently, the change in hydrate compositions has to be taken into account at the same time as its volume increases. Therefore, the growth needs to be discretized in the calculations. Once a new amount of hydrate is formed, it can remain in the system as metastable, and be omitted from the computations. Thus, these frameworks can be considered as successive flash calculations, or sequential flash simulations.

Unlike the kinetics approaches (see section 6.6), the calculations are based on thermodynamic equations for phase equilibria. Figure 6.10 illustrates the two main hypotheses for the hydrate formation.

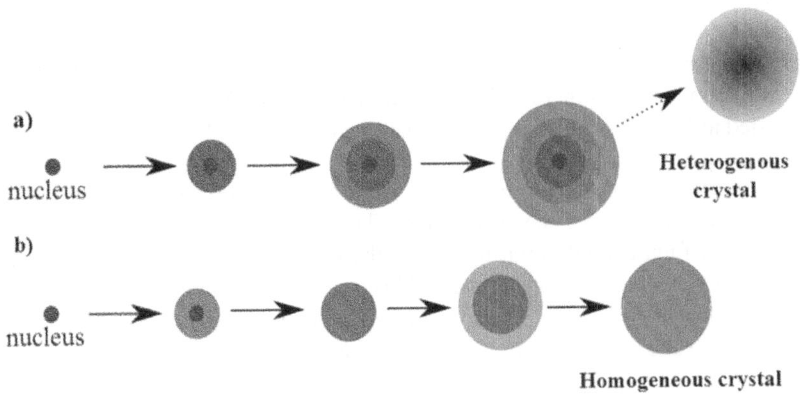

Figure 6.10. *Growth mechanism hypotheses for hydrate formation [BOU 16]*

The first mechanism (Figure 6.10(a)) represents a crystal growth in non-equilibrium conditions, with a varying composition of the hydrate phase. It is non-stoichiometric, as defined at the beginning of this chapter. As the hydrate crystals grow, the compositions of the gas phase changes. Even if there is a gas feed, its composition does not coincide with the gas composition after the onset of hydrate formation. The hydrate phase composition, predicted by assuming thermodynamic equilibrium at the surface of the growing crystals, changes as well. In the end, there is a continuous evolution of the hydrate cavity occupancy, from the core to the last layer (see Figure 6.11). On the aforementioned figure, the equilibrium curves in the T, P space move from the right to the left side of the diagram. This is not surprising since the gas molecules that stabilize the hydrate structure the most are enclathrated at first. Therefore, the hydrate stability zone is reduced and a higher pressure (or gas molecule solubility into the aqueous phase) is needed for hydrate growth, and crystallization to proceed. This also means that the core of the hydrate crystal is not at equilibrium with the surrounding liquid phase. It should dissociate in contact with this liquid. This situation corresponds to a process at high crystallization rate. The hydrate is only growing at local thermodynamic equilibrium, and is not rearranged since crystal growth is too quick for the most stable phase to appear. In the end, the whole hydrate is not at equilibrium, but only its last outermost layer which is in contact with the aqueous medium. This more likely represents the startup of a crystallization process (production in pipelines, gas capture/storage, etc.).

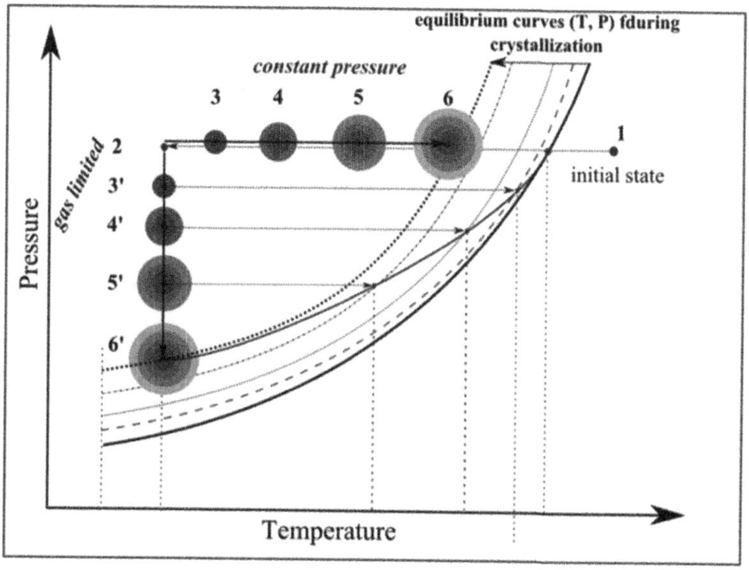

Figure 6.11. *Crystallization of non-stoichiometric hydrate at constant pressure or gas limited*

The second mechanism (Figure 6.10(b)) corresponds to a reorganization of the hydrate phase during crystallization. This is close to a conventional hydrate flash calculation, except that there is gas consumption, hence a hydrate composition that does not correspond to the initial gas composition. This is the approach of Tavasoli and Feyzi [TAV 15] for instance (flash TV). Therefore, the use of constant occupancy of the cavities is not feasible (like in standard flash PT simulations). This approach is presented in section 6.5.4.2.

6.5.4.2. *Algorithm for non-stoichiometric flash approach at constant pressure*

Figure 6.12 illustrates the framework of Tsuji *et al.* [TSU 05]. This approach concerns continuous and semi-batch operations for three or four phase equilibria (vapor, liquid, hydrate, and eventually large-molecule guest substance substance (LMGS)). The pressure is kept constant by a supply of gas (and large molecule guest substance for continuous operations).

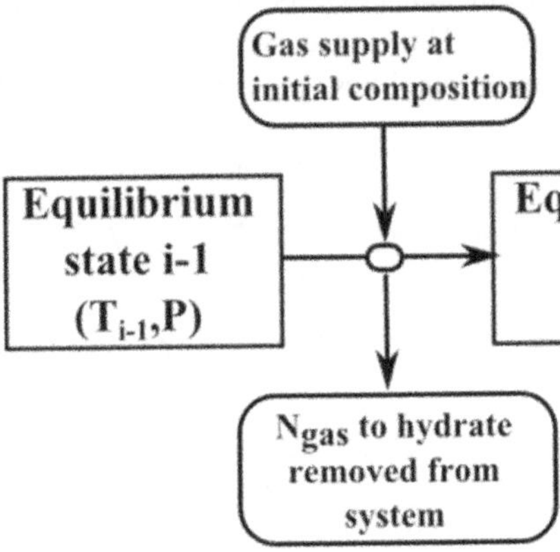

Figure 6.12. *Tsuji et al. Framework [TSU 05]*

Some assumptions are used in such a framework:

– the system contains three phases (or four phases with LMGS) instantly at thermodynamic equilibrium (vapor, aqueous and hydrate);

– the pressure is held constant by a feed of a gas mixture at given composition (continuous or semi-batch);

– water is not depleted so that there is an excess of water molecules;

– no influence of gas dissolution into water over the gas and hydrate compositions;

– amount of LMGS in the system is kept constant in the case of continuous processes, or decreased during hydrate formation for semi-batch processes;

– at each moment, only the most stable hydrate structure is formed (SI, SII or SH);

– the hydrate formed is instantly removed from the system so that it cannot interact with the new/future hydrates to be formed.

According to Figure 6.12, the mass balance at iteration i is written:

$$\begin{cases} \sum_{j\in gas} N_{in,i}^{j} = N_{hyd,i}^{j} \\ N_{in,i}^{LMGS} = N_{hyd,i}^{LMGS}\ (continuous) \\ N_{in,i}^{LMGS} = 0\ (semi-batch) \end{cases} \quad [6.21]$$

where the subscript "*in*" denotes the inflowing feed gas component needed to keep the pressure constant. This way, the mass flow for each sequence is determined. Then, the numbers $N_{hyd,i}^{j}$ need to be fully calculated. In their simulations, Tsuji *et al.* used CSMHYD for the LHE, but any thermodynamic model can be used.

The new number of moles at iteration $i+1$ is defined by

$$\begin{cases} \forall_{j\in gas} N_{i+1}^{j} = N_{i}^{j} - N_{hyd,i}^{j} + N_{in,i}^{j} \\ N_{i+1}^{LMGS} = N_{i}^{LMGS} - N_{hyd,i}^{LMGS} + N_{in,i}^{LMGS} \end{cases} \quad [6.22]$$

6.5.4.3. *Algorithm for non-stoichiometric gas limited flash approach at constant volume*

The algorithm used to compute a non-stoichiometric flash calculation at constant volume and given temperature is illustrated in Figure 6.13. This method provides a heterogeneous crystal. The constant volume suggests a gas limited process. It can be seen as a flash TV approach. Two articles can be found on this approach [KOB 07, BOU 16]. The first is a modification of previous efforts made by Tsu and coworkers [TSU 05], while the second is an original approach, although close to the first in its principles. In this section, the Bouillot and Herri algorithm [BOU 16] will be further presented and discussed.

The process of this isochoric algorithm starts in the non-hydrate zone (undersaturated liquid phase). Both the gas mixture and the water solution are "injected" into the system at constant volume. From an initial vapor liquid equilibrium (VLE), point A in Figure 6.13, flash TV calculations are performed until the VLE predicted pressure is equal to the LHE predicted pressure. This is achieved by decreasing the temperature step by step.

From point B, the crystallization begins, and successive flash calculations are performed. This corresponds to a discretization of the crystal growth into several steps, or several new amounts of different stoichiometric crystals in the system.

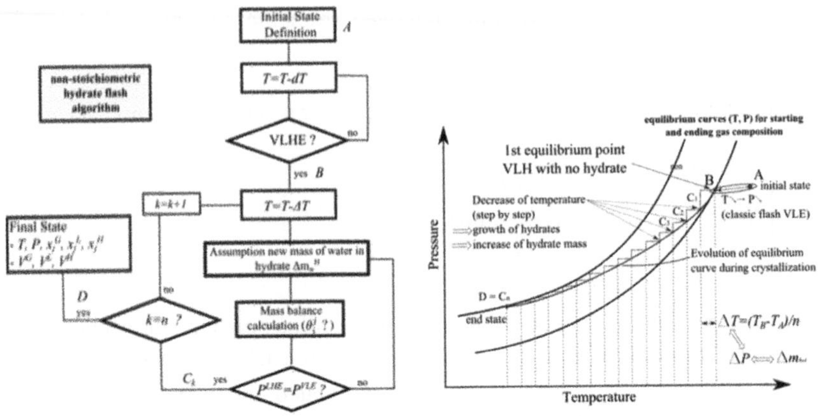

Figure 6.13. *Non-stoichiometric hydrate flash algorithm [BOU 16]*

For a given number of iterations n, the temperature step is $\Delta T = (\Delta T_B - \Delta T_D)/n$. This step corresponds to a new mass of water in the hydrate structure, at each iteration k : $\Delta m_{w,k}^H$. The last quantity is also the adjustment variable for the successive flash calculations. At each step, after choosing a value for $\Delta m_{w,k}^H$, a mass balance calculation is performed. To do so, the hydrate volume is determined from $\Delta m_{w,k}^H$:

$$V^H = \sum_k \Delta V_k^H = \sum_k \frac{\Delta m_{w,k}^H}{\rho^{H-\beta}} \qquad [6.23]$$

where $\rho^{H-\beta}$ is the density of the empty hydrate, that is to say, 790 kg/m^3 for SI, and 785 kg/m^3 for SII. These values can be retrieved from the formula given by Sloan and Koh [SLO 07]:

$$\rho^H = \frac{N_w MW_{H2O} + \sum_{j=1}^{C} \sum_{i=1}^{N} \theta_i^j v_i MW_j}{N_{Avo} V_{cell}} \qquad [6.24]$$

In this equation, N_w is the number of water molecules per unit cell, N_{Avo} the Avogadro's number, V_{cell} the unit cell volume, MW_j the molecular weight of compound j, θ_i^j the occupancy factor of j in cavity i, v_i the number of cavity i per unit cell, C the number of components, and N the number of cavities.

The amount of gas molecules in the hydrate phase is determined from the occupancy factor θ_i^j, as described in section 6.2.2. This occupancy factor can be established from the gas composition in the liquid phase at the beginning of the iteration (before the new amount of hydrate), or after. In the second case, the algorithm is a bit more complicated since the occupancy factor has to be the one at thermodynamic equilibrium, after the mass balance calculations. To avoid using this more complex algorithm, with more than 20 iterations, it suffices for the two approaches to converge. Hence, only the simpler first approach can be used.

Then, putting aside the hydrate phase, a VLE flash TV is performed. This provides an equilibrium pressure P^{VLE}. This pressure is compared to the one given by LHE, P^{LHE}. If the two pressures are equal, this means that the assumed new amount of crystallized water is correct.

Once this equality is obtained, the same calculations for the next iteration have to be performed, until the final temperature is reached. From all the mass balance calculations that were executed, the amount of molecules in each phase is completely determined. Also, the hydrate volume is known from equation [6.23]. The density of each fluid phase can be used to calculate its volume. A relevant equation of state, or correlation, is needed.

This kind of calculation can be time consuming. For the purpose of process simulation, this can be an obstacle without proper optimization. Also, the stability of the algorithm needs some improvements. A high temperature at the beginning of the algorithm (initial state, point *A* in Figure 6.13) is suitable to avoid starting in the hydrate zone.

6.5.4.4. *Case of stoichiometric hydrate*

From the previous algorithm (at constant volume and given temperature), it is also possible to converge to a stoichiometric hydrate. In this way, the algorithm approximates the one by Tavasoli and Feyzi [TAV 15]. To do so, the occupancy factor at the end of the process has to be the same as the one predicted by the thermodynamic equilibrium considering the surrounding liquid phase. Therefore, a loop on the occupancy factor has to be added. This is illustrated in Figure 6.14.

6.5.5. *Discussion*

Bouillot and Herri's [BOU 16] isochoric flash algorithm was compared to the experimental results of Le Quang *et al.* [LEQ 16] at quick and slow crystallization rates. They provided hydrate compositions, as well as water conversions and hydrate volumes, as a function of the initial quantity of mass in a given volume and a given final temperature. The results of the isochoric flash algorithm were compared to these experimental results. Figure 6.15 shows the thermodynamic path for $\{CO_2\text{-}CH_4\text{-}C_2H_6\}$ gas mixture at low crystallization rates, as suggested in Figure 6.13.

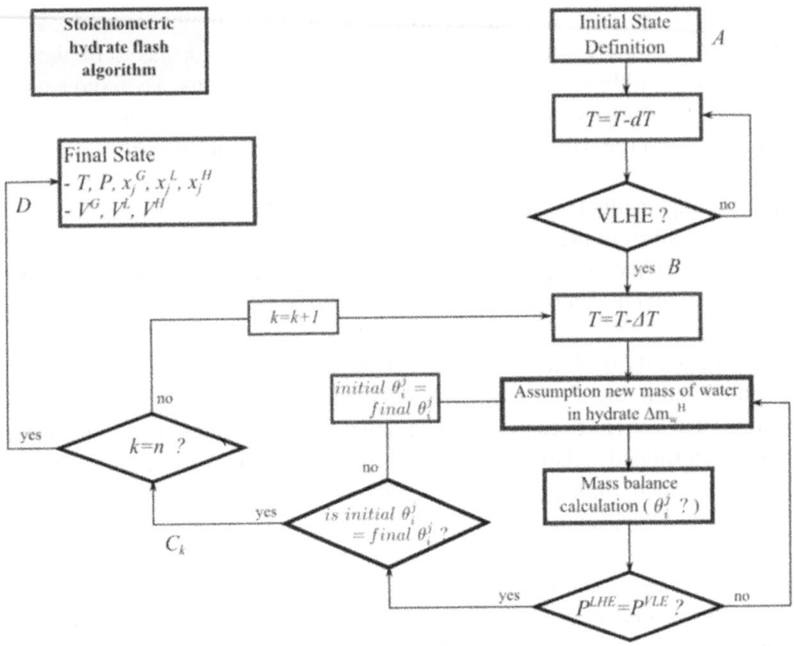

Figure 6.14. *Stoichiometric hydrate flash algorithm [BOU 16]*

Results at slow and quick crystallization rates were compared to heterogeneous and homogeneous frameworks (Figure 6.10). Of course, the number of iterations used has a significant impact on the results since it defines the crystallization discretization. The lower this number is, the more heterogeneous the hydrate phase. Above *20* iterations, Bouillot and Herri consider that the iteration number is less significant.

Bouillot and Herri observed a slightly better accordance of the non-stoichiometric framework with the experimental results at a quick crystallization rate. Therefore, stoichiometric simulations, i.e. homogeneous hydrates, were in better accordance with slow crystallization experiments (Table 6.1). This examination supports the initial guess: a more homogeneous hydrate phase is obtained when the driving force for the crystallization is low, and a more heterogeneous hydrate when this force is high. This is in accordance with the kinetic approach.

		Experimental		Non-stoichiometric		Non-stoichiometric		Stoichiometric	
				(N = 20)		(N = 5)			
		T_f	P_f	dev. P	dev. V^H	dev. P	dev. V^H	dev. P	dev. V^H
Gas		(±0.2°C)	(±0.1bar)	%	%	%	%	%	%
3	CO_2/ CH_4	3.4	33.3	5	21	5	21	5.9	22
4	CO_2/ CH_4	2.2	29.1	3.1	8.2	2.9	7.8	3.4	9.2
8	CO_2/ CH_4/ C_2H_6	2.75	35.4	4.3	22	4.3	22	8.8	28
12 *	CO_2/ CH_4/ C_2H_6	1.3	27.6	4.3	3.2	5.8	6.1	0.08	6.9

Table 6.1. *Simulations compared to the experimental results of Le Quang et al. [LEQ 16] for non-stoichiometric and stoichiometric hydrate flash algorithms [BOU 16] (* slow crystallization rate)*

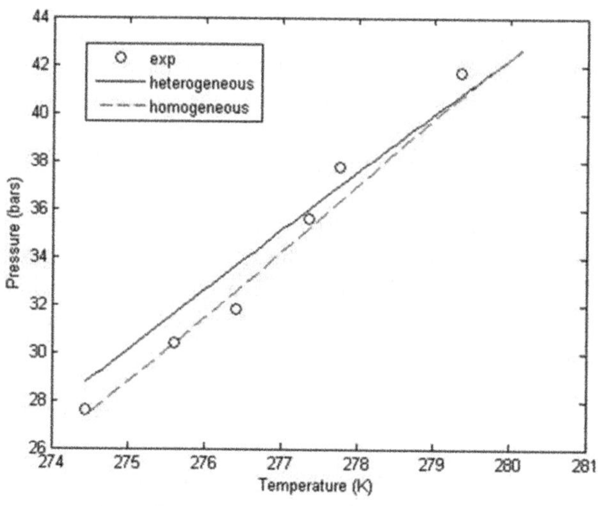

Figure 6.15. *Experimental and predicted thermodynamic path for non-stoichiometric and stoichiometric algorithms [BOU 16] according to Figures 6.13 and 6.14 for CO_2-CH_4-C_2H_6 mixture at slow crystallization rate [LEQ 16]*

In addition, Bouillot and Herri showed that the uncertainty on the Kihara parameters significantly affects the predicted pressure and hydrate volume. An uncertainty of 3% on the parameters (ε/k and σ) leads to a deviation of about 30% and 90% in the predicted pressure and hydrate volume, respectively.

6.6. A kinetic Langmuir based modeling approach

6.6.1. *Introduction to the kinetic approach of mixed hydrates*

The Kinetics of hydrate formation have been widely reviewed by Ribeiro and Lage [RIB 08]. They highlighted several approaches, such as:

– the former model of Englezos *et al.* [ENG 87a, ENG 87b];

– the Skovborg and Rusmussen model [SKO 94];

– the Herri *et al.* model [HER 99b];

– the Gnanendran and Amin model [GNA 04].

Other work is suggested in this review, but is regarded as limited compared to the four previous ones. For the purpose of this section, not all the models will be discussed since they usually do not account for the specificities of mixed hydrates and possible competition between the gas molecules being trapped.

The perspective of the kinetic modeling is to consider the driving force for the hydrate formation according to the speed of the transfer phenomena: gas-to-liquid mass transfer, the liquid diffusion in the bulk, the liquid-to-hydrate transfer, as shown in Figure 6.5. Equation [6.2] provides the driving force, depending mostly on the gas concentration in the liquid, or in this case at the liquid-hydrate interface.

Figure 6.16 sketches the different driving forces that could happen for different transfer rates to the liquid phase. Each of these "positive" driving forces ($\Delta\mu < 0$) leads to another hydrate, with different cage occupancy. The whole purpose of modeling the hydrate growth is to work on the kinetic of reaction causing the formation of the mixed hydrate as a solid solution.

Obviously, hydrate formation is crystallization. Like all crystallization processes, the solidification mechanism is usually divided into two parts: nucleation and growth. These concepts have been widely studied by Mullin [MUL 01] and Kashchiev [KAS 00]. In the specific case of gas hydrate, it is not surprising that several studies couple the mass transfer to crystallization.

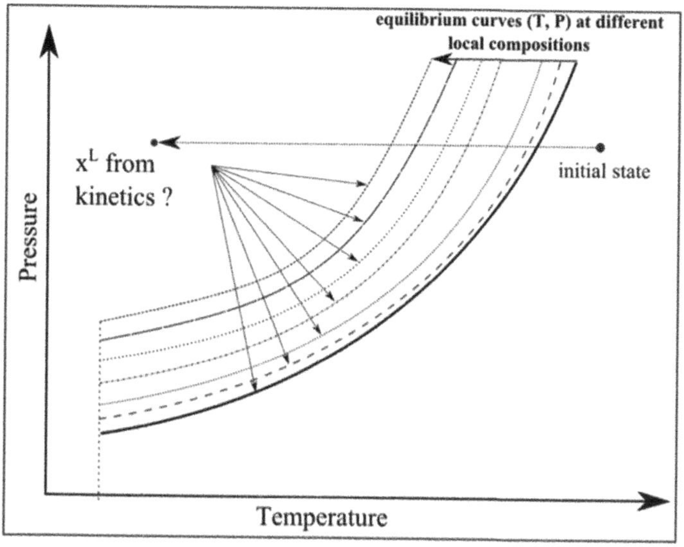

Figure 6.16. *Saturated metastable state regarding different gas composition in the liquid phase for a given hydrate structure*

6.6.1.1. *Former approach of Englezos et al.*

This early approach is based on crystallization theory: first the nucleation, then crystal growth. The nucleation is linked to the driving force for the crystallization. According to the two-films approach (see Figure 6.5), the hydrate formation process is divided into three parts:

– the gas-to-liquid transfer;

– the diffusion of the gas molecules into the bulk;

– the reaction at the hydrate particle interface.

Englezos *et al.* [ENG 87a, ENG 87b] described the hydrate growth as a function of the driving force (here the fugacity difference). Then, the authors suggested coupling the gas-to-liquid with the liquid to hydrate transfer rates at quasi-steady state conditions. To do this, they used the aforementioned two-film theory. Therefore, the rates in all regions/films are the same. The procedure is briefly explained hereafter.

At first the hydrate growth was studied. At that time, the theoretical model of Kashchiev and Firoozabadi [KAS 03] for clathrate hydrates did not exist yet. However, according to the definition of phase equilibrium [PRA 99], Englezos *et al.*

described the driving force as the fugacity difference between the experimental state, and the equilibrium (and not on temperature or pressure difference).

From this expression, the crystal growth per particle p is expressed as follows:

$$\left(\frac{dn}{dt}\right)_p = K^* A_p(f - f_{eq}) \qquad [6.25]$$

where f is the gas fugacity, f_{eq} the equilibrium fugacity, K a kinetic coefficient, and A_p the surface of the particle. Note that K corresponds to a combination of the gas molecule diffusion around the hydrate particle and the "reaction" rate of crystallization. In addition, since the particles are considered as spheres, their surface is given by their radius r. Then, there could be several formers for the crystallization reaction. In the approach of Englezos *et al.*, all the gas molecules are taken separately, and the gas consumption is the sum of all molecules consumed. Therefore:

$$\left(\frac{dn}{dt}\right)_p = \sum_{j\in gas}\left(\frac{dn_j}{dt}\right)_p = \sum_{j\in gas} K_j^* A_p\left(f_j - f_{j,eq}\right) \qquad [6.26]$$

Second is the coupling of the gas-to-liquid with the liquid to hydrate transfer rates at quasi steady state conditions. With the use of the two-film theory, the diffusion and reaction rates are similar, so that:

$$D_j \frac{d^2 c_j}{dy^2} = \sum_{j\in gas} K_j\left(f_j - f_{j,eq}\right) \qquad [6.27]$$

with D_j the diffusion coefficients, c_j the gas concentration in the bulk, and y the gas-liquid interface. Also K is calculated from K^* and the second moment of the particle size distribution (i.e. the average surface area per particle), so that $K_j = K_j^* 4\pi\langle A_p\rangle$. To simplify the model, they consider that the bulk is mixed at a high stirring rate so that the particle distribution is homogeneous in the liquid phase.

Of course, there is a need for a population balance in the model to determine the number of hydrate particles, and to know the surface area between the liquid and the crystals. The initial number of nuclei is determined from a mass balance at the turbidity point in experiments. The size of these nuclei is calculated from classic crystallization theory.

In this model, the fugacities of each molecule, as well as the equilibrium fugacities, have to be determined. Then, each former in the crystallization process is taken separately. Unfortunately, there was no discussion on the local hydrate stability. The gas molecule fugacity can be calculated from the occupancy of the cavities, like in the latter work of Al-Otaibi *et al.* [ALO 11], and this occupancy can be calculated from thermodynamics. Again, the mixed hydrate is supposed to grow at thermodynamic equilibrium; even if a local equilibrium could be achieved by another cage occupancy (non-equilibrium growth, questioned in Figure 6.16). Note: the work of Englezos *et al.* is based on the experimental set-up of Vysniauskas and Bishnoi [VYS 83]. These experiments were performed at constant temperature and pressure. Therefore, the driving force, expressed from the difference between the experimental conditions and the equilibrium ones, remains the same.

To conclude, this approach was successfully applied to model the incipient of gas consumption in a stirred batch reactor in the presence of binary gas mixtures containing methane and ethane.

6.6.1.2. *The Skovborg-Rasmussen model*

Skovborg and Rasmussen [SKO 94] suggested another approach for the hydrate growth for the mixtures investigated earlier by Englezos *et al.* [ENG 87a, ENG 87b]. Their model is in fact a simplification of the former approach. Indeed, they chose to represent the overall hydrate crystallization process by the simple gas-to-liquid mass transfer. This makes sense since some authors will show experimentally that the hydrate crystallization process is usually limited by the gas-to-liquid mass transfer [GAI 99, HER 99a]. In addition, according to the authors, the decrease in the gas consumption rate could be due to the reduction of the contact area between the gas phase and the liquid phase. In the approach of Englezos *et al.*, there is indeed a deviation between the experiments and the model after a while. Experimentally, there is a reduction in the gas consumption rate. Again, in the former approach, the gas consumption rate is proportional to the surface area of all the hydrate particles that grow over time. Hence, this consumption rate also grows over time. In the end, the idea of Skovborg and Rasmussen is to eliminate the particle size distribution component of the former approach, and to focus on the gas-to-liquid interface.

Therefore, equation [6.25] is changed into:

$$\left(\frac{dn}{dt}\right)_p = k_l A_{GL} c_{w0}(x_{int} - x_{bulk}) \quad [6.28]$$

where x_{int} is the mole fraction of the gas molecules at the gas/liquid interface (taken at thermodynamic equilibrium at P and T), x_{bulk} the gas mole fraction in the bulk, in equilibrium with the hydrate phase, and c_{w0} the initial concentration of water molecules. Similarly, equation [6.27] can be replaced by:

$$\left(\frac{dn}{dt}\right)_{total} = \sum_{j\in gas}\frac{dn_j}{dt} = c_{w0}\sum_{j\in gas} k_{l,j}A_{GL}\left(x_{j,int} - x_{j,bulk}\right) \quad [6.29]$$

A solvation of this system at constant temperature and pressure has been suggested. Similarly to the experiments performed by Englezos *et al.*, the pressure is kept constant by adding gas at initial composition to the reactor during the crystallization process, and the temperature is controlled. The algorithm is described in Figure 6.17.

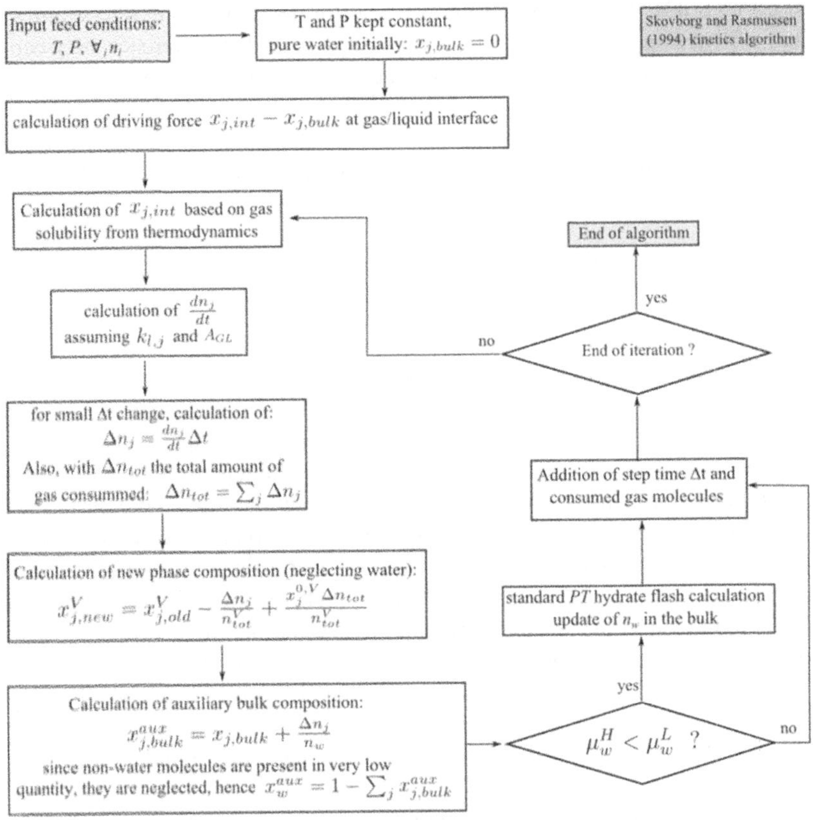

Figure 6.17. *Skovborg and Rasmussen algorithm for kinetic approach [SKO 94]*

In the calculation of the new phase composition, the second term of the equation $\left(\frac{\Delta n_j}{n^V_{tot}}\right)$ corresponds to the gas consumption, and the third $\left(\frac{x_j^{0,V}\Delta n_{tot}}{n^V_{tot}}\right)$ to the gas feed at initial composition $x_j^{0,V}$. It is considered that, to keep a constant pressure, the same amount of gas molecules needs to be added into the system after consumption.

n_{tot}^{V} is the total amount of moles in the vapor phase. Hence, the added amount of gas molecules is $x_j^{0,V}\Delta n_{tot}$.

Of course, this algorithm requires some parameters: mainly $k_{l,j}$ and A_{GL}. The latter has been estimated by Englezos *et al.* in the case of a batch reactor at different stirring rates (from 79.3 cm^2 at 300 rpm to 169.1 cm^2 at 450 rpm). Naturally, the numbers given are an estimation, and they have to be considered cautiously. In the case of a flow, surface area needs to be properly defined with an appropriate model. Then, there is the transfer coefficient $k_{l,j}$. This can be calculated from pure solubility measurements. But again, it is dependent on the experimental conditions.

Then, a flash calculation for the bulk, at T and P, is also needed, hence a procedure to compute the occupancy factor of the (mixed) hydrate as a function of the bulk composition. In their work, they used the Michelsen procedure [MIC 91]. Of course, the calculation of the LHE affects the estimation of the driving force $x_{j,int} - x_{j,bulk}$. Therefore, the accuracy of the methods also relies on the liquid-hydrate thermodynamic part. Skovborg and Rasmussen suggested taking into account this uncertainty in equation [6.29], resulting in an improvement of the method.

Ultimately, this method suggests modeling a kinetic equilibrium of the hydrate phase. Indeed, crystallization will occur at a concentration that is not the one predicted by the thermodynamic VLE. This is a kinetic flash framework at given T and P. However, the approach has a drawback, which was noticed by Gnanendran and Amin (or Krejbjerg and Sørensen [KRE 05]). Certainly, the explicit method detailed above can suffer from convergence issues, especially when the time step is too large, resulting in a significant super-saturation of the aqueous phase.

To eliminate the stability issue, Krejbjerg and Sørensen [KRE 05] suggested an implicit approach of the model. This implicit formulation is written as:

$$\Delta n_j = n|_{t+\Delta t} - n|_t = \Delta t\, k_l A_{GL} c_{wo} (x_{int} - x_{bulk})|_{t+\Delta t} \qquad [6.30]$$

In other words, the gas consumption is expressed as a function of the driving force at time step $t + \Delta t$ instead of the driving force at t. The mass balance equations are combined with equation [6.18] to properly write the system to solve. Note: the authors used fugacities instead of concentrations in their approach. According to them, this formulation eliminates stability problems. Besides, simultaneous solvation of all equilibrium and transport equations is needed. Since there is an implicit expression, there is also a need to perform more flash calculation in the aqueous-hydrate zone. In their article, Krejbjerg and Sørensen managed to perform kinetic flash calculations. Their simulation gave the phase fraction and phase compositions for H_2O-MeOH-CH_4-C_2H_6-*n*C8 mixtures. Unfortunately, it could not be compared to actual results.

Boesen *et al.* [BOE 14] extended this work to simulate hydrate crystallization under flow conditions using different specification variables (*P*, *T*, *V*, or *H*), and not only *P*-*T*.

6.6.1.3. *The Herri et al. model*

The experimental work of Herri [HER 96] also pointed out that the gas-to-liquid mass transfer is a limiting step in the global hydrate formation. However, the model of Herri *et al.* [HER 99b] looks at the hydrate formation as a crystallization process, like the former approach of Englezos *et al.*. Hence, there is again a coupling between the mass transfer and the crystallization equations. Of course, the particle size distribution comes back in the approach.

The authors chose to divide their approach in two parts:

– the mass and population balances;

– the effect of the stirring rate on the particle size distribution.

The first part concerns the mass transfer phenomena, from the gas phase to the hydrate phase. The two-film formalism is used (Figure 6.5), with the first zone describing the gas-to-liquid mass transfer, and the second one describing the bulk.

The second part concerns the influence of the stirring rate on the initial particle number and mean diameter. In their experiments, and in the experiments of Jones *et al.* [JON 92], they observed that an increase of the stirring rate had the consequence of increasing the mean diameter of the particle and a simultaneous decrease of the number of particles. In their analysis, they suggested that the primary nucleation mainly (but not only) occurs in the interface film layer between the gas and the liquid. Therefore, an increase of the stirring rate decreases the thickness of this layer, hence, the number of nuclei.

Then, the model proposes a quantitative approach. The mass balance (dissolution rate and crystal growth) is combined with the primary nucleation phenomenon. The mass balance is:

$$\frac{dC_{bulk}}{dt} = k_l A_{GL}(C_{int} - C_{bulk}) - \frac{4\pi}{v_m} G M_2 \qquad [6.31]$$

where M_2 is the second moment of the particle size distribution *f* (i.e. the volume). The particle size distribution is expressed as:

$$\frac{\partial f}{\partial t}+\frac{\partial f}{\partial R}=B'(R)-D'(R) \quad [6.32]$$

where B′ is the "birth" term, that is to say the nucleation/breakage contribution, and D′ is the "death" term, mainly due to agglomeration. Considering the "birth" term as a function of the primary nucleation only, this can be expressed as a function of the supersaturation according to classic nucleation theory [KAS 00], in the interface layer (BI,1), and in the bulk zone (BI,2):

$$\begin{cases} B_{I,1}=k_I h \frac{S}{V}\int_0^h \exp\left(-\frac{B}{log^2 S(x)}\right)dx \\ B_{I,1}=k_I \exp\left(-\frac{B}{log^2 S_{bulk}}\right) \end{cases} \quad [6.33]$$

where k_I and B are constants, h the thickness of the film region, S the supersaturation (gradient concentration in the film, and constant in the bulk).

Then, G is expressed from the growth rate kinetic constant k_g, such as:

$$G = k_g\left(C_{bulk}-C_{eq}\right) \quad [6.34]$$

Herri *et al.* gave an expression for the aforementioned constant.

In the end, the model was used to observe the trends on the mean diameter and the supersaturation. No quantitative data were compared. The simulation allowed to the effect of an additive in a high pressure reactor to be rapidly screened. Since the study was compared to pure methane hydrate crystallization, the non-equilibrium growth was not the point, although this method could (and has been, see section 6.6.1.4) extended to model a process including the crystal growth at a more precise scale around the hydrate particles.

6.6.1.4. *The Gnanendran and Amin model [GNA 04]*

This model again aims at predicting the hydrate formation at isothermal and isobaric conditions. It is in fact an extension of the work of Kashchiev and Firoozabadi [KAS 02a, KAS 02b, KAS 03]. In their effort, Gnanendra and Amin [GNA 04] consider gas mixtures instead of pure gas hydrates. The system is a semibatch spray reactor at constant temperature and pressure. Therefore, the approach is divided into two parts: the spray period, corresponding to a semi-batch process, and a stabilization period following the spraying period. Also, like Kashchiev and Firoozabadi [KAS 03], they aimed to take additives into account (para-toluenesulfonic acid as hydrate promoter in their work for example)

Classic theory for nucleation and growth is considered once again. At first, from a thermodynamic point of view, Gnanendran and Amin use the expression for the driving force described by Kashchiev and Firoozabadi [KAS 02a], see also equation [6.7], and calculate the fractional filling of the hydrate cavities from the Langmuir adsorption approach suggested by van der Waals and Platteeuw [VAN 59].

To begin with, the driving force is written for more than a single guest component, according to the expression of Kashchiev and Firoozabadi for isothermal regime:

$$\Delta\mu = kT\left(\sum_j x_j^H ln\left(\frac{\varphi(P,T)P}{\varphi(P_{eq},T)P_{eq}}\right)\right) + \Delta v_e(P - P_{eq}) \quad [6.35]$$

Of course, the former molecules for the crystallization process are here considered through a residual approach, involving fugacity coefficients. Δv_e is the volume difference between n_w molecules of water compared to the volume of a hydrate building unit at equilibrium pressure. Since the process is isothermal and isobaric, the use of such an expression leads to a constant driving force. In addition, the authors clearly consider that the gas mixture composition does not change, and so that the cavity occupation is a constant as well. There is therefore no discussion on the kinetics of enclathration of the guest compounds.

Then, the next step is to determine the amount of crystals in the solution. The amount of crystallized compounds is the fraction of the hydrate phase of the system ($\alpha(t)$). It is described by the mean of the rate of the crystallite nucleation rate $j(t)$, the growth rate $G(t)$, the initial volume of the solution V_0, the interface area A_S, and a shape factor m. According to Kashchiev and Firoozabadi [KAS 02a], this fraction is expressed as:

$$\alpha(t) = \left(\frac{bA_S}{V_0}\right)\int_0^t j(t')\left[\int_0^{t-t'} G(t'')dt''\right]^3 dt' \quad [6.36]$$

In the particular case of spray reactors, Gnanendran and Amin used the relevant expression for progressive nucleation (PN, i.e. continuous nucleation during the process), so that:

$$\alpha_{PN}(t) = \left[\frac{bG_0^{3m}J}{(1+3m)}\right]t^{1+3m} \quad [6.37]$$

where G_0 the growth constant, and J the stationary nucleation rate (in this precise case, $j(t) = J$). However, this expression is applicable for the post spray period (batch environment). During the spraying period, there is a slightly different approach. In this particular case, the expression becomes:

$$\alpha_{PN,spray}(t) = \left[\frac{bG_1^{3m}J_1}{(1+3m)(2+3m)}\right]t^{1+3m} \quad [6.38]$$

The growth constant G_l and nucleation rate J_l are also different. The complete expression can be found in Gnanendran and Amin.

In addition, the hydrate volume fraction as a function of time, as well as the hydrate formation rate $(d\alpha/dt)$, can be determined. This is then combined with a molar balance on the water molecules during the spray period and the batch environment. At last, the number of gas molecules into the hydrate structure is evaluated by the occupancy factor. Since the gas composition is supposed to be constant, the hydrate phase is homogeneous. Only the amount of water molecules is calculated in this approach.

6.6.1.5. *Conclusion*

Different approaches exist to model the hydrate crystal growth. Some of them, like the Skovborg and Rasmussen [SKO 94] approach, were developed to build kinetic flash algorithms. Usually, standard laws for mass transfer are used at the gas-to-liquid interface, or at the liquid-to-hydrate interface. When the liquid-to-hydrate mass transfer is taken into account, one sensitive point is the calculation of the occupancy of the hydrate phase during the crystallization. Usually, the Langmuir adsorption approach is used. However, this approach does not account for the kinetics of enclathration of the guest molecules. This is why another approach is suggested by Herri and Kwaterski [HER 12], of which the objective is to focus on the gas molecule integration into the hydrate structure. It can be combined with the typical mechanism described previously.

6.6.2. *Kinetic approach of enclathration*

6.6.2.1. *Introduction*

This approach has been published by Herri and Kwaterski [HER 12]. The main purpose is to model the whole process of the gas molecule enclathration into the hydrate structure, from the gas phase to the crystal phase. By this means, the crystal growth is modeled with a kinetic approach. It is influenced by the local composition around the crystal and the ability of the gas molecules to enter into the hydrate structure (characterized by kinetic constants). Ultimately, the crystal structure is non-stoichiometric, and the hydrate volume, as well as the growth speed are provided.

In the classic approach of van der Waals and Platteeuw, the hydrate phase is given by statistical thermodynamics while the liquid phase is expressed from standard thermodynamics (Gibbs-Duhem equation). The composition of the hydrate phase, that is to say the occupancy factor of the cavities, depends on the Langmuir constants and on the gas molecule fugacity (equation [6.10]). Usually, fugacities are calculated from VLE, and the Langmuir constants from a Kihara potential.

Thermodynamically, only the equality of the chemical potential is needed to represent the equilibrium. If the driving force $\Delta\mu$ (equation [6.3]) is lower than 0, the crystallization is possible. But, there is not a single possibility to consider the occupancy of the cavities. While $\Delta\mu < 0$, the hydrate crystal to be formed is stable. The main purpose of a kinetic approach is to consider that the mixed hydrate that will form may not be the most stable. Hence, this approach differs from a standard flash calculation that minimizes the Gibbs energy. Evidence of this metastable behavior has already been shown (see section 6.3).

In the model of Herri and Kwaterski, only the hydrate growth is considered, and not the overall process. Two layers are examined: the integration layer and the diffusion layer. The other parts (gas phase and liquid bulk) are supposed to be at thermodynamic equilibrium. Figure 6.18 illustrates the model.

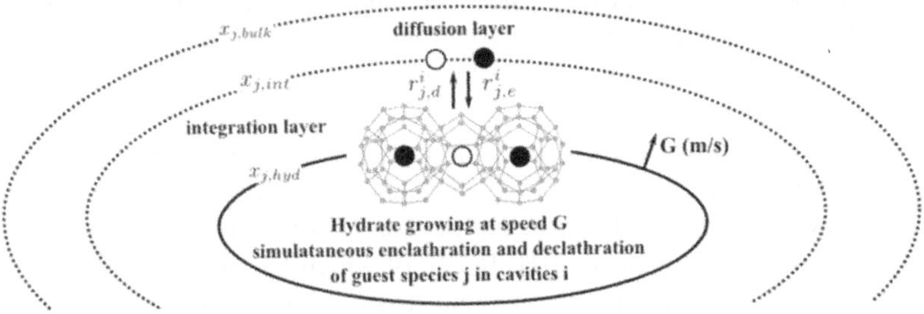

Figure 6.18. *Elementary steps for gas molecules integration at the vicinity of hydrate interface [HER 12]*

The diffusion layer corresponds to the liquid-to-hydrate layer, while the integration layer represents the solidification layer according to the approach of Svandal *et al.* [SVA 06].

The rate of enclathration and declathration of the gas molecule j into the cavity i into the hydrate structure are written respectively $r^i_{j,e}$ and $r^i_{j,d}$. They are expressed mathematically from kinetics constants, like any diffusion process in a boundary layer:

$$\begin{cases} r^i_{j,e} = k^i_{j,e} x_{j,int}(1-\theta_i) \\ r^i_{j,d} = k^i_{j,d}\theta^j_i \end{cases} \quad [6.39]$$

where $k^i_{j,e}$ are the kinetics constants of enclathrate and declathration respectively, and $(1-\theta_i)$ the fraction of empty cavities i, with $\theta_i = \sum_j \theta^j_i$. From these rates, two cases can be considered: the equilibrium (no driving force), and the dynamic system (during crystallization). Let us write F_j^i the difference between the two rates:

$$F_j^i = r_{j,e}^i - r_{j,d}^i \tag{6.40}$$

6.6.2.2. *Case of equilibrium*

At equilibrium, the rates of enclathration and declathration are the same, and $F_j^i = 0$, that is to say:

$$k_{j,d}^i \left(\frac{k_{j,e}^i}{k_{j,d}^i} x_{j,int}(1 - \theta_i) - \theta_i^j \right) = 0 \tag{6.41}$$

The ratio between the kinetic constant can be defined as the Langmuir constant of molecule j into the cavity i, hence:

$$C_{x,j}^i = \frac{k_{j,e}^i}{k_{j,d}^i} \tag{6.42}$$

Summing up the previous equations over all the species:

$$\theta_i = \frac{\sum_{j'} C_{x,j'}^i x_{j'}}{1 + \sum_{j'} C_{x,j'}^i x_{j'}} \Leftrightarrow (1 - \theta_i) = \frac{1}{1 + \sum_{j'} C_{x,j'}^i x_{j'}} \tag{6.43}$$

From equations [6.41]–[6.43], and since $\theta_i = \sum_j \theta_i^j$, the occupancy factor can finally be expressed as:

$$\theta_i^j = \frac{C_{x,j}^i x_j}{1 + \sum_{j'} C_{x,j'}^i x_{j'}} \tag{6.44}$$

This last form indeed looks like standard equation [6.10] in the van der Waals and Platteeuw model, with $C_{x,j}^i x_j = C_i^j f^j$. The subscript x indicates the formulation in mole fraction instead of fugacity.

In addition, equations [6.43] and [6.44] provide a relation between the Langmuir constants, i.e. the occupancy of the cavities, and the integration layer composition:

$$\theta_i^j = C_{x,j}^i x_j (1 - \theta_i) \tag{6.45}$$

Hence equation [6.40] can be written:

$$F_j^i = k_{j,d}^i C_{x,j}^i \left((x_{j,int} - x_j)(1 - \theta_i) \right) \quad [6.46]$$

In the case of the non-equilibrium, F_j^i is not equal to zero, and the hydrate crystals can grow ($F_j^i > 0$), or dissociate ($F_j^i < 0$).

6.6.2.3. *Crystal growth*

In the case of crystal growth, the rate of enclathration is higher than the rate of declathration. Equation [6.40] becomes the following inequality:

$$F_j^i \geq 0 \quad [6.47]$$

The growing surface A_s of the hydrate crystal is supposed to be surrounded by cavities. The surface density of cavity i around the hydrate is written Γ_i (mol.m^{-2}). Each cavity is exposed to the gas molecules at the rate F_j^i. In addition, the hydrate crystal is supposed to grow at speed G. From this starting point, the volume increase of the hydrate per dt element of time is:

$$\frac{dV}{dt} = GA_s \quad [6.48]$$

In this new amount of solid, there is a given number of water molecules, and a given number of cavities of type i hereafter written as c_{cav}^i. Considering each cavity i, the flow rate of enclathration R_i^j of the gas molecule is:

$$R_i^j = F_j^i \Gamma_i A_s \quad [6.49]$$

With equation [6.48], this rate can also be expressed as:

$$c_{cav}^i \theta_j^i \frac{dV}{dt} = c_{cav}^i \theta_j^i G A_s \quad [6.50]$$

Combination of equations [6.49] and [6.50], dividing by A_s, leads to:

$$F_j^i \Gamma^i = c_{cav}^i \theta_i^j G \quad [6.51]$$

or

$$F^i \Gamma^i = c_{cav}^i \theta_i G \quad [6.52]$$

with $F^i = \sum_j F_j^i$.

Furthermore, for two gas components j and j', the following general consideration can be described as:

$$\forall_{j\in gas}, \forall_{j'\in gas} \forall_{i\in cav}, \left(\frac{F_j^i}{F_{j'}^i} = \frac{\theta_i^j}{\theta_i^{j'}} \Leftrightarrow \frac{F_{j'}^i}{\theta_i^{j'}} = \frac{F_j^i}{\theta_i^j} \right) \qquad [6.53]$$

The right term of equation [6.53] states that there is a constant a^i so that:

$$\forall_{i\in cav}, \forall_{j\in gas}, a^i = \frac{F_j^i}{\theta_i^j} \qquad [6.54]$$

A summation over all the guest molecules can then lead to a simpler form such as $a^i = \frac{F^i}{\theta_i}$. In the end, equation [6.53] allows us to write:

$$a^i = \frac{c_{cav}^i}{\Gamma_i} G \qquad [6.55]$$

From equations [6.46] and [6.54], two expressions for the flow rate per moles of cavities (F_j^i) are suggested, so that:

$$\forall_{i\in cav}, \forall_{j\in gas}, a^i \theta_i^j = k_{j,d}^i C_{x,j}^i \left(\left(x_{j,int} - x_j \right)(1 - \theta_i) \right) \qquad [6.56]$$

Since $\theta_i^j = C_{x,j}^i x_j (1 - \theta_i)$ (see equation [6.45]), and

$$\forall_{i\in cav}, \forall_{j\in gas}, F_j^i = a^i C_{x,j}^i x_j (1 - \theta_i) = k_{j,d}^i C_{x,j}^i \left(\left(x_{j,int} - x_j \right)(1 - \theta_i) \right) [6.57]$$

Hence

$$\forall_{i\in cav}, \forall_{j\in gas}, x_j - x_{j,int} = \frac{x_{j,int}}{\left(1 + \frac{k_{j,d}^i}{a^i}\right)} \qquad [6.58]$$

This last expression provides the driving force for the gas molecules enclathration into the hydrate structure. It is independent of the cavities according to the left term. Therefore, the ratio $\frac{k_{j,d}^i}{a^i}$ is not i-dependent. Combined with equation [6.55], which gives the relation between the a^i constant, the speed of growth G, the cavity concentration c_i and cavity surface density Γ_i, an expression for the ratio $\frac{k_{j,d}^i}{a^i}$ can be suggested:

$$\frac{k_{j,d}^i}{a^i} = \frac{k_{j,d}^i \Gamma^i}{c_{cav}^i G} \qquad [6.59]$$

Herri and Kwaterski decided to define the kinetic constant of molecule *j* as:

$$k_j = \frac{k_{j,d}^i \Gamma^i}{c_{cav}^i} \qquad [6.60]$$

It can be considered as the intrinsic kinetic constant of component *j*. Consequently, this value needs to be determined. In the end, this expression can be injected into equation [6.59] so that:

$$\forall_{j \in gas}, x_j = x_{j,int} \frac{k_j/G}{\left(1+ k_j/G\right)} \qquad [6.61]$$

Thanks to this final expression, the chemical potential of the hydrate phase can be calculated from the classic Langmuir approach (equation [6.10]). Therefore, the occupancy factor is a function of the kinetic constants, the growth speed, and the composition at the interface of the diffusion and integration layers, hence:

$$\theta_i^j = \frac{\frac{k_j C_{x,j}^i x_{j,int}}{\left(1+ k_j/G\right)}}{1+\sum_{j' \in gas} \frac{k_{j'} C_{x,j'}^i x_{j',int}}{\left(1+ k_{j'}/G\right)}} \qquad [6.62]$$

One can remark that, if the kinetic constants are null, the occupancy factor expression is the same as the thermodynamic one (in equation [6.10]). Hence, in this case, the approach is identical to van der Waals and Platteeuw's.

From the equality of the rates at the interface of the diffusion and integration layers, the mass balance of the framework is:

$$\forall_{j \in gas}, G A_s \sum_{i \in cav} c_{cav}^i \theta_i^j = D_j^* A_s \left(x_{j,bulk} - x_{j,int}\right) \frac{\rho_w^0}{M_w} \qquad [6.63]$$

By assuming the growth speed and the kinetic constants, $x_{j,int}$ can be determined from the previous equation system, which can be rewritten as:

$$\forall_{j \in gas}, G \sum_{i \in cav} c_{cav}^i \frac{\frac{k_j C_{x,j}^i x_{j,int}}{\left(1+ k_j/G\right)}}{1+\sum_{j' \in gas} \frac{k_{j'} C_{x,j'}^i x_{j',int}}{\left(1+ k_{j'}/G\right)}} = D_j \left(x_{j,bulk} - x_{j,int}\right) \qquad [6.64]$$

with $D_j = D_j^* \frac{\rho_w^0}{M_w}$.

In the end, Herri and Kwaterski suggested an algorithm that combines the mass balance on equation [6.63] with the non-equilibrium thermodynamics inspired from standard equation [6.10] with equation [6.62]. This algorithm is illustrated in Figure 6.19.

Obviously, such an algorithm needs the knowledge of the intrinsic kinetic constants (k_j), the diffusivity of the gas in the solvents (D_j), and the Langmuir constants ($C^i_{x,j}$). The result is a kinetic flash algorithm, leading to a non-stoichiometric mixed hydrate. The growth rate is also determined.

However, this approach only focuses on the hydrate growth at the hydrate interface with the surrounding liquid phase. To model the hydrate formation process completely, the whole mass transfer of the gas molecules needs to be taken into account, as shown as an example in Figure 6.5.

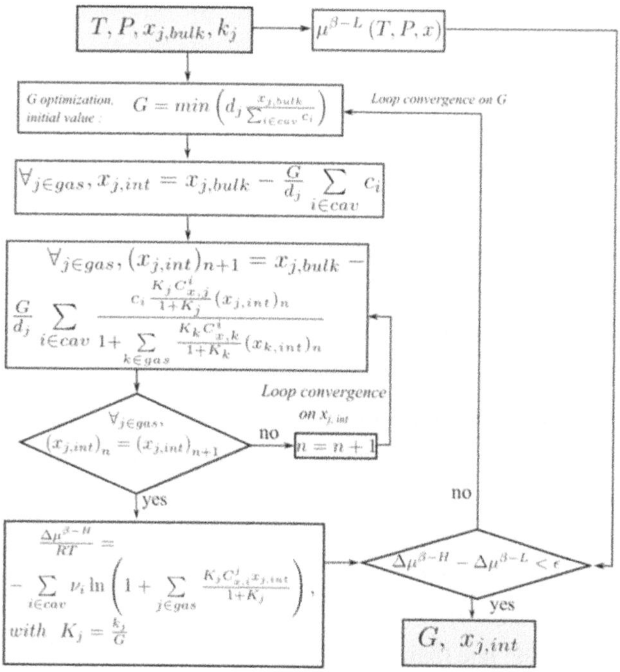

Figure 6.19. *Procedure of convergence to calculate the hydrate growth speed G and composition during crystallization $x_{j,int}$ [HER 12]*

6.7. Conclusion

Evidence of non-equilibrium formation of clathrate hydrates can be found in the literature, from standard laboratory scale experiments, to inclusion studies and molecular dynamics simulations. From this perspective, there is a wide open area toward kinetic understanding of clathrate crystallization. Many authors are already working on this novel subject, crucial for hydrate composition and volume predictive approaches. Both kinetic models and non-equilibrium thermodynamic models have been built. One important issue concerns the experimental studies. Among the thousands of pieces of data that can be found in the literature, few of them provide hydrate composition or volume/water conversion. Moreover, depending on the speed of crystallization, the final equilibrium could not be at thermodynamic equilibrium. Obtaining accurate and reliable experimental data still remains of prime importance.

6.8. Nomenclature

6.8.1. *Letters*

A_p: surface of particle (m^2)

A_{GL}: gas/liquid interface (m^2)

A_S: surface area of hydrate crystal (m^2)

C: molar concentration (mol/m^3)

C_i^j: Langmuir constant of molecule j in cavity i (Pa^{-1})

D: diffusion coefficient in section 6.3 ($mol/m^2/s$), or flow rate (mol/s)

f: fugacity (Pa), or particule size distribution

F: comparison between rates of enclathration and declatration (mol/s)

G: growth rate (m/s)

G_0: growth constant (m/s)

h: thickness of the film region (m)

j: nucleation rate ($particle/s$)

J: stationary nucleation rate ($particle/s$)

k_l: kinetic constant of mass transfer (m/s)

k_g: growth rate kinetic constant (m^4/s)

K: partition coefficient, or diffusion/reaction coefficient

LMGS: Large Molecule Guest Substance

m: mass (kg)

M: molar mass (*kg/mol*)

MW: molecular mass (*kg/molecule*)

M_2: second moment of particle size distribution (i.e. volume)

n *or* N: number of moles (*mol*)

N_w: number of water molecules per unit cell

P: pressure (*Pa*)

r: rate of enclathration/declathation (*mol/s*)

R: gas constant (*8.3145 J/mol/K*)

R_i^j: flow rate of enclatration of molecule j in cavity i (*mol/s*)

S: entropy (*J/K*)

T: temperature (*K*)

v_e: molar volume (m^3/mol)

V: volume (m^3)

w: interaction potential (*J*)

x: mole fraction

6.8.2. *Greek letters*

$\underline{\alpha}$: phase fraction

β: empty hydrate reference state

γ: activity coefficient

Γ_i: surface density of cavity I (*mol/m*^{2})

λ: stability variable

φ: fugacity coefficient

Φ: molar flux (*mol/m*^{2}*/s*)

μ: chemical potential (*J/mol*)

ν_j: number of cavity j in a crystal unit

π: pi number, or generic phase symbol

ρ: density (*kg/m*^{3})

θ: occupancy factor (matrix)

θ_i^j: occupancy of cavity i by molecule j

ρ: density (*kg/m*^{3})

6.8.3. *Subscript*

e: equilibrium state

in: inflowing feed gas component, in [TSU 07]

j: molecule index

L: large cavity

m: molar dimension

p: particle

S: small cavity

w: water

6.8.4. *Superscript*

0: reference state

B: basic hydrate according to Chen and Guo [CHE 96]

i: cavity index

j: molecule index

H: hydrate phase

L: liquid phase

6.9. Bibliography

[ADI 91] ADISASMITO S., FRANK III R.J., SLOAN E.D., "Hydrates of carbon dioxide and methane mixtures", *Journal of Chemical Engineering Data*, vol. 36, no. 1, pp. 68–71, 1991.

[ALO 11] AL-OTAIBI F., CLARKE M., BISHNOI P.R., "Kinetics of structure II gas hydrate formation for propane and ethane using an in-situ particle size analyzer and Raman spectrometer", *Chemical Engineering Science*, vol. 66, pp. 2468–2474, 2011.

[ARJ 05] ARJMANDI M., TOHIDI B., DANESH A. *et al.*, "Is subcooling the right driving force for testing low-dosage hydrate inhibitors?", *Chemical Engineering Science*, vol. 60, pp. 1313–1321, 2005.

[BAK 82] BAKER L.E., PIERCE A.C., LUKS K.D., "Gibbs energy analysis of phase equilibria", *Society of Petroleum Engineers Journal*, vol. 22, no. 5, pp. 731–742, 1982.

[BAL 02a] BALLARD A.L., SLOAN E.D., "The next generation of hydrate prediction: an overview", *Journal of Supramolecular Chemistry*, vol. 2, nos. 4–5, pp. 385–392, 2002.

[BAL 02b] BALLARD A.L., SLOAN E.D., "The next generation of hydrate prediction: Part I. Hydrate standard states and incorporation of spectroscopy", *Fluid Phase Equilibria*, vol. 194–197, pp. 371–383, 2002.

[BAL 04a] BALLARD A.L., SLOAN E.D., "The next generation of hydrate prediction: Part IV. A comparison of available hydrate prediction programs", *Fluid Phase Equilibria*, vol. 216, no. 2, pp. 257–270, 2004.

[BAL 04b] BALLARD A.L., SLOAN E.D., "The next generation of hydrate prediction: Part III. Gibbs energy minimization formalism", *Fluid Phase Equilibria*, vol. 218, no. 1, pp. 15–31, 2004.

[BIS 89] BISHNOI P.R., GUPTA A.K., ENGLEZOS P. *et al.*, "Multiphase equilibrium flash calculations for systems containing gas hydrates", *Fluid Phase Equilibria*, vol. 53, pp. 97–104, 1989.

[BOE 14] BOESEN R.R., SORENSEN H., PEDERSEN K.S., "New approach for hydrate flash calculations", *Proceedings of the 8th International Conference on Gas Hydrates – ICGH8*, Beijing, China, 2014.

[BOU 16] BOUILLOT B., HERRI J.M., "Framework for clathrate hydrate flash calculations and implications on the crystal structure and final equilibrium of mixed hydrates", *Fluid Phase Equilibria*, vol. 413, pp. 184–195, 2016.

[CHE 96] CHEN G.J., GUO T.M., "Thermodynamic modeling of hydrate formation based on new concepts", *Fluid Phase Equilibria*, vol. 122, pp. 43–65, 1996.

[CHE 98] CHEN G.J., GUO T.M., "New approach to gas hydrate modelling", *Chemical Engineering Journal*, vol. 71, pp. 145–151, 1998.

[COL 90] COLE W.A., GOODWIN S.P., "Flash calculations for gas hydrates: a rigorous approach", *Chemical Engineering Science*, vol. 45, no. 3, pp. 569–573, 1990.

[COL 79] COLLINS P.L.F., "Gas hydrates in CO_2-bearing fluid inclusions and the use of freezing data for estimation of salinity", *Economic Geology*, vol. 78, pp. 1435–1444, 1979.

[DAR 11] DARTOIS E.C.O., "Clathrate hydrate: near to mid-IR spectroscopic signatures", *Icarus,* vol. 212, pp. 950–956, 2011.

[DAV 87] DAVIDSON D.W., DESANDO M.A., GOUGH S.R. *et al.*, "A clathrate hydrate of carbon monoxide", *Nature*, vol. 328, pp. 418–419, 1987.

[DEH 13] DE HEMPTINE J.C., LEDANOIS J.M., MOUGIN P. *et al.*, *Select thermodynamic Models for Process Simulation*, Technip, Paris, 2013.

[DES 90] DESANDO M.A., HANDA Y.P., HAWKINS R.E. *et al.*, "Dielectric and 13C NRM studies of the carbon monoxide clathrate hydrate", *Journal of Inclusion Phenomena and Molecular Recognition in Chemistry*, vol. 8, nos. 1–2, pp. 3–16, 1990.

[ENG 87a] ENGLEZOS P., DHOLABHAI P., KALOGERAKIS N. *et al.*, "Kinetics of methane and ethane hydrate formation", *Chemical Engineering Science*, vol. 42, pp. 2647–2658, 1987.

[ENG 87b] ENGLEZOS P., DHOLABHAI P., KALOGERAKIS N. *et al.*, "Kinetics of gas hydrate formation from mixtures of methane and ethane", *Chemical Engineering Science*, vol. 42, pp. 2659–2666, 1987.

[ESL 12] ESLIMANESH A., MOHAMMADI A.H., RICHON D. *et al.*, "Application of gas hydrate formation in separation processes: a review of experimental studies", *Journal of Chemical Thermodynamics*, vol. 45, pp. 62–71, 2012.

[ESL 13] ESLIMANESH A., BABAEE S., GHARAGHEIZI F. *et al.*, "Assessment of clathrate hydrate phase equilibrium data for CO_2 + CH_4/N_2 + water system", *Fluid Phase Equilibria*, vol. 349, pp. 71–82, 2013.

[FIR 99] FIRROZABADI A., *Thermodynamics of Hydrocarbon Reservoirs*, McGraw-Hill, New York, 1999.

[GAI 99] GAILLARD C., MONFORT J.P., PEYTAVY J.L., "Investigation of methane hydrate formation in a recirculating flow loop: modelling of the kinetics and tests of efficiency of chemical additives on hydrate inhibition", *Oil and Gas Science and Technology*, vol. 54, pp. 365–374, 1999.

[GNA 04] GNANENDRAN N., AMIN R., "Modelling hydrate formation kinetics of a hydrate promoter–water–natural gas system in a semi-batch spray reactor", *Chemical Engineering Science*, vol. 59, pp. 3849–3863, 2004.

[GUP 90] GUPTA A.K., Steady state simulation of chemical processes, PhD Thesis, University of Calgary, Canada, 1990.

[GUP 91] GUPTA A.K., BISHNOI P.R., KALOGERAKIS N., "A method for the simultaneous phase equilibria and stability calculations for multiphase reacting and non-reacting systems", *Fluid Phase Equilibria*, vol. 63, pp. 65–89, 1991.

[HER 96] HERRI J.M., Etude de la formation de l'hydrate de méthane par turbidimétrie in situ, PhD Thesis, Université Paris VI, Paris, 1996.

[HER 99a] HERRI J.M., GRUY F., PIC J.S. *et al.*, "Interest of in situ turbidimetry for the characterization of methane hydrate crystallization: application to the study of kinetic inhibitors", *Chemical Engineering Science*, vol. 54, pp. 1849–1858, 1999.

[HER 99b] HERRI J.M., GRUY F., PIC J.S. *et al.*, "Methane hydrate crystallization mechanism from in-situ particle sizing", *AIChE Journal*, vol. 45, no. 3, pp. 590–602, 1999.

[HER 11] HERRI J.M., BOUCHEMOUA A., KWATERSKI M. *et al.*, "Gas hydrate equilibria for CO_2–N_2 and CO_2–CH_4 gas mixtures—experimental studies and thermodynamic modelling", *Fluid Phase Equilibria*, vol. 301, pp. 171–190, 2011.

[HER 12] HERRI J.M., KWATERSKI M., "Derivation of a Langmuir type of model to describe the intrinsic growth rate of gas hydrates during crystallization from gas mixtures", *Chemical Engineering Science*, vol. 81, pp. 28–37, 2012.

[JAC 10] JACOBSON L.C., HUJO W., MOLINERO V., "Amorphous precursors in the nucleation of clathrate hydrates", *Journal of American Chemical Society*, vol. 132, pp. 11806–11811, 2010.

[JON 92] JONES A.G., HOSTOMSKI J., LI Z., *Chemical Engineering Science*, vol. 47, p. 3817, 1992.

[KAS 00] KASHCHIEV D., *Nucleation*, Butterworth-Heinemann, Oxford, 2000.

[KAS 02a] KASHCHIEV D., FIROOZABADI A., "Driving force for crystallization of gas hydrates", *Journal of Crystal Growth*, vol. 241, pp. 220–230, 2002.

[KAS 02b] KASHCHIEV D., FIROOZABADI A., "Nucleation of gas hydrates", *Journal of Crystal Growth*, vol. 243, pp. 476–489, 2002.

[KAS 03] KASHCHIEV D., FIROOZABADI A., "Induction time in crystallization of gas hydrates", *Journal of Crystal Growth*, vol. 250, pp. 499–515, 2003.

[KOB 07] KOBAYASHI T., MORI Y.H., "Thermodynamic simulations of hydrate formation from gas mixtures in batch operations", *Energy Conversion and Management*, vol. 48, pp. 242–250, 2007.

[KRE 05] KREJBJERG K., SØRENSEN H., "A solution algorithm for simulating hydrate growth", *Proceedings of the Fifth International Conference on Gas Hydrates – ICGH5*, Trondheim, Norway, 2005.

[KRO 09] KROLEIN K., MUZNY C.D., KAZAKOV A. *et al.*, Clathrate hydrate physical property database, NIST Standard Reference Database # 156, available at: http://gashydrates.nist.gov/HydrateViewer/, 2009.

[KVA 14] KVAMME B., KUZNETSOVA T., JENSEN B. *et al.*, "Consequences of CO_2 solubility for hydrate formation from carbon dioxide containing water and other impurities", *Physical Chemistry Chemical Physics*, vol. 16, pp. 8623–8638, 2014.

[LAU 14] LAURICELLA M., MELONI S., ENGLISH N.J. *et al.*, "Methane clathrate hydrate nucleation mechanism by advanced molecular simulations", *Journal of Physical Chemistry B*, vol. 118, pp. 22847–22857, 2014.

[LAU 15] LAURICELLA M., MELONI S., LIANG S. *et al.*, "Clathrate structure-type recognition: application to hydrate nucleation and crystallization", *Journal of Chemical Physics*, vol. 142, p. 244503, 2015.

[LEQ 16] LE QUANG D., LE QUANG D., BOUILLOT B. *et al.*, "Experimental procedure and results to measure the composition of gas hydrate, during crystallization and at equilibrium, from N_2-CO_2-CH_4-C_2H_6-C_3H_8-C_4H_{10} gas mixtures", *Fluid Phase Equilibria*, vol. 413, pp. 10–21, 2016.

[LIA 11] LIANG S., KUSALIK P.G., "Exploring nucleation of H2S hydrates", *Chemical Science*, vol. 2, no. 7, pp. 1286–1292, 2011.

[MA 13a] MA Q.L., CHEN G.J., SUN C.Y., "Vapor–liquid–liquid–hydrate phase equilibrium calculation for multicomponent systems containing hydrogen", *Fluid Phase Equilibria*, vol. 338, pp. 87–94, 2013.

[MA 13b] MA Q.L., CHEN G.J., "Prediction of vapor-liquid-liquid-hydrate phase equilibrium for multicomponent systems containing tetrahydrofuran", *Science China Chemistry*, vol. 56, no. 12, pp. 1800–1810, 2013.

[MAH 16] MAHABADIAN M.A., CHAPOY A., BURGASS R. *et al.*, "Development of a multiphase flash in presence of hydrates: experimental measurements and validation with the CPA equation of state", *Fluid Phase Equilibria*, vol. 414, pp. 117–132, 2016.

[MAK 10] MAKOGON Y.F., "Natural gas hydrates – a promising source of energy", *Journal of Natural Gas Science and Engineering*, vol. 2, no. 1, pp. 49–59, 2010.

[MCK 63] MCKOY V., SINAGOĞLU O.J., "Theory of dissociation pressures of some gas hydrates", *Journal of Chemical Physics*, vol. 38, pp. 2946–2956, 1963.

[MIC 82a] MICHELSEN M.L., "The isothermal flash problem. Part I. Stability", *Fluid Phase Equilibria*, vol. 9, pp. 1–19, 1982.

[MIC 82b] MICHELSEN M.L., "The isothermal flash problem. Part II. Phase-split calculations", *Fluid Phase Equilibria*, vol. 9, pp. 21–40, 1982.

[MIC 91] MICHELSEN M.L., "Calculation of hydrate fugacities", *Chemical Engineering Science*, vol. 46, pp. 1192–1993, 1991.

[MIC 94] MICHELSEN M.L., "Calculation of multiphase equilibrium", *Computer & Chemical Engineering,* vol. 18, pp. 545–550, 1994.

[MIL 85] MILLER S.L., ""Clathrate hydrates in the Solar System" in KLINGER J., BENEST D., DOLLFUD A. (eds), *Ice in the Solar System*, vol. 59, Reidel, 1985.

[MOH 05] MOHAMMADI A.H., ANDERSON R., TOHIDI B., "Carbon monoxide clathrate hydrates: equilibrium data and thermodynamic modeling", *AIChE Journal*, vol. 51, pp. 2825–2833, 2005.

[MOH 10] MOHAMMADI A.H., RICHON D., "Ice-clathrate hydrate-gas phase equilibria for air, oxygen, nitrogen, carbon monoxide, methane, or ethane + water system", *Industrial & Engineering Chemistry Research*, vol. 49, pp. 3976–3979, 2010.

[MOH 09] MOHAMMADI-MANESH H., ALAVI S., WOO T.K., *et al.*, "Molecular dynamics simulation of 13C NMR powder lineshapeds of CO in structure I clathrate hydrate", *Physical Chemistry Chemical Physics*, vol. 11, pp. 8821–8828, 2009.

[MUL 01] MULLIN J.W., *Crystallization*, 4th ed., Butterworth-Heinemann, Oxford, 2001.

[MUR 95] MURPHY P.J., ROBERTS S., "Laser Raman spectroscopy of differential partitioning in mixed-gas clathrates in H_2O-CO_2-N_2-CH_4 fluid inclusions: implications for microthermometry", *Geochimica et Cosmochimica Acta*, vol. 59, no. 23, pp. 4809–4824, 1995.

[MUR 97] MURPHY P.J., ROBERTS S., "Melting and nucleation behavior of Clathrate in multivolatile fluid inclusions: evidence of thermodynamic disequilibrium", *Chemical Geology*, vol. 135, pp. 1–20, 1997.

[NGU 12] NGUYEN A.H., JACOBSON L.C., MOLINERO V., "Structure of the clathrate/solution interface and mechanism of cross-nucleation of clathrate hydrates", *Journal of Physical Chemistry C*, vol. 116, pp. 19828–19838, 2012.

[NGU 13] NGUYEN A.H., MOLINERO V., "Stability and metastability of bromine clathrate polymorphs", *Journal of Physical Chemistry B*, vol. 117, pp. 6330–6338, 2013.

[NGU 14] NGUYEN A.H., MOLINERO V., "Structure of the clathrate/solution interface and mechanism of cross-nucleation of clathrate hydrates", *Journal of Chemical Physics*, vol. 140, p. 084506, 2014.

[OHM 04] OHMURA R., SHIMADA W., UCHIDA T. *et al.*, "Clathrate hydrate crystal growth in liquid water saturated with a hydrate-forming substance: variations in crystal morphology", *Philosophical Magazine*, vol. 84, no. 1, pp. 1–16, 2004.

[PRA 99] PRAUSNITZ J.M., LICHTENTHALER R.N., DE AZEVEDO E.G., *Molecular Thermodynamics of Fluid-Phase Equilibria*, 3rd ed., Prentice Hall, Upper Saddle River, New Jersey, 1999.

[RIB 08] RIBEIRO C., LAGE P.L.C., "Modelling of hydrate formation kinetics: state-of-the-art and future directions", *Chemical Engineering Science*, vol. 63, pp. 2007–2034, 2008.

[SCH 04] SCHICKS J., RIPMEESTER J.A., "The coexistence of two different methane hydrate phase under moderate pressure and temperature conditions: kinetic versus thermodynamic products", *Angewandte Chemie International Edition*, vol. 43, pp. 3310–3313, 2004.

[SCH 13] SCHICKS J., LUZI-HELBING M., "Cage occupancy and structural changes during hydrate formation from initial stages to resulting hydrate phase", *Spectrochimica Acta Part A: Molecular and Biomolecular Spectroscopy*, vol. 115, pp. 528–536, 2013.

[SCH 15] SCHICKS J., LUZI-HELBING M., "Kinetic and thermodynamic aspects of clathrate hydrate nucleation and growth", *Journal of Chemical & Engineering Data*, vol. 60, no. 2, pp. 269–277, 2015.

[SEG 16] SEGTOVITCH I.S.V., BARRETO A.G., TAVARES F.W., "Simultaneous multiple flash and stability analysis calculations including hydrates", *Fluid Phase Equilibria*, vol. 413, pp. 196–208, 2016.

[SHI 12] SHIN W., PARK S., RO H. *et al.*, "Spectroscopic confirmation of metastable structure formation occurring in natural gas hydrates", *Chemistry Asian Journal*, vol. 7, pp. 2235–2238, 2012.

[SKO 94] SKOVBORG P., RASMUSSEN P., "A mass transport limited model for the growth of methane and ethane gas hydrates", *Chemical Engineering Science*, vol. 49, no. 8, pp. 1131–1143, 1994.

[SLO 04] SLOAN E.D., "A changing hydrate paradigm – from apprehension to avoidance to risk management", *Fluid Phase Equilibria*, vol. 228, pp. 67–74, 2004.

[SLO 07] SLOAN E.D., KOH C.A., *Clathrate Hydrates of Natural Gases*, 3rd ed., CRC Press, Boca Raton, 2007.

[SUB 99] SUBRAMANIAN S., SLOAN E.D., "Molecular measurements of methane hydrate formation", *Fluid Phase Equilibria*, vol. 158, pp. 813–820, 1999.

[SUB 00] SUBRAMANIAN S., KINI R.A., DEC S.F. *et al.*, "Evidence of structure II hydrate formation from methane + ethane mixtures", *Chemical Engineering Science*, vol. 55, pp. 1981–1999, 2000.

[SUB 02] SUBRAMANIAN S., SLOAN E.D., "Solubility effects on growth and dissolution of methane hydrate needles", *Proceedings from the fourth International Conference on Gas Hydrates*, Yokohama, Japan, pp. 856–861, 2000.

[SUM 11] SUM A.K., WU D.T., YASUOKA K., "Energy science of clathrate hydrates: simulation-based advances", *MRS Bulletin*, vol. 36, pp. 205–210, 2011.

[SVA 06] SVANDAL A., KVAMME B., GRANAZY L. *et al.*, "The phase-field theory applied to CO_2 and CH_4 hydrate", *Journal of Crystal Growth*, vol. 287, pp. 486–490, 2006.

[TAV 15] TAVASOLI H., FEYZI F., "Compositional data calculation of vapor–aqueous–hydrate systems in batch operations by a new algorithm", *Journal of Natural Gas Science and Engineering*, vol. 24, pp. 474–488, 2015.

[TSU 05] TSUJI H., KOBAYASHI T., OKANO Y. *et al.*, "Thermodynamic simulations of isobaric hydrate-forming operations: formulation of computational scheme and its application to hydrate formation from a methane + ethane + propane mixture", *Energy & Fuels*, vol. 19, pp. 1587–1597, 2005.

[VAN 59] VAN DER WAALS J.H., PLATTEEUW J.C., "Clathrate solutions", *Advances in Chemical Physics*, vol. 2, pp. 1–57, 1959.

[VAT 06] VATAMANU J., KUSALIK P.G., "Molecular insights into the heterogeneous crystal growth of sI methane hydrate", *The Journal of Physical Chemistry B*, vol. 110, pp. 15896–15904, 2006.

[VYS 83] VYSNIAUSKAS A., BISHNOI P.R., "A kinetic study of methane hydrate formation", *Chemical Engineering Science*, vol. 38, pp. 1061–1072, 1983.

[WAL 09] WALSH M.R., KOH C.A., SLOAN E.D. *et al.*, "Microsecond simulations of spontaneous methane hydrate nucleation and growth", *Science*, vol. 326, pp. 1095–1098, 2009.

[ZHU 14] ZHU J., DU S., YU X. *et al.*, "Encaspulation kinetics and dynamics of carbon monoxide in clathrate hydrate", *Nature Communications*, vol. 5, no. 4128, 2014.

List of Authors

Laura BEDOURET
Institut Laue Langevin
Grenoble
and University of Bordeaux
France

Baptiste BOUILLOT
SPIN
Ecole des Mines de Saint-Etienne
France

Patrick BOURIAT
University of Pau and
Pays de l'Adour
France

Daniel BROSETA
University of Pau and
Pays de l'Adour
France

Ross BROWN
CNRS – University of
Pau and Pays de l'Adour
France

Bertrand CHAZALLON
PhLAM CNRS
University of Lille
France

Didier DALMAZZONE
ENSTA ParisTech
University Paris-Saclay
France

Anthony DELAHAYE
IRSTEA-GPAN
Antony
France

Arnaud DESMEDT
CNRS – University of Bordeaux
France

Laurence FOURNAISON
IRSTEA-GPAN
Antony
France

Jean-Michel HERRI
SPIN
Ecole des Mines de Saint-Etienne
France

Nelly HOBEIKA
University of Pau and
Pays de l'Adour
France

Maria Lourdes MARTINEZ DE BAÑOS
University of Pau and
Pays de l'Adour
France

Jennifer A. NOBLE
CNRS – University of Bordeaux
France

Jacques OLLIVIER
Institut Laue Langevin
Grenoble
France

Patrice PARICAUD
ENSTA ParisTech
University Paris-Saclay
France

Claire PETUYA
CNRS – University of Bordeaux
France

Livio RUFFINE
Ifremer
Brest
France

Luiz Paulo SALES SILVA
ENSTA ParisTech
University Paris-Saclay
France

Index

P, R

S, T, V

Other titles from

iSTE

in

Energy

2017

LA SCALA Massimo
From Smart Grids to Smart Cities: New Challenges in Optimizing Energy Grids

SIMON Patrice, BROUSSE Thierry, FAVIER Frédéric
Supercapacitors Based on Carbon or Pseudocapacitive Materials (Energy Storage – Batteries, Supercapacitors Set – Volume 3)

2016

ALLARD Bruno
Power Systems-on-Chip: Practical Aspects of Design

ANDRE Michel, SAMARAS Zissis
Energy and Environment

DUFOUR Anthony
Thermochemical Conversion of Biomass for the Production of Energy and Chemicals

2015

CROGUENNEC Laurence, MONCONDUIT Laure, DEDRYVÈRE Rémi
Electrodes for Li-ion Batteries
(Energy Storage – Batteries, Supercapacitors Set – Volume 2)

LEPRINCE-WANG Yamin
Piezoelectric ZnO Nanostructure for Energy Harvesting
(Nanotechnologies for Energy Recovery Set – Volume 1)

ROBYNS Benoît, FRANÇOIS Bruno, DELILLE Gauthier, SAUDEMONT Christophe
Energy Storage in Electric Power Grids

ROSSI Carole
Al-based Energetic Nanomaterials
(Nanotechnologies for Energy Recovery Set – Volume 2)

TARASCON Jean-Marie, SIMON Patrice
Electrochemical Energy Storage
(Energy Storage – Batteries, Supercapacitors Set – Volume 1)

2013

LALOUI Lyesse and DI DONNA Alice
Energy Geostructures: Innovation in Underground Engineering

2012

BECKERS Benoit
Solar Energy at Urban Scale

ROBYNS Benoît, DAVIGNY Arnaud, FRANÇOIS Bruno, HENNETON Antoine, SPROOTEN Jonathan
Electricity Production from Renewable Energies

2011

GAO Fei, BLUNIER Benjamin, MIRAOUI Abdellatif
Proton Exchange Membrane Fuel Cell Modeling

MULTON Bernard
Marine Renewable Energy Handbook

2010

BRUNET Yves
Energy Storage

2009

SABONNADIERE Jean-Claude
Low Emission Power Generation Technologies and Energy Management

SABONNADIERE Jean-Claude
Renewable Energy Technologies

Printed and bound by CPI Group (UK) Ltd, Croydon, CR0 4YY

13/09/2026

14961024-0005